199
Advances in Polymer Science

Editorial Board:
A. Abe · A.-C. Albertsson · R. Duncan · K. Dušek · W. H. de Jeu
J.-F. Joanny · H.-H. Kausch · S. Kobayashi · K.-S. Lee · L. Leibler
T. E. Long · I. Manners · M. Möller · O. Nuyken · E. M. Terentjev
B. Voit · G. Wegner · U. Wiesner

Advances in Polymer Science
Recently Published and Forthcoming Volumes

Polymers for Regenerative Medicine
Volume Editor: Werner, C.
Vol. 203, 2006

Peptide Hybrid Polymers
Volume Editors: Klok, H.-A., Schlaad, H.
Vol. 202, 2006

**Supramolecular Polymers ·
Polymeric Betains**
Vol. 201, 2006

Ordered Polymeric Nanostructures at Surfaces
Volume Editor: Vansco, G. J.
Vol. 200, 2006

Emissive Materials · Nanomaterials
Vol. 199, 2006

Surface-Initiated Polymerization II
Volume Editor: Jordan, R.
Vol. 198, 2006

Surface-Initiated Polymerization I
Volume Editor: Jordan, R.
Vol. 197, 2006

Conformation-Dependent Design of Sequences in Copolymers II
Volume Editor: Khokhlov, A. R.
Vol. 196, 2006

Conformation-Dependent Design of Sequences in Copolymers I
Volume Editor: Khokhlov, A. R.
Vol. 195, 2006

Enzyme-Catalyzed Synthesis of Polymers
Volume Editors: Kobayashi, S., Ritter, H., Kaplan, D.
Vol. 194, 2006

Polymer Therapeutics II
Polymers as Drugs, Conjugates and Gene Delivery Systems
Volume Editors: Satchi-Fainaro, R., Duncan, R.
Vol. 193, 2006

Polymer Therapeutics I
Polymers as Drugs, Conjugates and Gene Delivery Systems
Volume Editors: Satchi-Fainaro, R., Duncan, R.
Vol. 192, 2006

Interphases and Mesophases in Polymer Crystallization III
Volume Editor: Allegra, G.
Vol. 191, 2005

Block Copolymers II
Volume Editor: Abetz, V.
Vol. 190, 2005

Block Copolymers I
Volume Editor: Abetz, V.
Vol. 189, 2005

Intrinsic Molecular Mobility and Toughness of Polymers II
Volume Editor: Kausch, H.-H.
Vol. 188, 2005

Intrinsic Molecular Mobility and Toughness of Polymers I
Volume Editor: Kausch, H.-H.
Vol. 187, 2005

Polysaccharides I
Structure, Characterization and Use
Volume Editor: Heinze, T.
Vol. 186, 2005

Advanced Computer Simulation Approaches for Soft Matter Sciences II
Volume Editors: Holm, C., Kremer, K.
Vol. 185, 2005

Emissive Materials · Nanomaterials

With contributions by
T. Fuhrmann-Lieker · A. C. Grimsdale · J. Jang · M. Kaneko
K. Müllen · R. Pudzich · J. Salbeck · M. Yagi

The series *Advances in Polymer Science* presents critical reviews of the present and future trends in polymer and biopolymer science including chemistry, physical chemistry, physics and material science. It is adressed to all scientists at universities and in industry who wish to keep abreast of advances in the topics covered.
As a rule, contributions are specially commissioned. The editors and publishers will, however, always be pleased to receive suggestions and supplementary information. Papers are accepted for *Advances in Polymer Science* in English.
In references *Advances in Polymer Science* is abbreviated *Adv Polym Sci* and is cited as a journal.

Springer WWW home page: http://www.springer.com
Visit the APS content at http://www.springerlink.com/

Library of Congress Control Number: 2006921179

ISSN 0065-3195
ISBN-10 3-540-31250-1 Springer Berlin Heidelberg New York
ISBN-13 978-3-540-31250-5 Springer Berlin Heidelberg New York
DOI 10.1007/11611967

This work is subject to copyright. All rights are reserved, whether the whole or part of the material is concerned, specifically the rights of translation, reprinting, reuse of illustrations, recitation, broadcasting, reproduction on microfilm or in any other way, and storage in data banks. Duplication of this publication or parts thereof is permitted only under the provisions of the German Copyright Law of September 9, 1965, in its current version, and permission for use must always be obtained from Springer. Violations are liable for prosecution under the German Copyright Law.

Springer is a part of Springer Science+Business Media

springer.com

© Springer-Verlag Berlin Heidelberg 2006
Printed in Germany

The use of registered names, trademarks, etc. in this publication does not imply, even in the absence of a specific statement, that such names are exempt from the relevant protective laws and regulations and therefore free for general use.

Cover design: *Design & Production* GmbH, Heidelberg
Typesetting and Production: LE-TEX Jelonek, Schmidt & Vöckler GbR, Leipzig

Printed on acid-free paper   02/3100 YL – 5 4 3 2 1 0

Editorial Board

Prof. Akihiro Abe
Department of Industrial Chemistry
Tokyo Institute of Polytechnics
1583 Iiyama, Atsugi-shi 243-02, Japan
aabe@chem.t-kougei.ac.jp

Prof. A.-C. Albertsson
Department of Polymer Technology
The Royal Institute of Technology
10044 Stockholm, Sweden
aila@polymer.kth.se

Prof. Ruth Duncan
Welsh School of Pharmacy
Cardiff University
Redwood Building
King Edward VII Avenue
Cardiff CF 10 3XF, UK
DuncanR@cf.ac.uk

Prof. Karel Dušek
Institute of Macromolecular Chemistry,
Czech
Academy of Sciences of the Czech Republic
Heyrovský Sq. 2
16206 Prague 6, Czech Republic
dusek@imc.cas.cz

Prof. W. H. de Jeu
FOM-Institute AMOLF
Kruislaan 407
1098 SJ Amsterdam, The Netherlands
dejeu@amolf.nl
and Dutch Polymer Institute
Eindhoven University of Technology
PO Box 513
5600 MB Eindhoven, The Netherlands

Prof. Jean-François Joanny
Physicochimie Curie
Institut Curie section recherche
26 rue d'Ulm
75248 Paris cedex 05, France
jean-francois.joanny@curie.fr

Prof. Hans-Henning Kausch
Ecole Polytechnique Fédérale de Lausanne
Science de Base
Station 6
1015 Lausanne, Switzerland
kausch.cully@bluewin.ch

Prof. Shiro Kobayashi
R & D Center for Bio-based Materials
Kyoto Institute of Technology
Matsugasaki, Sakyo-ku
Kyoto 606-8585, Japan
kobayash@kit.ac.jp

Prof. Kwang-Sup Lee
Department of Polymer Science &
Engineering
Hannam University
133 Ojung-Dong
Daejeon 306-791, Korea
kslee@hannam.ac.kr

Prof. L. Leibler
Matière Molle et Chimie
Ecole Supérieure de Physique
et Chimie Industrielles (ESPCI)
10 rue Vauquelin
75231 Paris Cedex 05, France
ludwik.leibler@espci.fr

Prof. Timothy E. Long
Department of Chemistry
and Research Institute
Virginia Tech
2110 Hahn Hall (0344)
Blacksburg, VA 24061, USA
telong@vt.edu

Prof. Ian Manners
School of Chemistry
University of Bristol
Cantock's Close
BS8 1TS Bristol, UK
ian.manners@bristol.ac.uk

Prof. Martin Möller
Deutsches Wollforschungsinstitut
an der RWTH Aachen e.V.
Pauwelsstraße 8
52056 Aachen, Germany
moeller@dwi.rwth-aachen.de

Prof. Oskar Nuyken
Lehrstuhl für Makromolekulare Stoffe
TU München
Lichtenbergstr. 4
85747 Garching, Germany
oskar.nuyken@ch.tum.de

Prof. E. M. Terentjev
Cavendish Laboratory
Madingley Road
Cambridge CB 3 OHE, UK
emt1000@cam.ac.uk

Prof. Brigitte Voit
Institut für Polymerforschung Dresden
Hohe Straße 6
01069 Dresden, Germany
voit@ipfdd.de

Prof. Gerhard Wegner
Max-Planck-Institut
für Polymerforschung
Ackermannweg 10
Postfach 3148
55128 Mainz, Germany
wegner@mpip-mainz.mpg.de

Prof. Ulrich Wiesner
Materials Science & Engineering
Cornell University
329 Bard Hall
Ithaca, NY 14853, USA
ubw1@cornell.edu

Advances in Polymer Science
Also Available Electronically

For all customers who have a standing order to Advances in Polymer Science, we offer the electronic version via SpringerLink free of charge. Please contact your librarian who can receive a password or free access to the full articles by registering at:

springerlink.com

If you do not have a subscription, you can still view the tables of contents of the volumes and the abstract of each article by going to the SpringerLink Homepage, clicking on "Browse by Online Libraries", then "Chemical Sciences", and finally choose Advances in Polymer Science.

You will find information about the

– Editorial Board
– Aims and Scope
– Instructions for Authors
– Sample Contribution

at springer.com using the search function.

Contents

Polyphenylene-type Emissive Materials: Poly(*para*-phenylene)s,
Polyfluorenes, and Ladder Polymers
A. C. Grimsdale · K. Müllen 1

Spiro Compounds for Organic Electroluminescence and Related Applications
R. Pudzich · T. Fuhrmann-Lieker · J. Salbeck 83

Charge Transport and Catalysis by Molecules Confined in Polymeric Materials
and Application to Future Nanodevices for Energy Conversion
M. Yagi · M. Kaneko 143

Conducting Polymer Nanomaterials and Their Applications
J. Jang 189

Author Index Volumes 101–199 261

Subject Index 287

Polyphenylene-type Emissive Materials: Poly(*para*-phenylene)s, Polyfluorenes, and Ladder Polymers

Andrew C. Grimsdale · Klaus Müllen (✉)

Max-Planck-Institut für Polymerforschung, Ackermannweg 10, 55128 Mainz, Germany
muellen@mpip-mainz.mpg.de

1	Introduction	3
2	**Poly(*para*-phenylene)s**	6
2.1	Poly(*para*-phenylene) (PPP)	6
2.1.1	PPP by Coupling of Benzene Rings	6
2.1.2	Precursor Routes to PPP	8
2.2	Poly(*para*-phenylene)s with Solubilising Substituents	9
2.2.1	Synthetic Routes to Substituted PPPs	9
2.2.2	Luminescent Properties of Soluble PPPs	12
2.2.3	Copolymers with Partial Substitution	16
2.2.4	Blends of PPPs with Other Polymers	18
3	**Ladder-type Poly(*para*-phenylene)s**	19
3.1	Ladder-type PPPs with Methine Bridges	20
3.2	Blends of LPPPs with Other Polymers	24
3.3	Defect Emission from LPPPs	25
3.4	Ladder-type PPPs with Two-atom Bridges	26
3.5	Copolymers of LPPP and PPP – 'Stepladder' Copolymers	29
4	**Stepladder-type Poly(*para*-phenylene)s**	30
4.1	Polyfluorenes	30
4.1.1	Synthetic Routes to Polyfluorenes	31
4.1.2	Optical Properties of PDAFs	32
4.1.3	PFs with Substituted Alkyl Side Chains	34
4.1.4	Defect Emission from PDAFs	35
4.1.5	Colour Control by Copolymerisation	39
4.1.6	Dendronised Polyfluorenes – Towards Stable Blue Emission	43
4.1.7	Polyfluorenes with Improved Charge Injection	48
4.1.8	Alternating Copolymers of Fluorenes and Other Arylenes	54
4.1.9	Polymers Containing Spirobifluorene Units	56
4.2	Poly(indenofluorene)s	58
4.3	Poly(ladder-type pentaphenylene)s	61
4.4	Other 'Stepladder'-type Polyphenylenes	62
4.4.1	Poly(tetrahydropyrene)s	62
4.4.2	Polycarbazoles	63

5	Polyphenylene-based Block Copolymers	65
5.1	Rod–Coil Block Copolymers	66
5.2	Rod–Rod Block Copolymers	72
6	Conclusion	73
	References	74

Abstract This chapter reviews the synthesis of the various classes of polyphenylene-based materials that have been investigated as active materials in light-emitting applications. In particular, it is shown how the electronic properties may be controlled by synthetic design. Insoluble poly(*para*-phenylene) can be made by a variety of precursor routes. Attachment of solubilising side chains gives soluble polyphenylenes in which a high degree of torsion between adjacent benzene rings produced by steric interactions between the substituents strongly reduces their electronic interaction. The phenylene units can be made coplanar by bridging them with methine or other bridges to produce ladder-type polymers, which show excellent photophysical properties, but strong intermolecular interactions lead to problems in obtaining blue emission. Similar problems are seen for 'stepladder' polymers such as polyfluorenes with only partial bridging of the phenylene rings. These interactions may be controlled by introduction of bulky substituents. The electroluminescence efficiency of these materials can also be enhanced by use of charge-transporting substituents. Copolymerisation with lower-band-gap units enables tuning of the emission colour across the entire visible range.

Keywords Poly(*para*-phenylene)s · Ladder polymers · Photoluminescence · Electroluminescence · Light-emitting diodes

Abbreviations

Ac	acetyl
AIBN	azabis(isobutyronitrile)
Boc	*tert*-butoxycarbonyl
Bu	butyl
sec-Bu	*sec*-butyl
*t*Bu	*tert*-butyl
Cp	cyclopentadienyl
DMSO	dimethyl sulfoxide
ECL	effective conjugation length
EL	electroluminescence
eV	electron volt
HOMO	highest occupied molecular orbital
IR	infrared
ITO	indium tin oxide
LB	Langmuir–Blodgett
LEC	light-emitting electrochemical cell
LED	light-emitting diode
LPPP	ladder-type poly(*para*-phenylene)
LUMO	lowest unoccupied molecular orbital
MALDI-TOF	matrix assisted laser desorption ionisation–time of flight mass spectrometry

Me	methyl
Me-LPPP	methyl-substituted LPPP
M_n	number-averaged molecular weight
M_w	weight-averaged molecular weight
NMR	nuclear magnetic resonance spectroscopy
PANI	polyaniline
PDAF	poly(dialkylfluorene)
PEDOT	poly(3,4-ethylenedioxythiophene)
PEO	poly(ethylene oxide)
PF	polyfluorene
Ph	phenyl
Ph-LPPP	phenyl-substituted LPPP
PIF	polyindenofluorene
PMMA	poly(methyl methacrylate)
PPP	poly(*para*-phenylene)
PPV	poly(*para*-phenylene vinylene)
PS	polystyrene
PSS	poly(styrene sulfonate)
PVK	poly(*N*-vinyl carbazole)
TGA	thermal gravimetric analysis
UV-VIS	ultraviolet–visible

1
Introduction

Polyphenylenes are one of the most important classes of conjugated polymers and have been the subject of extensive research, particularly as active materials for use in light-emitting diodes (LEDs) [1, 2] and polymer lasers [3]. These materials have been of particular interest as potential blue emitters in such devices. The discovery of stable blue-light-emitting materials is a major goal of research into luminescent polymers [4]. In this chapter we review the synthesis of polyphenylene-based materials for light-emitting applications, with a particular emphasis on how the properties of the materials may be tuned by synthetic design. The classes of material that will be covered in this review (Scheme 1) are polyphenylenes including poly(*para*-phenylene) (PPP, **1**) and soluble PPPs (**2**), in which there is only a single bond between each adjacent pair of phenylene units, ladder-type PPPs (LPPPs, **3**), in which all the phenylene units are tied together in a coplanar fashion by methine bridges, and so-called 'stepladder' polymers such as polyfluorenes (PFs, **4**) and polyindenofluorenes (PIFs, **5**), in which only some of the phenylenes are linked by methine bridges. In this introduction we present an overview of the important properties of these materials that need to be controlled in order to obtain efficient LEDs, of the general principles of how these properties may be controlled by synthetic or other methods, and of the general synthetic methods available for the preparation of these polymers. We will then present a more

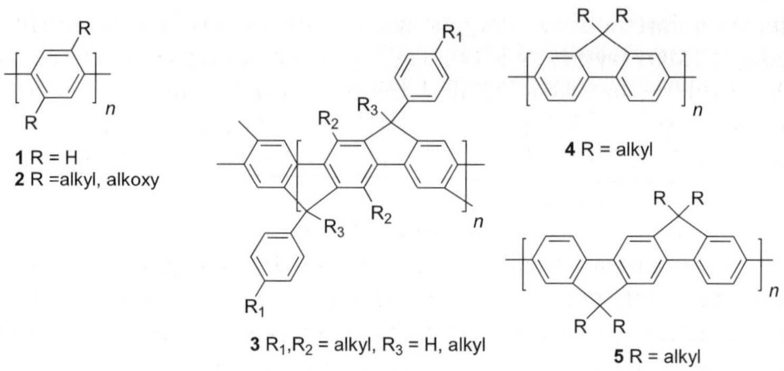

Scheme 1 Typical polyphenylene-based conjugated polymers

detailed discussion of the synthesis and optimisation of polymer properties for each class of material.

There are several properties of luminescent materials that need to be controlled in order to make efficient LEDs and lasers. The first is the colour of the emission, which is primarily determined by the energy difference (band gap) between the highest occupied molecular orbital (HOMO) and the lowest unoccupied molecular orbital (LUMO), but in the solid state is also affected by interactions between the molecules or polymer chains which can lead to red shifts in the emission due to formation of aggregates. This can be controlled by manipulating both the polymer backbone and the substituents. Polyphenylenes are intrinsically blue-emitting materials with large HOMO-LUMO gaps but, as we will show, by copolymerisation with other materials it is possible to tune the emission colour across the entire visible spectrum. Even without incorporation of comonomers it is possible to tune the emission colour over a substantial range by controlling the conjugation length through restriction of the torsion of adjacent phenylene rings. Thus substitution of PPP (**1**) with solubilising groups to give soluble PPPs (**2**) causes a blue shift in the emission as the steric interactions between the side chains induce increased out-of-plane twisting of the phenylene units, while the enforced coplanarity produced by the methine bridges in LPPP (**3**) results in a marked red shift in the emission. The 'stepladder' polymers such as PFs (**4**) and PIFs (**5**) show emission colours intermediate between those of **2** and **3**. As already mentioned, solid-state interactions between polymer chains can cause red shifts in the emission from these materials, which can be suppressed by suitable choice of substituents. This can however adversely affect the charge-transporting properties of the material (see below).

The second critical property to be tuned is the efficiency of charge injection, which is determined by the energy barrier between the HOMO and the anode (for hole injection) and between the LUMO and the cathode (for electron injection), and of charge transport which is controlled by intermolecular or interchain interactions. Polyphenylenes are materials with intrinsically low-lying HOMOs (typically 5.8–6.0 eV), which creates a large barrier to hole injection from the most widely used anode material, indium tin oxide (ITO), which has a work function of 4.8–5.0 eV. The LUMO values are typically around 2.2–2.5 eV, which makes electron injection from air-stable metals like aluminium (work function 4.3 eV) difficult, thus requiring the use of more electropositive metals such as calcium (work function 2.9 eV) as cathodes. Obviously, one can improve the charge injection by raising the HOMO and/or lowering the LUMO energy of the polymer, but in doing so one reduces the size of the energy gap and so red shifts the emission colour. As a result, obtaining efficient blue emission is a particular problem.

The efficiency of devices can be increased by the incorporation of layers of charge-injecting (or level-matching) materials which have energy levels intermediate between that of the emissive layer and the work function of the electrode, but the use of such layers has the disadvantage of increasing the device thickness which increases the driving voltage, and also complicates device fabrication as successive layers have to be deposited in ways such that the lower layers are not disturbed by deposition of the upper ones. Blending charge- transporting materials into the emissive layer leads to problems with phase separation giving unstable device performance. Incorporation of charge-accepting units into the polymer backbone or onto the side chains avoids both these problems. This approach has been used to successfully improve both the hole- and the electron-accepting properties of phenylene-based polymers. Good charge transport in the solid state requires close packing of polymer chains permitting rapid and efficient hopping of charges between chains. As mentioned above, strong interchain interactions can cause undesirable red shifts in the emission spectrum, so that it is sometimes necessary to compromise the charge-transport properties of the material to obtain the desired emission colour.

Other desirable properties are the ability to form defect-free films, preferably by solution-processing techniques, a high solid-state photoluminescence (PL) quantum efficiency, and good stability towards oxygen and light. For some applications the ability to obtain polarised light is also desirable. All of these properties are to some extent controllable by design of the structure and the synthetic pathway. The formation of a thick, uniform defect-free film by spin coating or similar solution-processing methods is dependent upon many factors. The first is the molecular mass of the compound, as low molar mass materials tend to be crystalline and so do not form high-quality amorphous films, while very high molar mass polymers are difficult to dissolve. The second is the solubility. To obtain a good film, the material must be reasonably soluble as too dilute solutions give too thin films, and must not form aggregates in solution as these will tend to lower the film quality (uneven morphology) and may produce red shifts in the emission. The exact PL efficiency of a given material is as yet not predictable, but the removal of fluorescence quenching defects, such as halide atoms or carbonyl groups, and the suppression of interchain interactions leading to non-radiative decay pathways, are known to assist in improving solid-state quantum efficiency of materials. The stability of a polymer towards photo-oxidation can be improved by avoiding susceptible functional groups, e.g. benzyl protons. Polarised emission is obtained by alignment of the polymer chains, which seems to be easiest for polymers which possess a liquid-crystalline mesophase [5]. Circularly polarised emission has been obtained by using chiral side chains.

The synthesis of polyphenylenes has been reviewed most recently by Kaeriyama [6] and Scherf [7]. There exist three general methods for the synthesis of polyphenylene-based materials: (a) aromatisation of poly(cyclohexa-1,3-diene) precursor polymers, which is used only to make poly(*para*-phenylene (PPP, 1) and rod–coil copolymers containing PPP rods; (b) oxidative coupling of monomers, which is of strictly limited synthetic utility as only low molecular mass materials are obtained from such methods; and (c) transition-mediated polycondensations of substituted aromatic compounds and/or aryl organometallic compounds. This last is the main method for preparing phenylene-based polymers. The two main polycondensation methods used are the Suzuki polycondensation of aryl halides with arylboronic acids [8] and the Yamamoto polymerisation of aryl dihalides using nickel(0) reagents [9]. Generally speaking, the Suzuki method gives higher molecular masses than the Yamamoto procedure, but is synthetically more demanding. A more detailed comparison of the relative merits of the two methods will be given in the discussion of the synthesis of soluble PPPs (Sect. 2.2, below). Other coupling reactions, e.g. Stille coupling of aryl halides and aryl tin reagents [10] or Kumada coupling of aryl halides with aryl Grignard reagents [11], have also been used to make polyphenylenes, but generally give lower molecular masses than the Suzuki or Yamamoto methods.

2
Poly(*para*-phenylene)s

2.1
Poly(*para*-phenylene) (PPP)

2.1.1
PPP by Coupling of Benzene Rings

Unsubstituted PPP (1) is insoluble, and so direct synthesis by oxidation of benzene by the Kovacic method [12] or by nickel(0) coupling of 1,4-dihalobenzenes [9] (Scheme 2) gives insoluble and intractable powders with degrees of polymerisation of about 20 as determined by IR absorption meas-

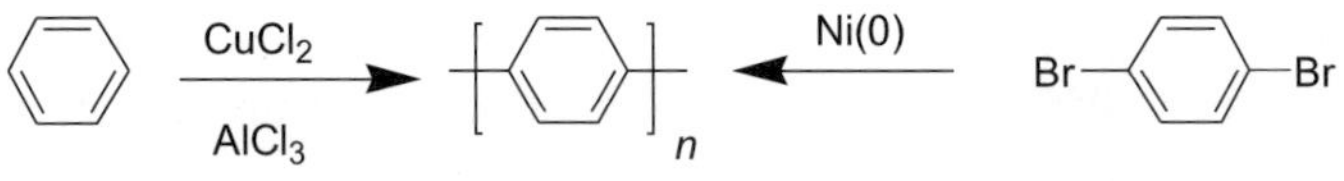

Scheme 2 Direct routes to PPP

urements [13]. The material obtained by the Kovacic method contains a large amount of defects due to 1,2-coupling and/or formation of condensed polyaromatic units, while the Yamamoto method gives only 1,4-coupling.

Films of oligophenylenes with an average of about nine benzene rings (**1**, $n = 9$) can be prepared from these powders by vacuum deposition [13, 14]. Due to the difficulty of obtaining pure oligomers of PPP with more than six benzene rings, the effective conjugation lengths (ECLs) for absorption and emission of PPP (i.e. the lengths beyond which increasing the length of the polymer chain produces no further red shift in absorption or emission) have not been determined. A study of oligomers with solubilising substituents (**2**, $n = 3-17$) determined the ECLs for such compounds to be 11 phenyl rings for absorption and seven rings for emission [15]. As substituted PPPs (**2**) have shorter conjugation lengths than PPP (**1**) due to the steric interactions between the substituents leading to greater out-of-plane twisting of the conjugated backbone, the ECLs for **1** can be expected to be longer. Related polymers such as ladder-type PPPs (**3**, Sect. 3.1), polyfluorenes (**4**, Sect. 4.1), and polyindenofluorenes (**5**, Sect. 4.2) in which the torsion angles between adjacent units are similar to or smaller than in PPP all show ECLs for emission of between 11 and 15 phenyl rings. The EL emission maximum from devices using vacuum-deposited films of PPP oligomers ($n \sim 9$) occurs at 446 nm [16–18]. Comparison of the EL maxima from these devices and those using high molecular weight PPP made by a precursor route (see below) suggests that the ECL for emission from **1** is greater than 10 benzene rings. Insertion of a poly(N-vinylcarbazole) (PVK) charge-transporting layer improves the efficiency tenfold, but at low voltages there is considerable emission ($\lambda_{max} = 550$ nm) from an excimer between PVK and the oligophenylenes [18]. As the voltage increases this decreases as the recombination zone shifts into the bulk of the oligophenylene layer, thus giving voltage-tunable EL. Leising and coworkers have used well-defined films of a hexaphenyl oligomer (**1**, $n = 6$, sexiphenyl) to make blue-emitting devices, with an emission maximum at 425 nm [19–26]. Comparison of the EL spectra of sexiphenyl films with the orientation of the oligomers variously parallel or perpendicular to the substrate shows that the latter gives brighter emission with a narrower spectrum [27]. Polarised emission has been reported from oriented films of sexiphenyl [28, 29]. The Leising group have made red- and green-emitting devices using thin films of sexiphenyl covered with appropriate dyes to convert its blue emission [23, 30–35]. White emission is obtainable by appropriate colour mixing [35].

2.1.2
Precursor Routes to PPP

Films of PPP have to be prepared via precursor routes (previously reviewed by Gin and Conticello [36]). The route most often used to prepare films of PPP (**1**) is one developed at ICI (Scheme 3) [37, 38]. This starts with a microbial oxidation of benzene to cyclohexadienediol **6**. Radical-initiated polymerisation of the diacetate **7** gives the precursor polymer **8**, which is then thermally converted to **1**. However, the material is not stereoregular as it contains about 10–15% of 1,2-linkages. This material has been used by Leising et al. to prepare blue-emitting LEDs (λ_{max} = 459 nm) with efficiencies of up to 0.05% [39–41].

A number of other diesters of **6** have been polymerised by the same method (again with 10–15% of 1,2-linkages in the polymer) and thermally converted to PPP, with the highest molecular weights being obtained with dimethyl carbonate and dipivaloyl esters [42]. Copolymers with blue-shifted PL spectra can be prepared by copolymerisation of **7** and 10 mol % of vinylbiphenyl or *N*-vinylcarbazole [43, 44]. The blue shift is due to interruption of the conjugation by the non-aromatic units.

Totally stereoregular PPP (i.e. with all 1,4-linkages) has been prepared by Grubbs and coworkers (Scheme 4) by a stereospecific nickel-catalysed polymerisation of a cyclohexadienediol disilylether **9**, followed by conversion of the resulting polymer **10** to **1** via the acetoxy-precursor **8** [45–48]. However,

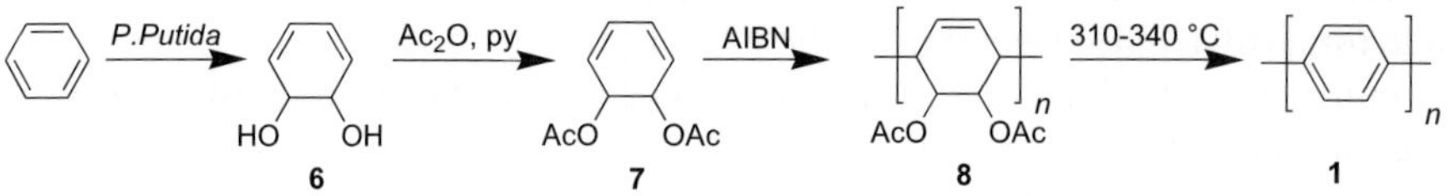

Scheme 3 ICI route to PPP

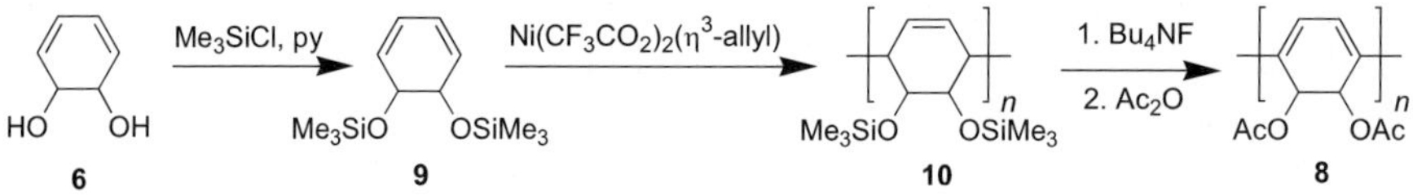

Scheme 4 Grubbs route to regioregular PPP

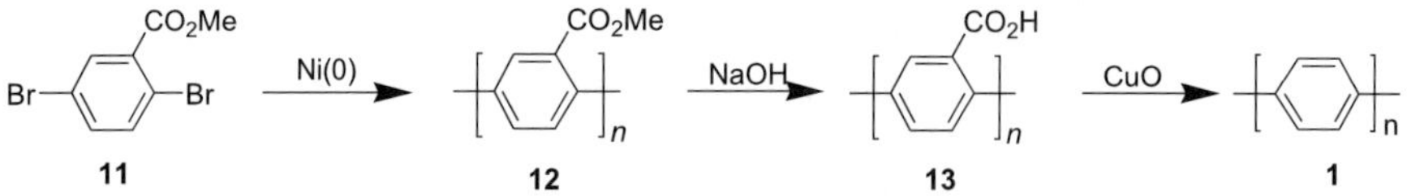

Scheme 5 Kaeriyama precursor route to PPP

to assist processing they used an acid catalyst in the final step, which badly contaminated their product, making it unsuitable for use in LEDs.

Another method for preparing all 1,4-linked PPP is that of Kaeriyama and coworkers (Scheme 5) [49, 50]. Yamamoto polycondensation of methyl 2,5-bromobenzoate 11 produced a soluble all-para polymer 12, which was hydrolysed to the polyacid 13 and then decarboxylated to give PPP. However, the decarboxylation cannot be satisfactorily performed in the solid state, so this method is unsuitable for preparing high-quality films.

PPP films can also be prepared by electropolymerisation under either reductive or oxidative conditions, but the EL properties have been found to be highly dependent on the polymerisation conditions [51]. A study of the PL efficiency of PPP thin films of varying chain length concluded that for highly ordered PPP films a chain length of 25–30 units was optimal [52]. Oriented films of PPP have been prepared by a friction deposition method and found to show highly polarised fluorescence [53].

2.2
Poly(*para*-phenylene)s with Solubilising Substituents

2.2.1
Synthetic Routes to Substituted PPPs

Yamamoto and coworkers prepared a PPP derivative 14 by treatment of 1 (prepared by the Yamamoto method [54]) with perfluorpropylperoxide (Scheme 6). From NMR and other analysis they estimated that the average chain length was 13 benzene units with an average of two perfluorpropyl units per molecule. This material showed blue PL (λ_{max} = 450 nm), and was used to construct a device whose emission colour was found to shift from green to blue with increasing applied voltage [55].

More generally, PPP derivatives 15 with solubilising side chains, most commonly alkyl or alkyoxy groups, are prepared by transition metal catalysed polycondensations [56]. The main methods used are the Suzuki and Yamamoto polycondensations. Two variations of the Suzuki polycondensation can be used. The simpler, so-called 'AB coupling' (Scheme 7), involves conversion of a dibromobenzene 16 into a bromoarylboronic acid 17, which is then homocoupled to give 15. Random copolymers can be made by using a mixture of monomers.

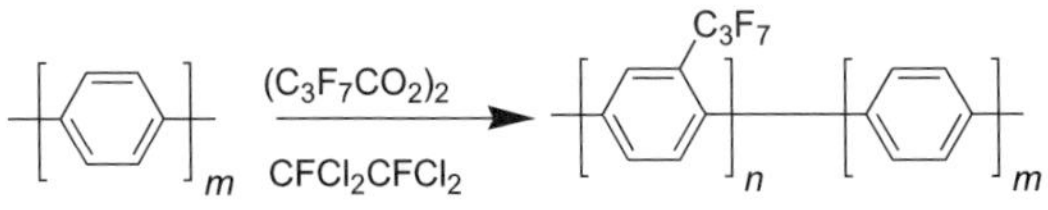

Scheme 6 Perfluorpropylation of PPP

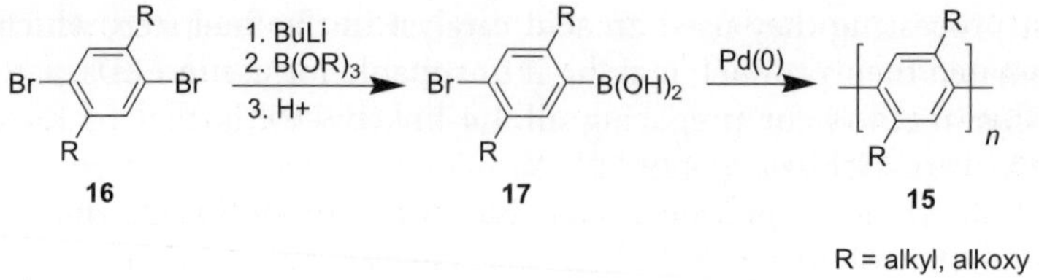

Scheme 7 Soluble PPPs by AB-type Suzuki polycondensation

In the other 'AA-BB coupling' method (Scheme 8) a phenylbisboronic acid **18** is coupled with a dihalobenzene **16**. This method has the advantage of permitting the preparation of alternating copolymers **19**, but there can be experimental difficulties as exactly equimolar amounts of **16** and **18** are required for optimal polymerisation. In particular, boronic acids are hygroscopic, may be contaminated with significant amounts of the anhydrides, and are often difficult to purify. As a result boronic esters are frequently preferred as reagents, even though their preparation involves an extra step, as they are usually easier to purify and handle.

Considerable work in optimising the Suzuki polycondensation methods for making polyphenylenes has been done, particularly by the Wegner and Schlüter groups [57–64]. For a fuller description of the scope and problems of Suzuki polycondensation, the reader is referred to the recent review by Schlüter [8].

The Yamamoto method of condensing dihalobenzenes **16** with nickel(0) (Scheme 9) has the advantage of experimental simplicity, but is limited to preparation of homopolymers and random copolymers, and requires stoichiometric amounts of expensive nickel(0) reagents. These can be generated in situ by the reduction of nickel(II) salts in the presence of suitable ligands,

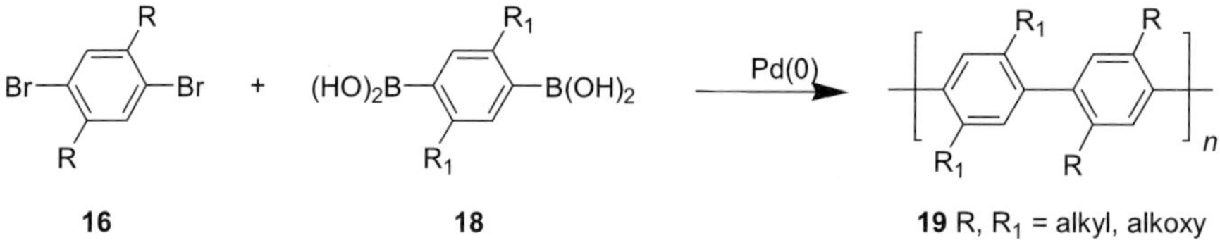

Scheme 8 Soluble PPPs by AA-BB-type Suzuki polycondensation

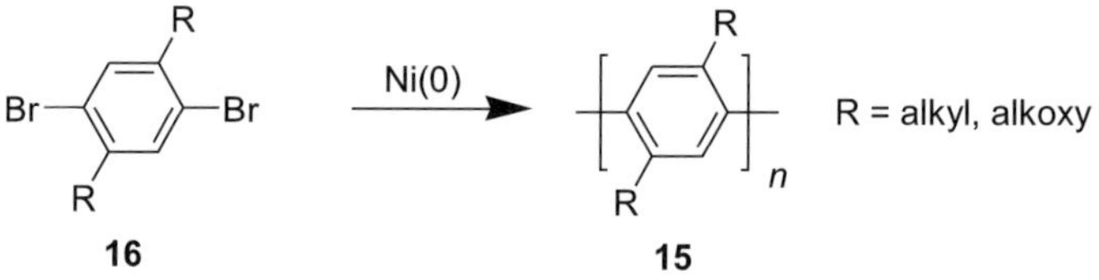

Scheme 9 Soluble PPPs by Yamamoto polycondensation

but the results from such polymerisations tend to be inferior than from the use of commercially available nickel(0) reagents. For a more comprehensive discussion of the reagents and conditions for this method, the reader is referred to the review by Yamamoto [9].

Percec and coworkers have developed a variation of this method in which hydroquinone bistriflates **20** [65–67] or bismesylates **21** [68, 69] are coupled to give alkyl, aryl, or ester functionalised PPPs **22** (Scheme 10). An advantage of this method is that the monomers are easily prepared from hydroquinone.

Poly(2,5-dialkoxy-1,4-phenylene)s can also be made by oxidation of *para*-dialkoxybenzenes (Scheme 11). Thus, 1,4-dimethoxybenzene (**23**) can be polymerised with aluminium chloride and copper(II) chloride or iron(III) chloride. The polymer **24** is only soluble in sulfuric acid, however, and so not usable in LEDs [7]. Oxidation of 1,4-dibutoxybenzene (**25**) with iron(III) chloride by contrast gives a polymer **26** which is soluble in organic solvents [70, 71].

Of these methods it is reported that the Suzuki method gives the highest degrees of polymerisation in alkyl or alkoxy PPPs [72]. In addition to alkyl and alkoxy groups, a wide range of other substituents have been used to solubilise PPPs. In polymers where the rings have two identical substituents, the question of polymer regioregularity does not arise, but in other cases there is the possibility of the units coupling in either a 'head-to-head' or 'head-to-tail' fashion (Scheme 12). The different steric interactions between the substituents on adjacent units may cause different degrees of out-of-plane twisting in the two cases, which will affect the conjugation length, while the differences in the interactions between the chains may cause differences in chain packing, which influences the morphology of the polymer film and the rates of charge-carrier and exciton migration between chains. To date

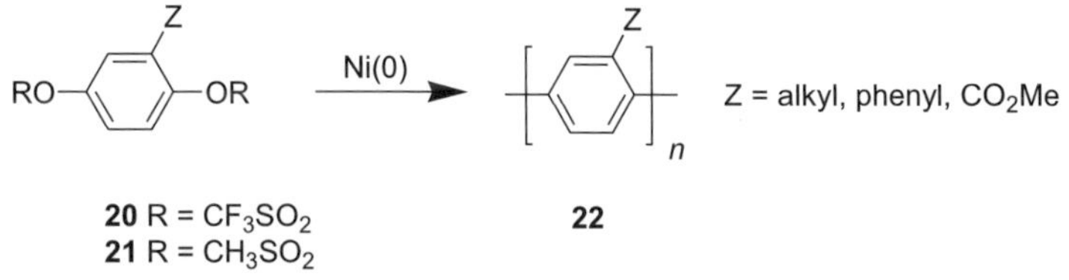

20 R = CF_3SO_2
21 R = CH_3SO_2

22 Z = alkyl, phenyl, CO_2Me

Scheme 10 Percec route to soluble PPPs

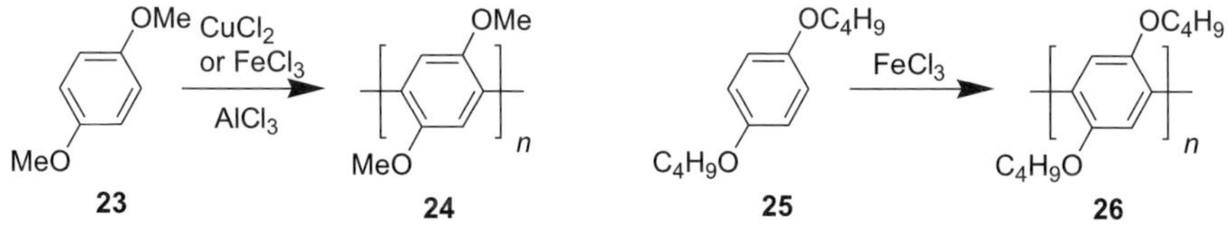

23, **24**, **25**, **26**

Scheme 11 Alkoxy-substituted PPPs by oxidative polymerisation

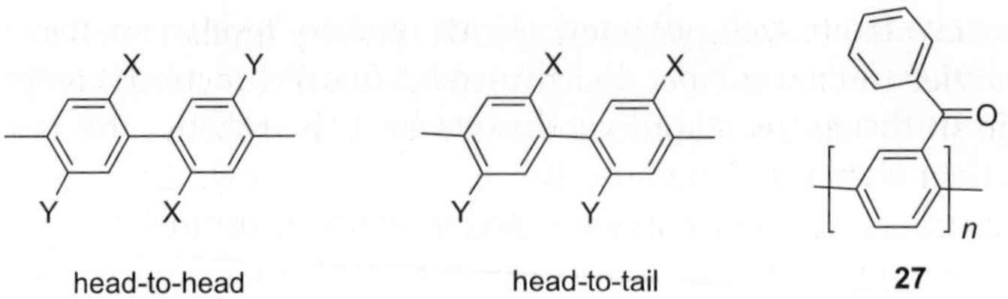

Scheme 12 Regioisomerism in substituted PPPs

the only report of the synthetic control of the regioregularity of a substituted PPP influencing the properties comes from the synthesis of poly(2-benzoyl-1,4-phenylene) (**27**) by Yamamoto polycondensation [73]. When the polycondensation was performed with 2,2-bipyridyl as a ligand, the resulting polymer was found to have a sharply red-shifted UV absorption maximum (λ_{max} = 352 nm versus 328 nm) indicating a much longer conjugation length, which was attributed to the material being much more head-to-tail regioregular.

2.2.2
Luminescent Properties of Soluble PPPs

These polymers show blue PL emission and many of them have been used by various groups to make LEDs (Scheme 13) [71, 74–89]. The EL spectra are often red shifted compared to the PL spectra and, due to formation of aggregates which show longer-wavelength, usually yellow, emission the overall EL emission colour is not always blue, but may be green or even white, particularly after operation of the device for some time. Efficiencies from single-layer devices are about 0.05% with aluminium cathodes (work function 4.3 eV), and up to 1.8% with more electropositive calcium cathodes (work function 2.9 eV), due to the smaller energy barrier between the LUMO ($\sim$ 2.2–2.5 eV) and the electrode. By using a hole-injecting layer of PVK (work function 5.5 eV), efficiencies of up to 3% have been obtained using calcium cathodes, and of up to 0.8% using other, more air-stable metal cathodes. This suggests that the main limiting factor for emission efficiency in these materials is

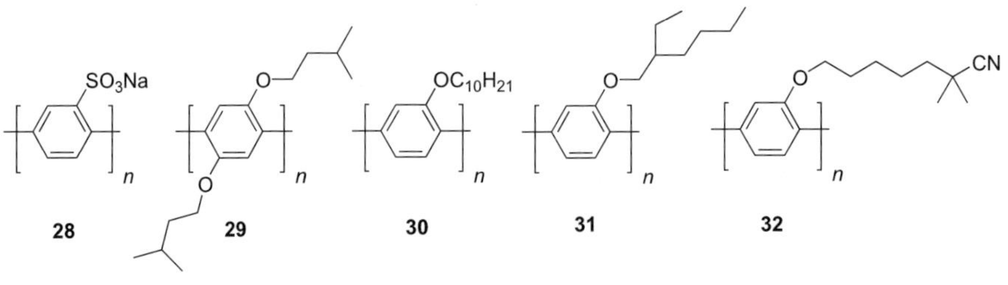

Scheme 13 Typical substituted PPPs used in blue-emitting devices

hole injection due to their very low HOMO energy levels (~ 6.0 eV) creating a large barrier to charge injection from ITO (work function 4.8–5.0 eV). No systematic study has been done on the effects of the substituents upon the EL efficiency, but Neher and coworkers reported that single-layer devices using the sulfonium-substituted polymer **28** gave efficiencies of 0.5–0.8% with an aluminium cathode, which is an order of magnitude higher than for the dialkoxysubstituted polymer **29** in identical devices [80]. Heeger and coworkers [77] have compared the PL and EL efficiencies of the mono-substituted polymers **30–32**. They showed identical PL efficiencies in solution (85%), but in the solid state the polymers **31** and **32** with branched side chains showed higher efficiencies (40% and 46%, respectively) than the polymer **30**, which has straight alkyl chains (35%). This may reflect less efficient polymer chain packing in the former, due to the bulkier substituents, leading to less exciton migration and non-radiative decay. Conversely, the EL efficiency of double-layer devices using **30** (3.0%) was higher than for **31** (2.0%) or **32** (1.4%).

Use of sulfonium or other ionic substituents also gives solubility in very polar solvents including water [90–93]. Though water is not a desirable solvent for processing materials for use in electronic devices due to the danger of corrosion of the electrodes, the ability to process from highly polar solvents such as ethanol can be advantageous in constructing multi-layer devices as many organic electronic materials have very limited solubility in such solvents, thus allowing an ethanol-soluble material to be deposited from solution on top of such a material. Use of a polyaniline (PANI) anode or blending of the PPP derivative with a hole-transporting material lowers the operating voltage [77, 78].

Polarised EL emission has been obtained from a device using polymer **29** deposited as a Langmuir–Blodgett (LB) film [79, 81]. In contrast to the blue emission obtained from spin-coated films, the emission from the LB films is mainly yellow due to the formation of aggregates, with the emission parallel to the dipping direction ($\lambda_{max} = 536$ nm) slightly red shifted compared to the perpendicular emission ($\lambda_{max} = 524$ nm). Polarised EL emission has also been obtained by rubbing alignment of a film of **33** (Scheme 14) [94]. Another way to induce alignment is to incorporate mesogenic units in the side chains. Thus, polarised PL has been obtained from films of polymer **34** containing liquid-crystalline cyanobiphenyl group substituents [95–97].

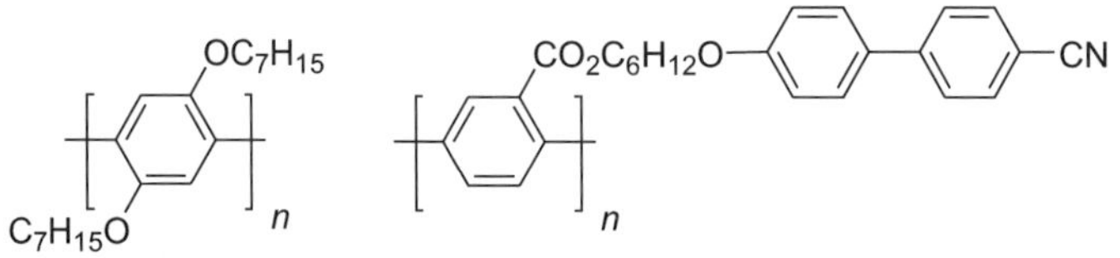

Scheme 14 PPPs from which polarised emission has been obtained

Circularly polarised emission is possible from polymers containing chiral groups. Scherf and coworkers have prepared a cyclophane-substituted PPP by the Suzuki route using the dibromocyclophane **35** and the corresponding bisboronic acid **36** (Scheme 15) [98, 99]. If racemic monomers were used the resulting polymer **37** was not chiral with the cyclophanes randomly distributed on either face of the polymer (atactic). If resolved enantio-pure monomers were used, then the stereoregular isotactic **38** or syndiotactic **39** polymers could be obtained depending upon which enantiomer of each monomer was used. The isotactic polymer is chiral and both enantiomers have been prepared.

Not all the substituents need to be chiral to achieve overall chirality in the polymer. Thus, the copolymer **40** (Scheme 16) containing only 5% chiral units shows a Cotton effect in the circular dichroism spectrum [100]. Circularly polarised PL has been obtained from a copolymer **41** containing both mesogenic and chiral side chains [101].

Complexation of a PPP **42** having carboxylate substituents (Scheme 17) with cyclodextrin gives a fluorescent polyrotaxane which shows a PL

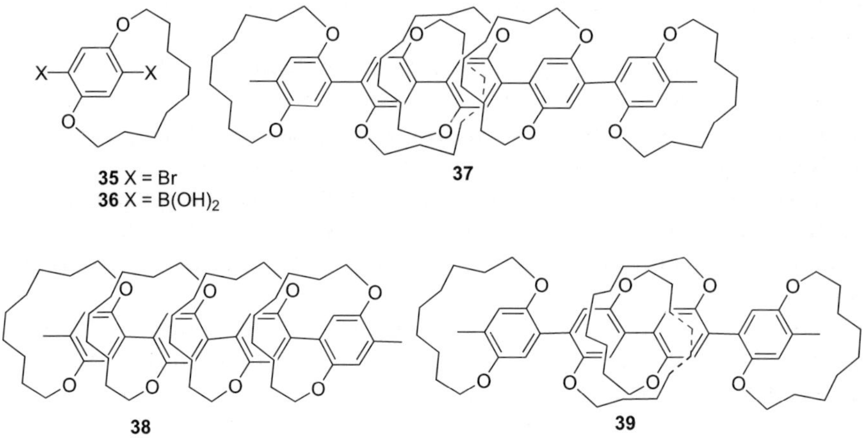

Scheme 15 Stereoisomeric cyclophane-substituted PPPs

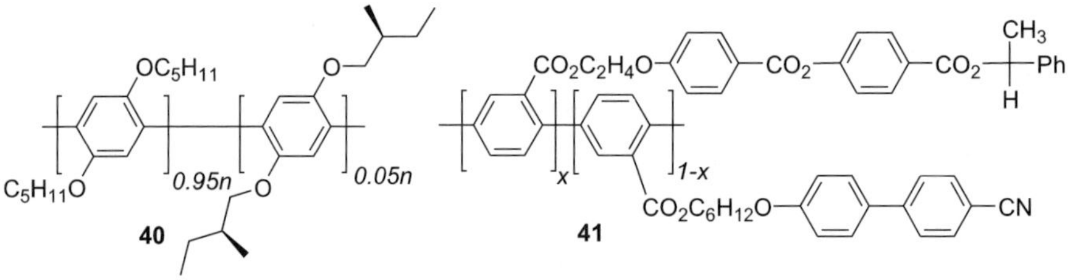

Scheme 16 Other PPPs producing circularly polarised emission

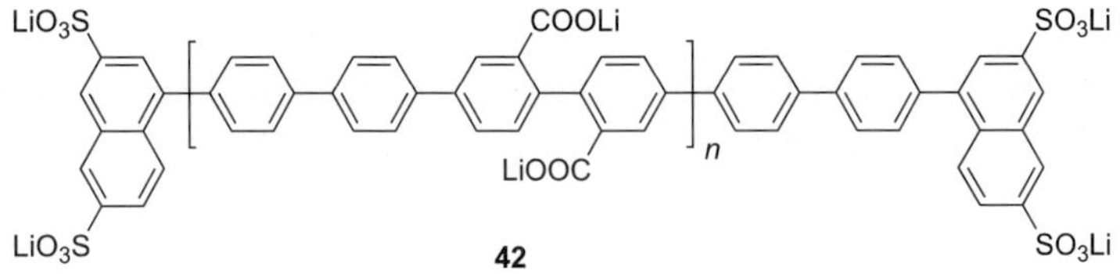

Scheme 17 Polycarboxylate used to make fluorescent polyrotaxane

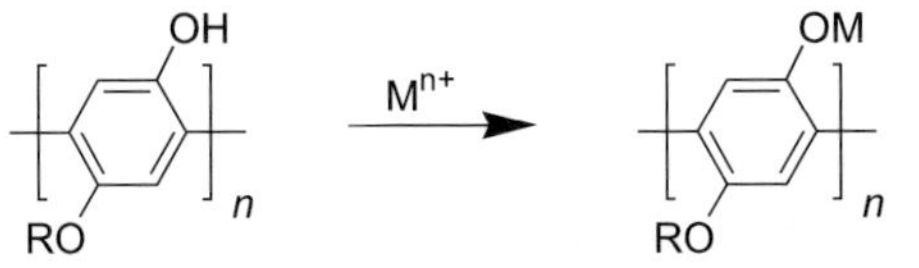

43 R = $C_{12}H_{25}$
44 R = $C_{16}H_{33}$
45 R = $C_{18}H_{37}$

M = Na^+, Co^{2+}, Cu^{2+}, Fe^{3+}

Scheme 18 Complexation of polyphenols with metal ions

maximum at $\lambda_{max} = 410$ nm compared to 430 nm for the uncomplexed chain [102]. The EL for the rotaxane is reported to be slightly red shifted compared with the PL [103].

A red shift in the PL emission from the polyphenols **43–45** has been observed upon complexation with metal ions (Scheme 18) [104]. The uncomplexed polymers show violet PL ($\lambda_{max} = 402$ nm) in THF solution. Treatment with one equivalent of sodium hydroxide solution produces a marked red shift to give blue-green emission, together with a large drop in the PL intensity. The degree of red shifting depends upon the chain length of the alkoxy substituents, with the red shifts observed for **43** ($\lambda_{max} = 474$ nm) and **44** ($\lambda_{max} = 479$ nm) being much larger than for **45** ($\lambda_{max} = 461$ nm). Similar effects are seen upon addition of methanolic solutions of metal ions to solutions of the polymers in THF. Thus, complexation of **43** with cobalt(II) or copper(II) shifts the PL maximum into the blue ($\lambda_{max} = 436$ nm), while the iron(III) complex emits in the green ($\lambda_{max} = 509$ nm). Here the length of the side chain has an even bigger effect on the size of the red shift, as complexation of **44** with copper produces blue-green ($\lambda_{max} = 471$ nm) PL, with the cobalt and iron complexes emitting in the green ($\lambda_{max} = 471$ nm) and the yellow ($\lambda_{max} = 551$ nm) regions of the spectrum, respectively. Polymer **45** shows smaller red shifts than **44** upon complexation with copper ($\lambda_{max} = 499$ nm) or iron ($\lambda_{max} = 519$ nm), but a greater shift with cobalt ($\lambda_{max} = 489$ nm). Presumably these effects reflect either changes in the planarity of the polymer chain or emission from aggregates brought about by the complexation. There is no report of whether similar effects are seen in the solid-state spectra of these materials, but this approach is an excellent example of how control-

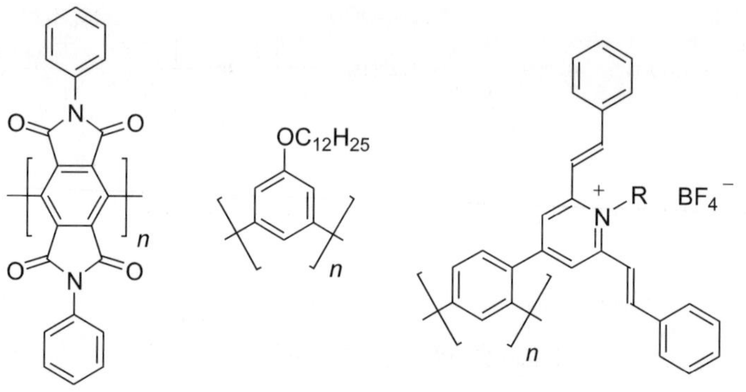

Scheme 19 Soluble PPPs with very short conjugation lengths

ling interchain interactions can profoundly influence the optical properties of a conjugated polymer.

A major feature of these materials is that steric interactions between the solubilising substituents lead to an increased phenylene–phenylene torsion from 23° in PPP to around 60–80° in polymers with substituents in the 2- and 5- positions [105]. As a result, the conjugation along the backbone is much reduced, so that their absorption and emission are blue shifted compared with PPP. Their PL emission is thus largely in the violet with a maximum typically between 400 and 420 nm. As mentioned in Sect. 2.1 above, the effective conjugation lengths for the poly(dialkoxyphenylene) **29** have been determined to be 11 phenyl rings for absorption and seven rings for emission, which are somewhat lower than for other polyphenylene-based materials [15].

An exception is the bisimide **46** (Scheme 19), which shows green PL (λ_{max} = 553 nm) that is similar to that of the monomer, suggesting that the emission comes from isolated monomer units [106]. A soluble poly(*meta*-phenylene) **47**, made by Reddinger and Reynolds by the Yamamoto method, which has an even shorter conjugation length, emits mainly in the ultraviolet (λ_{max} = 346 nm) [107]. Other *meta*-linked polymers **48**, however, show violet-green PL (λ_{max} = 445–532 nm) as the emission comes from the substituents [108].

2.2.3
Copolymers with Partial Substitution

One way to red shift the emission is to make copolymers with only partial substitution. Holmes and coworkers prepared the random copolymer **49** with 33% of unsubstituted phenylene units ($m:n = 2:1$) by copolymerisation of the substituted and unsubstituted bromobenzene boronic acids (Scheme 20) [75]. The PL emission was blue ($\lambda_{max} \sim 420$ nm), but the EL was

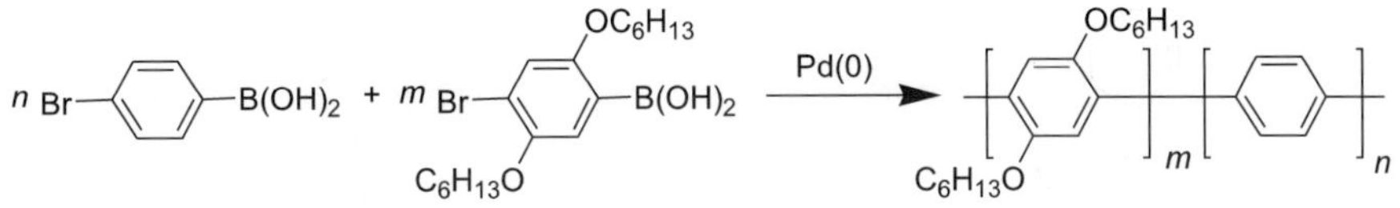

49

Scheme 20 Partially substituted PPP random copolymers by Suzuki polycondensation

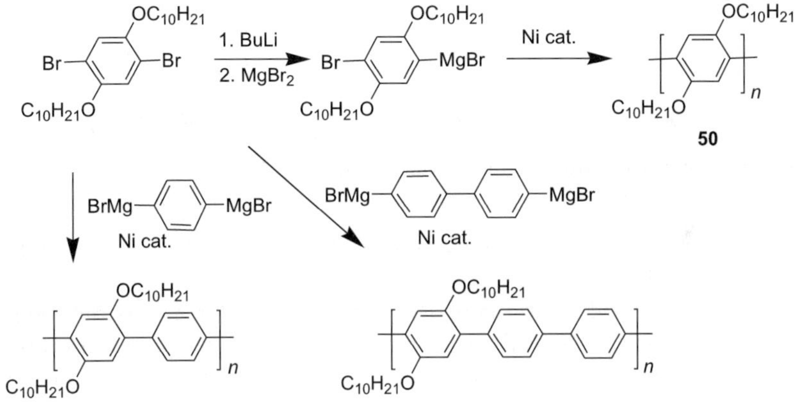

51 52

Scheme 21 PPPs with varying degrees of substitution by Grignard coupling

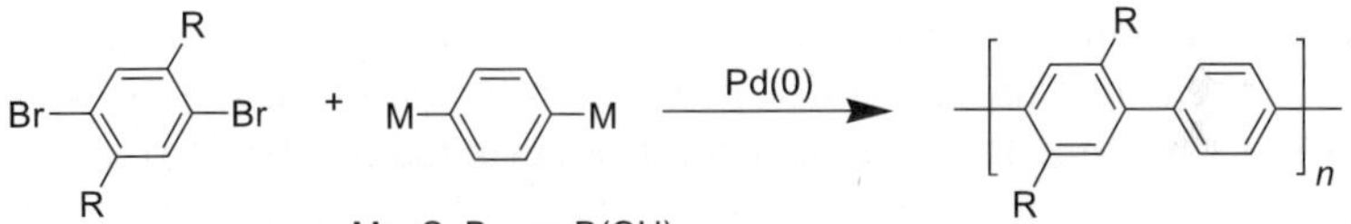

M = $SnBu_3$ or $B(OH)_2$

53 R = alkyl, alkoxy

Scheme 22 Partially substituted PPP alternating copolymers by other routes

white with the emission maximum being red shifted by about 70 nm, and a broad featureless tail up to 800 nm being seen, which was attributed to emission from excimers.

Fu has prepared the homopolymer **50** and the copolymers **51–52** by Grignard coupling (Scheme 21) [88]. While **50** shows violet emission (λ_{max} = 415 nm), the PL from **51** and **52** is blue-green (λ_{max} = 450, 500 nm).

Copolymers **53** with alternating substituted and unsubstituted phenylenes have also been made by Stille [109, 110] or Suzuki coupling (Scheme 22) [72, 111, 112]. These copolymers show violet to blue (λ_{max} = 370–425 nm) fluorescence. The emission from the dialkoxy-substituted polymers is red shifted by ~ 50 nm compared with their dialkyl analogues [89, 109–111]. These copoly-

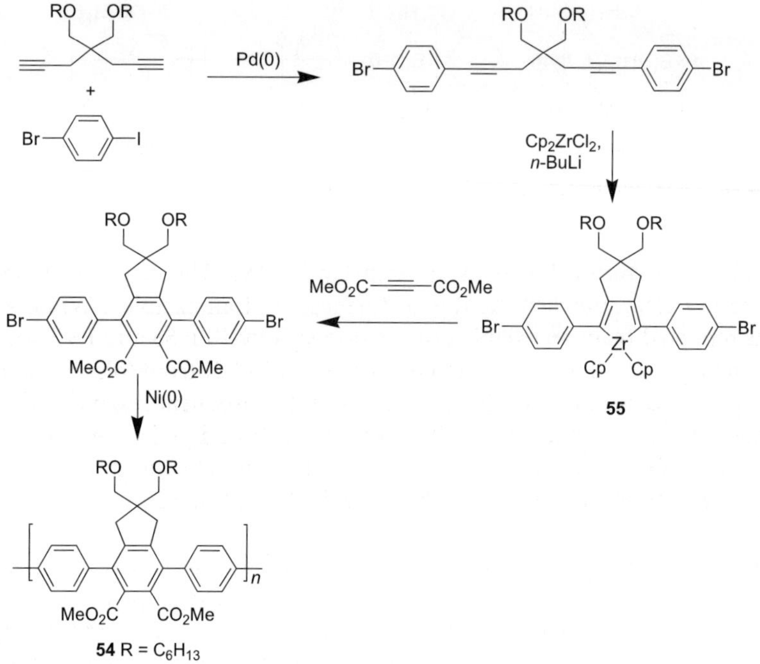

Scheme 23 Zirconocene-precursor route to substituted PPPs

mers also show less red-shifted emission due to aggregates than the corresponding homopolymers [89].

The triphenylene monomer for polymer **54** is made via a zirconcene precursor **55** (Scheme 23). This is a versatile intermediate as the zirconium can be displaced by a variety of reagents, thus permitting the synthesis of a range of monomers. The polymer shows blue emission in THF solution with a maximum in the violet at 376 nm, suggesting that there is particularly large torsion between the substituted and unsubstituted rings, and a long tail into the green, perhaps due to aggregates [113].

2.2.4
Blends of PPPs with Other Polymers

Blending also provides a method for tuning the emission from substituted PPPs. Thus, Salaneck and coworkers found that blending the violet-blue-emitting copolymer **56** (λ_{max} = 389, 443 nm) with the blue-green emitter **57** (λ_{max} = 479 nm) (Scheme 24) gave rise to blue EL (λ_{max} = 460 nm) with an optimal efficiency of 1.9% for a blend containing 10 wt % of **57** [114–116].

Edwards et al. reported that blending **27** (Scheme 12) with PVK produced a red shift in the EL with the maximum moving from λ_{max} = 433 nm to 446 nm [117]. A similar red shift in the EL emission (λ_{max} = 448 nm) has also

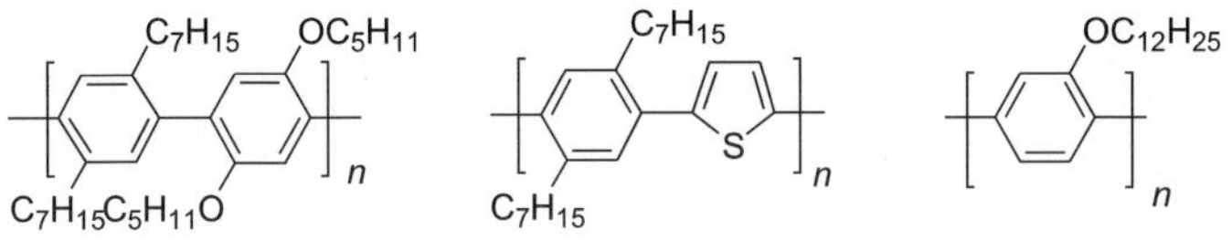

Scheme 24 Phenylene-based polymers used in blends in LEDs

been obtained from a blend of **15** with PVK [118]. The cause of these red shifts is not clear, but may be due to formation of an exciplex. White emission is reported from an exciplex of the dodecyloxy-PPP **58** and PVK formed when the layer of **58** is spin coated onto a layer of PVK from toluene solution due to partial dissolution of the lower layer and consequent mixing of the two polymers at the interface [85]. The spectrum is broad and covers the range from 400 to 700 nm, with maxima at λ_{max} = 495 and 533 nm. If the PPP layer is deposited instead from hexane, in which PVK is insoluble, the EL emission is blue (λ_{max} = 412 nm).

Whereas the above examples show a red shift in the emission by blending with a lower-band-gap material, a blue shift from λ_{max} = 430 nm to λ_{max} = 400 nm has been obtained in the EL emission of **29** (Scheme 13) by blending it with poly(methylphenylsilane)s, whose PL emission is in the near ultraviolet [89]. This is because the polysilanes prevent aggregation of the emitting polymer chains, and so suppress the long-wavelength emission seen from films of pure **29**. The emission colour is also stabler with none of the red shifting seen for the pristine polymer during device operation. A second effect of this better confinement of the excitons is that the EL efficiency is increased by up to 30 times due to suppression of the non-radiative decay pathways.

3
Ladder-type Poly(*para*-phenylene)s

An obvious approach to overcoming the problem of phenylene–phenylene torsion in substituted PPPs is to tether adjacent rings together with short alkyl bridges to make a ladder-type polymer. Ladder-type polymers are of considerable scientific interest as they are intermediate between linear and three-dimensional materials [119]. They can be prepared in two ways: (a) by iterative multi-centre condensation or addition (e.g. Diels–Alder cycloaddition) reactions; (b) by polymer-analogous conversion of suitably functionalised single-stranded precursors. A major feature of the second method is that the polymer-analogous reactions must proceed quantitatively to avoid formation of defects in the final polymer. Though ribbon-like polyacenes can be prepared by polycycloaddition methods [119], linear ladder-type PPPs are only accessible through the conversion of single-stranded PPPs. If methine

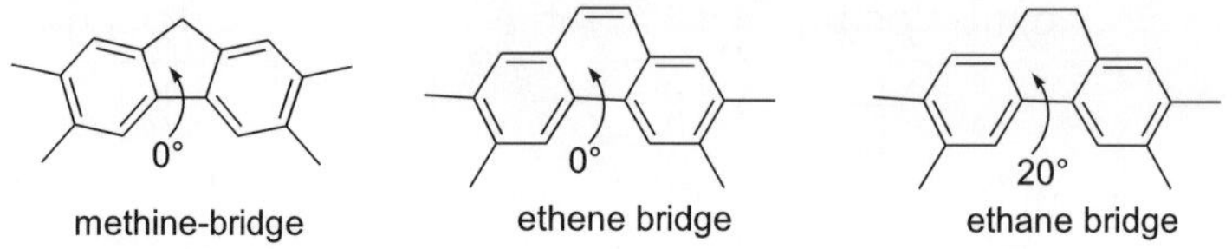

Scheme 25 Bridges and associated torsion angles in ladder polymers

or ethene bridges are used the phenylene backbone is forced to be coplanar, but use of ethane or longer alkyne bridges allows some torsion between adjacent phenylene rings (Scheme 25). For an ethane bridge the torsion angle is predicted to be about 20° [120]. Thus, some degree of control over the optical properties can be achieved by varying the type of bridge(s) used, as the more coplanar the polymer the greater the expected degree of conjugation, and thus the longer the wavelengths of the absorption and emission maxima. The methine and ethene bridges also impart greater rigidity to the structure and thus reduce the Stokes shift between the absorption and emission maxima.

3.1
Ladder-type PPPs with Methine Bridges

Scherf and Müllen prepared (Scheme 26) the ladder-type polyphenylene (LPPP, **3**) with methine bridges [121–124], via a poly(diacylphenylene-*co*-phenylene) precursor copolymer **59** obtained by an AA-BB-type Suzuki polycondensation. The key step is the polymer-analogous Friedel–Crafts ring-closing reaction on the polyalcohol **60**, obtained by the reduction of **59**. This was found to proceed quickly and smoothly upon addition of boron trifluoride to a solution of **60** in dichloromethane. The reaction appeared to be complete by both NMR and MALDI-TOF analysis, indicating the presence of less than 1% of defects due to incomplete ring closure. LPPPs with number-average molecular weights (M_n) of up to 50 000 g/mol have been obtained corresponding to about 150 phenylene rings.

A chiral LPPP **61** (Scheme 27) containing cyclophane units has been prepared by using the resolved cyclophane bisboronic acid **36** (Scheme 15) [98, 99, 125]. This is a potential candidate for obtaining circularly polarised EL.

In order to determine the effective conjugation length of **3**, oligomers with between three (**62**) and seven (**63**) benzene rings (Scheme 28) were prepared by adding a suitable amount of a monofunctional end-capping reagent to the Suzuki polycondensation shown in Scheme 26, separating the resulting oligophenylene precursors, and then performing the reduction and ring closing on them [126]. By extrapolation of a plot of their UV-VIS absorption maxima (in eV) against the reciprocal number of benzene rings, the effective conjugation length for absorption in LPPPs was estimated to be about 11 phenyl rings [120].

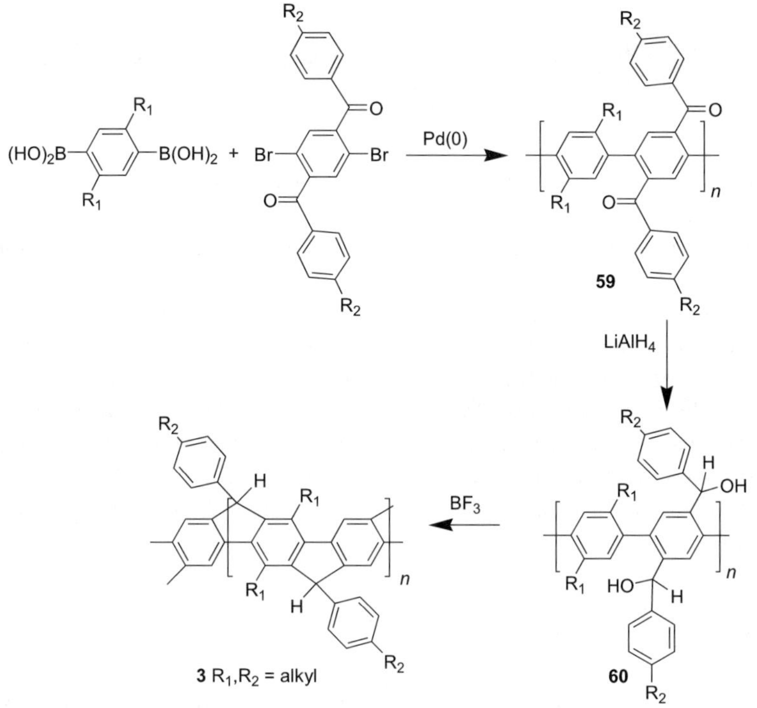

Scheme 26 Synthesis of ladder-type PPP

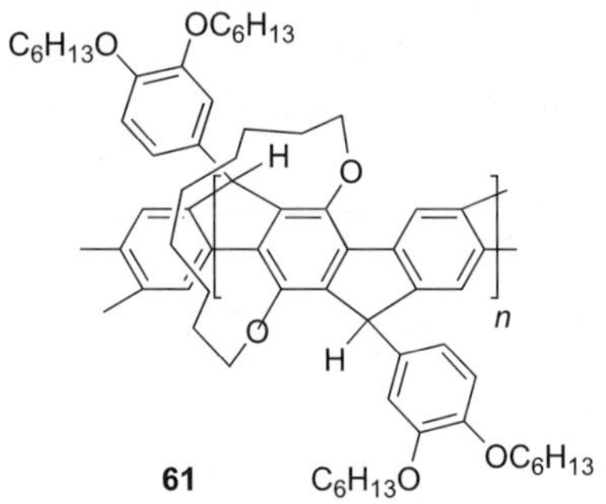

Scheme 27 Chiral ladder-type PPP

By comparison, the effective conjugation length for absorption in poly-(tetrahydropyrene)s **64** (see Sect. 4.3, below), which have an estimated 20° torsion angle between adjacent phenylene rings, was found to be about 19 phenyl rings. Thus, contrary to expectation it has been found that increased planarisation of the aromatic π-system leads to a decrease and not an increase in the effective conjugation length for absorption in PPP derivatives [120].

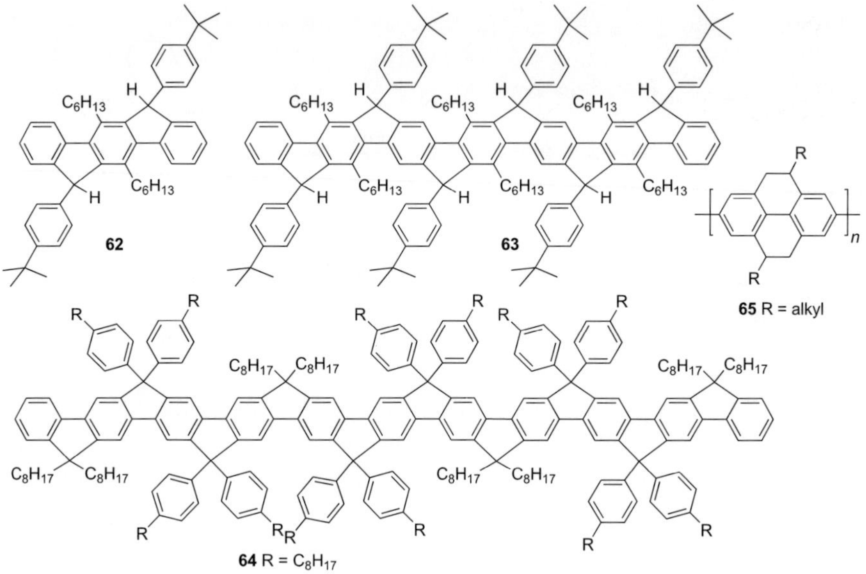

Scheme 28 Oligomers of LPPP used to determine the effective conjugation length

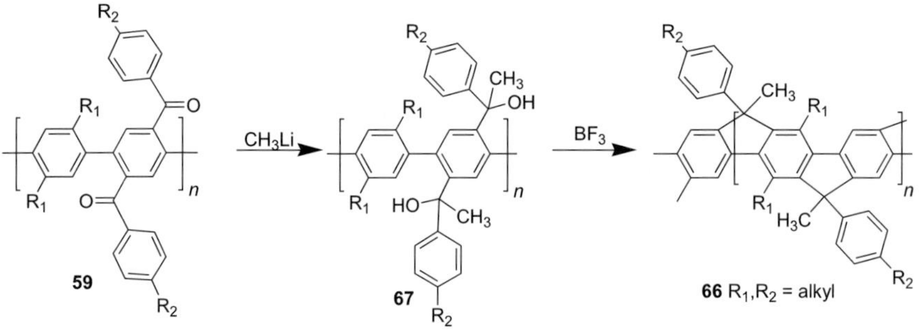

Scheme 29 Synthesis of Me-LPPP

An oligomer **65** with 11 benzene rings has been prepared [127], and its emission spectrum was found to be only slightly blue shifted compared with Me-LPPP (**66**, Scheme 29). Single molecule spectroscopy studies of **65** and **66** showed that the emission from the former at 451 nm matched that from the smallest emissive chromophores observed in the latter, but that the emission maximum from the polymer at around 460 nm comes from longer segments containing probably 14–15 benzene rings. The effective conjugation length for emission in LPPPs is thus similar to the values seen for polyfluorenes and polyindenofluorenes (see 4.1 and 4.2, below).

The planarisation of the PPP backbone in LPPP (**3**) has been found to lead to better vibrational resolution in both absorption and emission spectra and to a much smaller Stokes shift [19]. The absorption maximum is at 440–450 nm, which is considerably bathochromically shifted with regard to single-stranded PPP. The absorption band also shows an unusually sharp absorption edge. The PL of **3** is an intense blue colour in solution with a maximum at 450–460 nm. The Stokes shift is thus only about 150 cm^{-1}. Such a small value is a clear indication of how the rigidity of the polymer hinders deformation in going from the ground to the excited state. A further result of this rigidity is that the PL quantum efficiencies in solution of LPPPs are very high (up to 90%), as non-radiative decay pathways are seriously reduced [128].

While the PL from solutions of LPPP is blue, in thin films the emission is dominated by a broad, featureless band in the yellow (λ_{max} = 600 nm) [20, 40, 129, 130]. The relative intensity ratio of the blue and yellow bands is strongly dependent upon the method used to prepare the films and varies with solvent and film thickness. The blue band disappears completely upon annealing a film of **3** at 150 °C. As a result, LEDs using **3** show yellow EL [131]. The efficiencies from single-layer devices are 0.4% with calcium cathodes and 0.02% with aluminium cathodes. Double-layer devices using PPV as a hole-transporting layer show 0.6% and 0.04% efficiencies with calcium and aluminium cathodes, respectively. Blue emission (λ_{max} = 450–460 nm) has been observed from LEDs using **3** but has been found to be unstable, with the yellow band rapidly appearing [129].

Originally this yellow emission band was attributed to excimers from aggregates formed by π-stacking of the polymer chains. Evidence supporting this came from photophysical experiments, including site-selective excitation experiments [132], time-resolved PL measurements [132], and photovoltaic experiments [133]. Also consistent with this hypothesis is the obtaining of pure blue EL from blends of **3** (1 wt %) in PVK [131]. The efficiency was 0.15% with a calcium cathode. The emission was found to turn white after only a few tens of minutes of device operation, which was attributed to formation of excimers due to the Joule heat produced by passing of electricity through the device. The EL efficiency of devices using such blends can be improved 2–5 times by use of an oxadiazole electron-transporting layer [134].

The stability of emission from LPPPs can be substantially enhanced by replacement of the hydrogen at the methine bridges with a methyl group to give Me-LPPP (**66**) (Scheme 29) [135]. This is achieved by treating the precursor polymer **59** with methyl lithium, followed by ring closure of the resulting polyalcohol **67** with boron trifluoride as in the preparation of **3**.

The emission from **66** is slightly red shifted compared with **3**, with an emission maximum at λ_{max} = 461 nm and a secondary peak at λ_{max} = 491 nm, so that the emission colour is blue-green. Unlike **3**, Me-LPPP shows almost identical emission from films and solutions, with PL quantum efficiencies of over 90% in solution and up to 60% in the solid state. There is a broad emission band centred at 560 nm, which has been attributed to emission from aggregates [136]. This band is much weaker than the yellow band from LPPP (**3**), and the emission does not change upon annealing. LEDs using **66** produce blue-green emission with EL efficiencies of up to 4% [21, 23, 24, 137, 138]. These high emission efficiencies make Me-LPPP a particularly promising material for use in organic solid-state lasers [3]. Optically pumped lasing has been observed from films of **66** in both waveguide and 'distributed feedback' configurations by the groups of Leising and Lemmer [33, 35, 139–141].

3.2
Blends of LPPPs with Other Polymers

Blends of Me-LPPP, poly(ethylene oxide) (PEO), and lithium triflate have been used as the emissive layer in a light-emitting electrochemical cell (LEC) [142]. The emission efficiency with an aluminium cathode was 0.3%, which is somewhat lower than for the corresponding LED (1%), but the onset voltage was only 2.7 V compared with 12 V for an LED. The emission colour changed rapidly from blue to green due to an increase in the intensity of peaks at 530 and 560 nm. The change in the emission was slower in devices with very low or very high amounts of PEO, which is attributed to the decrease in interaction between **65** and PEO in such blends.

Blends of Me-LPPP (**66**) and the red-emitting polymer **68** (Scheme 30) show predominant emission from the latter even at concentrations of only 0.2% **68**. The emission is orange at 0.2% **68** with an EL efficiency of 1.6% (cf. 1% for **66** and 0.01% for **68**), but drops rapidly at higher concentrations of **68**. The PL efficiency is optimal (41% versus 30–60% for **66** and 11% for **68**) with 0.7% **68** and again drops as the concentration of **68** increases [35, 143, 144]. A blend with 0.05% **68** shows high (0.8%) efficiency white EL emission due to emission from both polymers [33, 35, 144]. Blending the polymers with poly (methyl methacrylate) (PMMA) leads to separation of the emissive polymer chains and less efficient energy transfer so that 0.08% of **68** is required for obtaining white emission, but the EL efficiency is increased to 1.2% [35, 145].

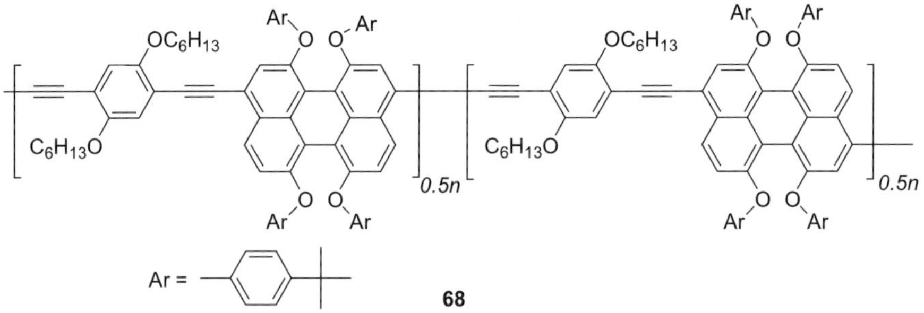

Scheme 30 Perylene-based polymer used in blends with Me-LPPP

3.3
Defect Emission from LPPPs

Lupton [146] has reported that the broad emission feature centred at 560 nm could be detected in the delayed fluorescence from both films and dilute solutions of Me-LPPP. This is not consistent with the suggestion that the long-wavelength emission from ladder-type PPPs comes from aggregates, as dilute solutions should not contain any aggregates. He therefore proposed that the long wavelength emission band originates from defects on the polymer chains.

Convincing evidence has been produced that long-wavelength emission from polyfluorenes is due to fluorenone (**69**) units (see Sect. 4.1, below), so a probable structure for the defects in LPPPs is a ketone as in **70** (Scheme 31). As the ketone in **70** has more extended conjugation than in fluorenone (**69**) one would expect the emission from it to be bathochromically shifted, which is consistent with the long-wavelength emission from LPPPs occurring at 560–600 nm, and that from polyfluorenes at about 530 nm. This is supported by comparison of experimental measurements of **3** and **66** with theoretical calculations of the properties of potential defect structures [147].

The ketone **70** presumably arises from oxidation of the methine bridge by oxygen from the air. The difference between the emission spectra in solution and the solid state would then reflect the more efficient energy transfer to the defect sites in the latter due to the increased intermolecular interactions. The much lower intensity of the defect band and the greater emission stability of Me-LPPP over LPPP can be explained as being due to the greater difficulty in oxidising the methyl-substituted methine bridge, producing a much lower level of defect sites. It is reported that when 9,9-dialkylfluorene bisboronates are substituted for benzene bisboronates in Scheme 28, the resulting polymers display blue-green emission (maxima at 460 nm), which is much stabler than that from **3** [148]. It is suggested that this is because fewer bridges are being formed in the final polymer-analogous reaction, and thus there is less chance of a defect arising from incomplete ring closure.

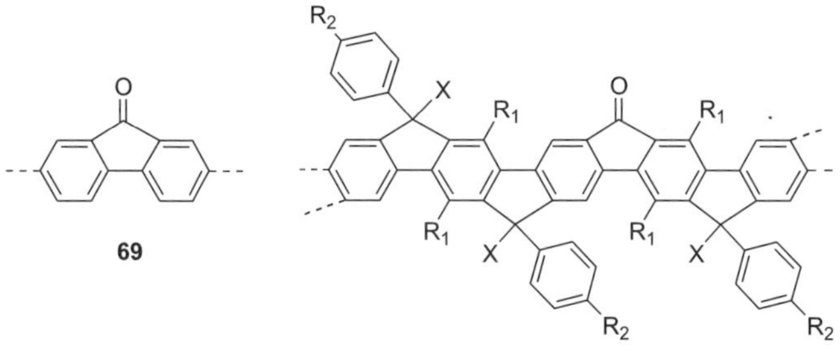

70 R_1, R_2 = alkyl, X = H, CH_3

Scheme 31 Proposed emissive defects in polyfluorenes and ladder-type PPPs

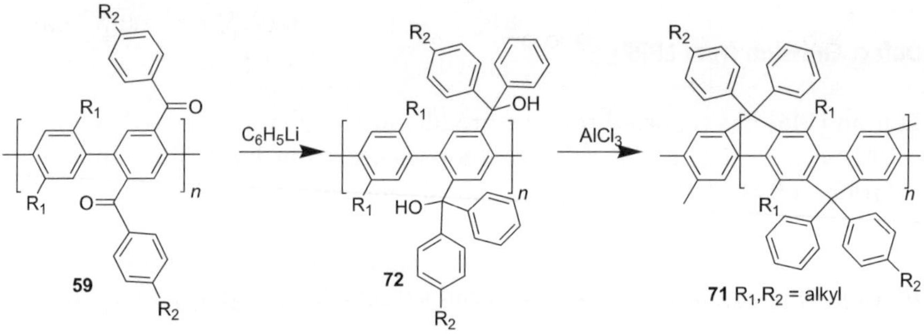

Scheme 32 Synthesis of Ph-LPPP

That even low levels of defects can produce strong emission is exemplified by the case of Ph-LPPP (**71**). The synthesis (Scheme 32) is similar to that of Me-LPPP (**66**), except that complete ring closure of the polyalcohol **71** could not be obtained using boron trifluoride [119]. As a result, other reagents had to be tested and it was found that complete ring closure could be obtained by using aluminium chloride [149].

The PL emission from **71** is very similar to that from **66** with maxima at 460 and 490 nm. However, the EL spectrum shows an additional long-wavelength band. This is not a broad featureless band as seen for the defect emission from **3** or **66**, but one with well-resolved maxima at 600 and 650 nm. Photophysical investigation of this emission showed the feature at 600 nm to be emission from a triplet exciton (phosphorescence) with a vibronic shoulder at 650 nm [150]. Elemental analysis of the polymer showed that it contained 80 ppm of palladium (cf. < 2 ppm in **66**). It was therefore proposed that residues of the palladium catalyst used to make the precursor polymer **59** reacted with the phenyl lithium and the polymer to introduce covalently bound palladium centres onto the polymer chain. These then act as sites for phosphorescent emission.

3.4
Ladder-type PPPs with Two-atom Bridges

The polymers discussed above have methine bridges. A ladder polymer **73** with dihydroxyethane bridges has been made by Forster and Scherf (Scheme 33) [151]. Yamamoto polycondensation of a dibromodibenzoylbenzene **74** gave the poly(diacylphenylene) **75**, which was coupled with samarium(II) iodide coupling to give **73**. This polymer shows strong blue-green fluorescence in solution (λ_{max} = 459 nm) and the solid state (λ_{max} = 482 nm). The red shift in the solid-state PL indicates that **73** is less rigid than the LPPPs with methine bridges, but still shows only weak geometrical changes in going

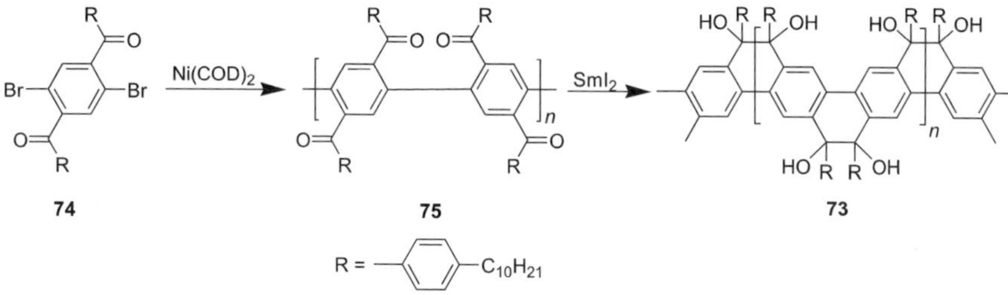

Scheme 33 Synthesis of a ladder-type polyphenylene with dihydroxyethane bridges

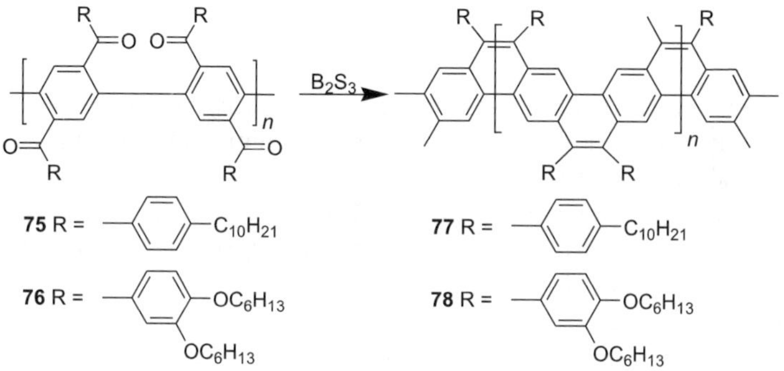

Scheme 34 Synthesis of ladder-type polyphenylene with ethene bridges

from the ground to the excited state. No long wavelength emission band is seen in the PL spectrum of **73**.

Treatment of the poly(dibenzoylphenylene)s **75** and **76** with boron sulfide gives polymers **77** and **78** with ethene bridges (Scheme 34) [152, 153]. These also show blue-green emission (λ_{max} = 478 nm and 484 nm, respectively) with some long-wavelength emission in the solid state which has been attributed to aggregates. Their EL efficiency is reported to be very low (< 0.1%) [154].

A similar polymer **79** was prepared by Goldfinger and Swager by an acid-catalysed cyclisation of a PPP precursor **80** with alkyne side chains (Scheme 35) [155]. There is no report of the emission from this material, but the absorption edge was reported to be at 478 nm, suggesting it should be a blue-green or green emitter.

Tour and Lambda [156, 157] have prepared the aza-ladder polymers **81** from the alternating copolymers **82** (Scheme 36) by treatment with acid to remove the Boc protecting groups and induce imine formation. Unfortunately these materials are only soluble in strongly protic solvents, and there is no report of their luminescent properties. The absorption maxima in protic solvents are around 400 nm, with secondary bands between 510 and 550 nm, while in the solid state maxima between 460 and 490 nm were seen, indicating that protonation has a major effect on the conjugation length.

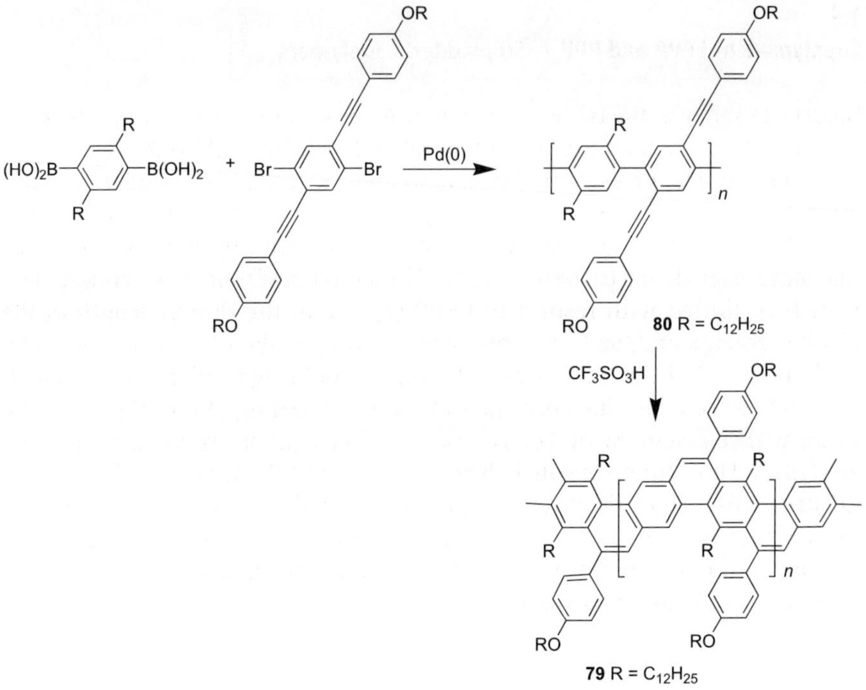

Scheme 35 Ladder-type polyphenylene by alkyne cyclisation route

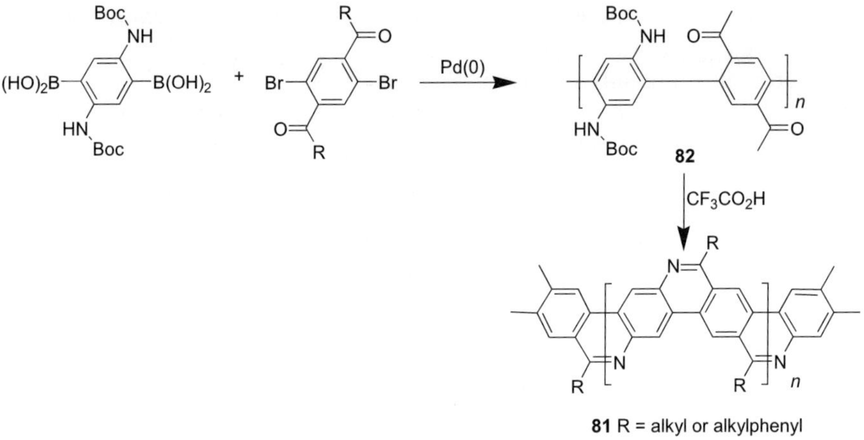

Scheme 36 Synthesis of an aza-bridged ladder polymer

3.5
Copolymers of LPPP and PPP – 'Stepladder' Copolymers

Incorporation of a dialkyldibromo comonomer into the Suzuki copolymerisation used to make the LPPP precursor **59** (Scheme 26), followed by reduction and ring closure, gives statistical copolymers **83** called 'stepladder' copolymers (Scheme 37). These contain oligo-LPPP units connected via twisted phenylene spacers. Copolymers with 40–70% of unbridged phenylene content have been been prepared [129]. The emission from these copolymers **83** is blue shifted with respect to LPPP (**3**) due to the shorter length of the LPPP segments and/or to out-of-plane twisting of the phenylene rings [40, 158]. Films of the copolymers with higher unbridged phenylene content ($m = 50\%+$) remain blue emitting even after annealing [129]. Blue-emitting LEDs with efficiencies of nearly 1% have been made using these materials [159]. This difference in behaviour from LPPP cannot be due to any greater resistance of the copolymers towards oxidation to form defects, but must reflect less efficient exciton diffusion to defect sites due to less close packing of the polymer chains brought about by the introduction of the random twisted phenylene groups.

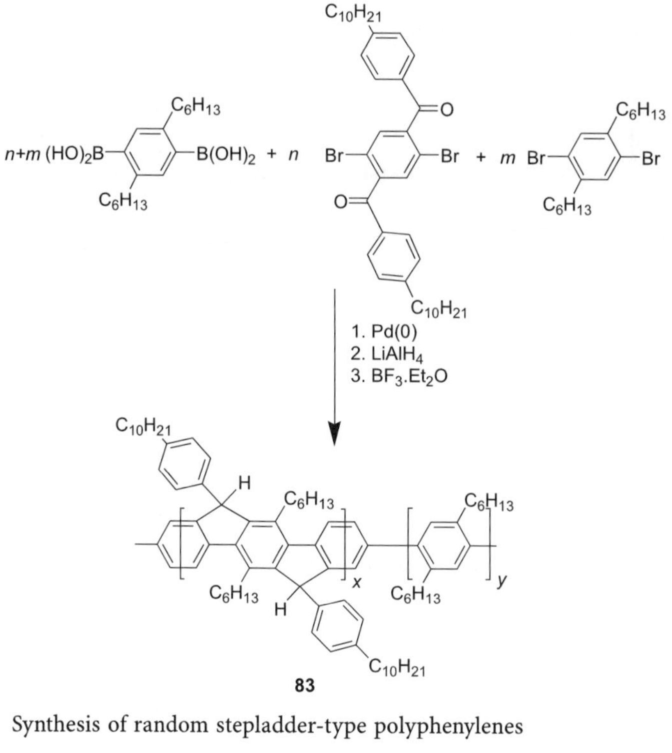

Scheme 37 Synthesis of random stepladder-type polyphenylenes

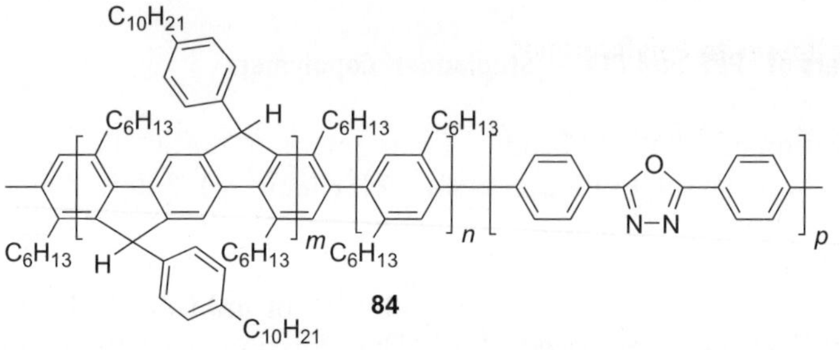

Scheme 38 A stepladder copolymer incorporating charge-transport moieties

A copolymer **84** (Scheme 38) containing dialkyl PPP, ladder-type PPP, and electron-transporting diaryloxadiazole segments ($m:n:p = 4:3:3$) has been prepared by a similar copolymerisation with 2,5-bis(4-bromophenyl)-1,3,4-oxadiazole added as a comonomer [160]. It emits blue light ($\lambda_{max} = 410$, 480 nm) with an efficiency of 0.4% when aluminium cathodes are used [161]. This is twice the efficiency obtained for similar devices using the copolymers **83** without the oxadiazole units.

4
Stepladder-type Poly(*para*-phenylene)s

In addition to the random stepladder polymers described in Sect. 3 above, in which the bridged and unbridged phenylene units are statistically distributed along the polymer chain, there exist regular stepladder-type polyphenylenes. Such polymers are intermediate in structure between PPP and LPPP with a defined number of bridged phenyl rings connected by single-bond linkages. Such materials have been looked at as blue-emitting materials, with the aim of achieving a balance between the excellent photophysical properties of LPPP (small Stokes shift and well-resolved vibronic structure in the emission spectrum) and synthetic accessibility.

4.1
Polyfluorenes

Polyfluorenes (PFs), the simplest regular stepladder-type polyphenylenes, in which only every second ring is bridged, have been much studied in recent years due to their large PL quantum efficiencies and excellent chemical and thermal stability, as evidenced by the number of recent reviews [162–164]. A further attractive feature of poly(9,9-dialkylfluorene)s (PDAFs) is the synthetic accessibility of the monomers, as alkylation and halogenation of fluorene proceed smoothly and in high yields.

4.1.1
Synthetic Routes to Polyfluorenes

Polyfluorenes can be made by oxidative coupling of the monomer with iron(III) chloride. Poly(9,9-dihexyl-2,7-fluorene) (**85**) with a molecular mass of 5000 ($n \sim 20$) was made in this way by Yoshino and coworkers (Scheme 39) [165, 166], and used to make low-efficiency blue-emitting ($\lambda_{max} = 470$ nm) devices [167–169]. These were the first blue-emitting LEDs reported using a phenylene-based polymer.

The disadvantages of this method are that the degree of polymerisation is low, and the high level of defects produced due to coupling other than at the 2- and 7-positions. As a result this method is generally not used to make PFs. A recent exception is the synthesis of the polymer **86** with a cyanoalkyl substituent (Scheme 39) [170]. Here, transition-metal-mediated coupling methods such as Suzuki or Yamamoto polycondensation could not be used as the nitrile deactivates the metal catalysts by binding to the metal.

Advincula et al. oxidised the polyionene **87** with iron(III) chloride to obtain an insoluble polyfluorene network **88** (Scheme 40) [171]. This shows violet PL ($\lambda_{max} \sim 410$ nm), suggesting that the polyfluorene segments are rather short. Due to its insolubility this material cannot be used to make good-quality films for use in LEDs, but the incorporation of a conjugated backbone within a network is one possible way to obtain isolated chains and so avoid the problems associated with interchain interactions, e.g. excimer formation.

Most syntheses of PFs use Suzuki polycondensations or Yamamoto Ni(0) couplings of dibromomonomers. A group at Dow [172–177] have developed a Suzuki cross-coupling route leading to high molecular weight PDAFs (Scheme 41), e.g. poly(9,9-dioctylfluorene) (**89**) with $M_n > 100\,000$ g/mol was obtained after less than 24-h reaction time. Bradley and coworkers [178] used **89** made by this method to make a blue ($\lambda_{max} = 436$ nm) LED with relatively high (0.2%) efficiency being obtained when a hole-transporting layer was used.

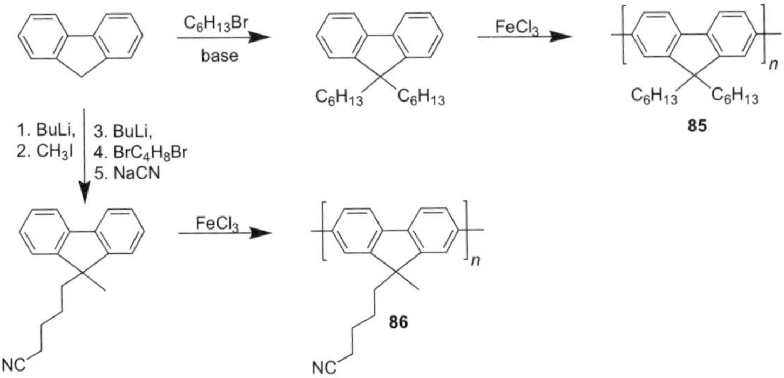

Scheme 39 Synthesis of polyfluorenes by oxidative polymerisation

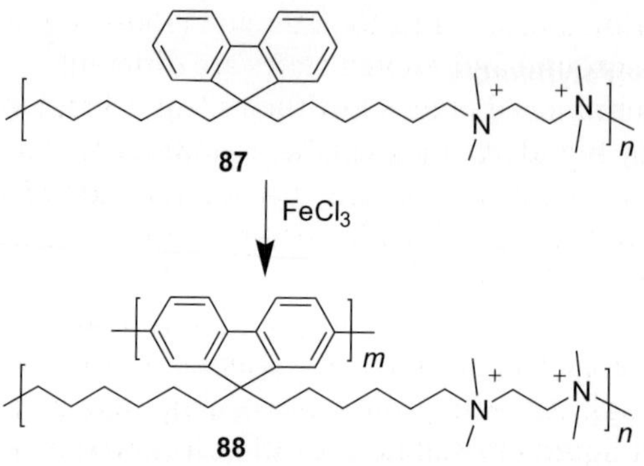

Scheme 40 Synthesis of a polyfluorene network by oxidative coupling

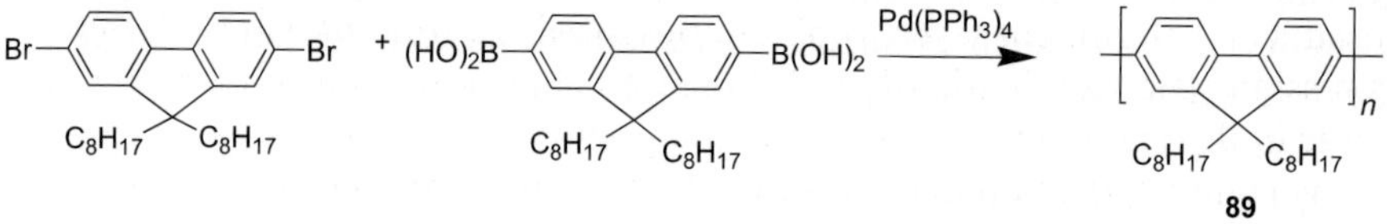

Scheme 41 Dow route to high molecular weight polyfluorenes

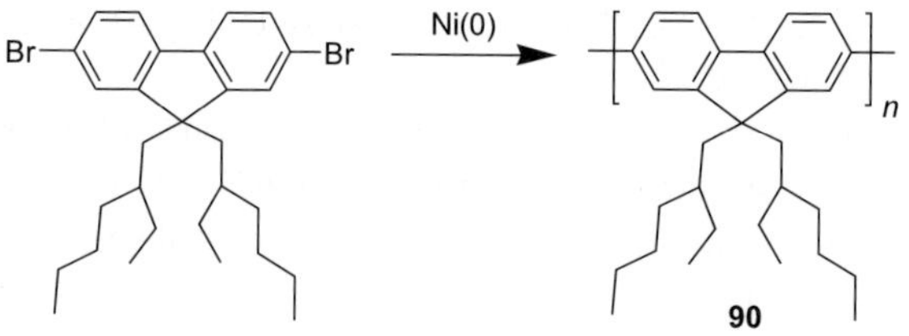

Scheme 42 PDAFS by Yamamoto polycondensation

Yamamoto-style polycondensations have also been used to make high molecular weight PFs. For example, Scherf and coworkers prepared poly[9,9-bis(2-ethylhexyl)fluorene] (**90**) with M_n of over 100 000 g/mol by coupling the dibromofluorene monomer with bis(cycloocta-1,7-dienyl)nickel(0) and bipyridine (Scheme 42) [179].

4.1.2
Optical Properties of PDAFs

The emission from PDAFs is violet-blue with a primary emission maximum at about 425 nm, and a secondary peak at about 445 nm. Well-defined oligomers of **85** have been prepared, and as a result the effective conjugation length for PDAFs has been determined to be about 12 units (24 phenyl rings) for absorption and six units for emission [180–182], indicating that the geometries of the ground and excited states are different.

PDAFs with unbranched alkyl substituents, e.g. **85** and **89**, show two thermotropic nematic liquid crystalline phases [183, 184]. Longer alkyl chains lower the transition temperatures, so that whereas for **85** the phases occur at 162–213 °C and 222–246 °C, with isotropisation occurring at 290–300 °C, for **89** the nematic phases are between 80–103 °C and 108–157 °C with isotropisation at 278–283 °C, and for the dodecyl-substituted polymer **91** (Scheme 43) they are observed at 62–77 °C and 83–116 °C with isotropisation at 116–118 °C [184]. Annealing films of the polymers at a temperature just above the second liquid crystalline transition followed by rapid cooling preserves the liquid-crystalline order [183, 184]. Deposition of such a film of **89** upon a rubbing-aligned polyimide alignment layer has been used to obtain polarised PL and EL [5, 185]. A rubbing-aligned layer of PPV has also been used as an alignment layer to obtain polarised EL from films of **89**, but the emission spectrum is slightly red shifted due to some absorption by the PPV layer in the region of the main emission peak at 433 nm [186–188].

By contrast, the polymer **90** with branched alkyl chains shows only one nematic phase. Thus, **90** shows a phase transition at 167 °C upon heating with the reverse transition upon cooling being seen at 132 °C. Annealing of a film of **90** at 185 °C upon a rubbing-aligned polyimide layer followed by rapid cooling produced a film which showed polarised EL with a dichroic ratio of up to 20 [179, 185, 189, 190].

Circularly polarised PL and EL emission has been obtained from polymers, e.g. **92** and **93** (Scheme 43), bearing chiral side chains [191–194].

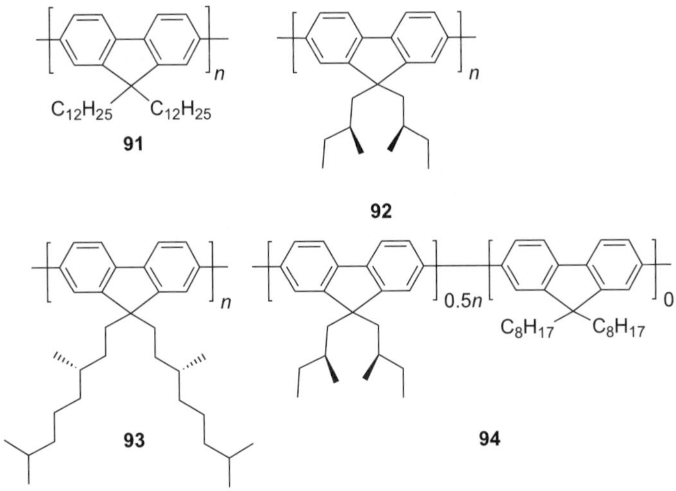

Scheme 43 Polyfluorenes with linear and branched alkyl chains

The copolymer **94** also showed circularly polarised emission, but the degree of dissymmetry was much reduced [191–193]. This was attributed to the dioctylfluorene units preferring a planar conformation of the backbone while the chiral-substituted units adopt a helical structure. A combination of a helical backbone conformation and liquid crystallinity is thought essential for obtaining a high degree of circular polarisation.

4.1.3
PFs with Substituted Alkyl Side Chains

PDAFs are highly soluble in non-polar solvents, e.g. toluene, but insoluble in highly polar solvents, e.g. ethanol. Solubility in such solvents can be obtained by introduction of charged groups onto the side chain. The polycation **95** made by reduction of the polyamide **96** (prepared by Yamamoto polycondensation of the dibromomonomer) and quaternisation of the resulting polyamine **97** (Scheme 44) is soluble in ethanol, methanol, and water, but insoluble in non-polar solvents such as THF [195].

Hydrolysis of the polyester **98** with potassium *tert*-butoxide and potassium hydroxide in THF to make the polyacid **99** (Scheme 45) gave a material which was insoluble in water, even at high pH, but soluble in pyridine. ^{1}H NMR performed in d_5-pyridine showed that complete hydrolysis had occurred [195].

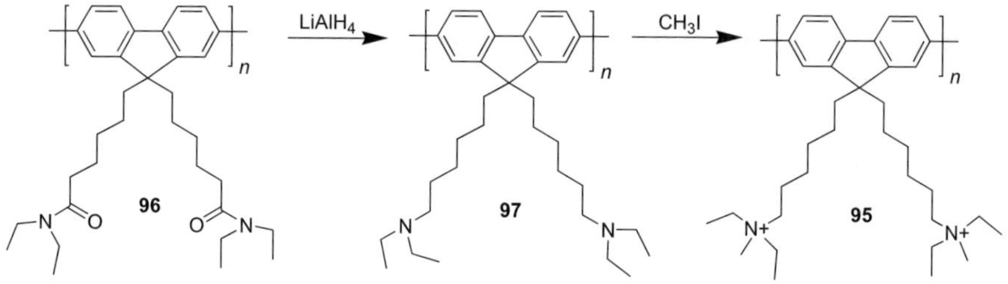

Scheme 44 Synthesis of a water-soluble polyfluorene

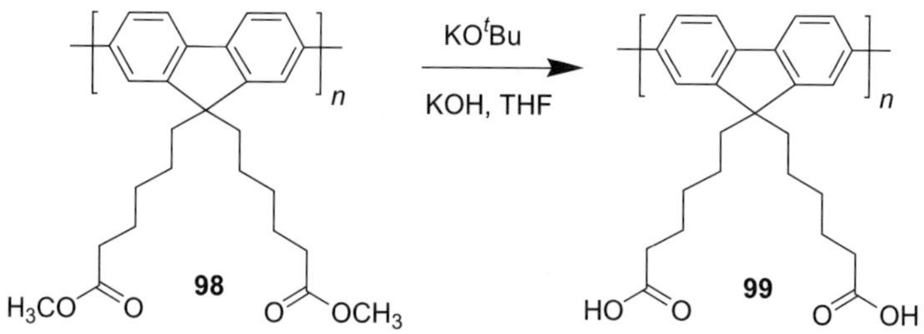

Scheme 45 Preparation of a fluorene-based polyacid

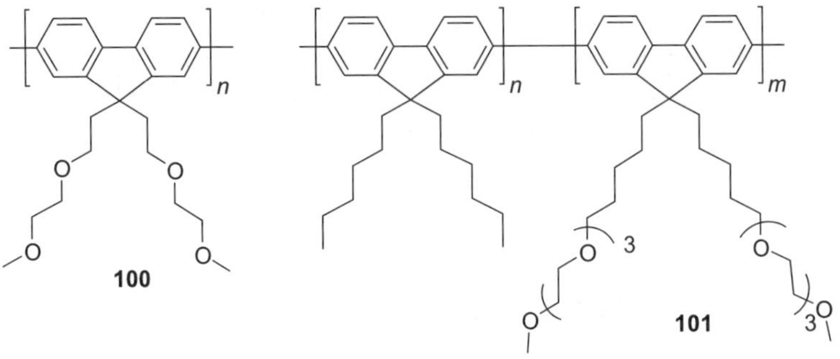

Scheme 46 Polyfluorenes with ethylene oxide side chains

Polymers **100** [196] and **101** [197] (Scheme 46) with ethylene oxide side chains have been prepared by Ni(0) coupling. The oxygenated side chains enable them to bind with metal ions, making them suitable as emissive materials for blue-emitting LECs as well as LEDs. Their emission in both LED and LEC devices rapidly changes from blue (λ_{max} = 430 nm) to blue-green due to the appearance of long-wavelength emission centred at 530 nm. When **100** is blended with PEO in a LEC, phase separation occurs, resulting in highly efficient (2.4%) white emission with an emission maximum at λ_{max} = 550 nm [198]. By use of colour filters, pure red, green, or blue emission can be obtained.

4.1.4
Defect Emission from PDAFs

As with LPPP the blue emission from PDAFs is unstable, with the appearance of a strong emission band around 530 nm after annealing or upon running an EL device [163, 164, 199, 200]. Initially this long wavelength emission band was believed to be due to emission from excimers [201, 202]. In support of this, absorption and PL studies of **85**, **89**, and **91** showed that annealing resulted in a red shift in the absorption edge, which was greatest for **91**, which was attributed to the formation of a more planar extended ground state conformation [184].

However, it was also noted that polymers end-capped with fluorenone units showed a similar emission band at around 530 nm to that seen from annealed films of PDAFs [201], suggesting that emission from fluorenone defect sites **69** was an alternative explanation for this emission. Consistent with this, the emission from the fluorenone-containing polymer **102** (Scheme 47) made by Bunz and coworkers is blue (λ_{max} = 428, 447 nm) in solution, but in the solid state is orange (λ_{max} = 533 nm), with all the emission coming from the fluorenone units [203]. Furthermore, it has been observed that when fluorene groups were used as end-capping groups for PFs, they were readily oxidised to

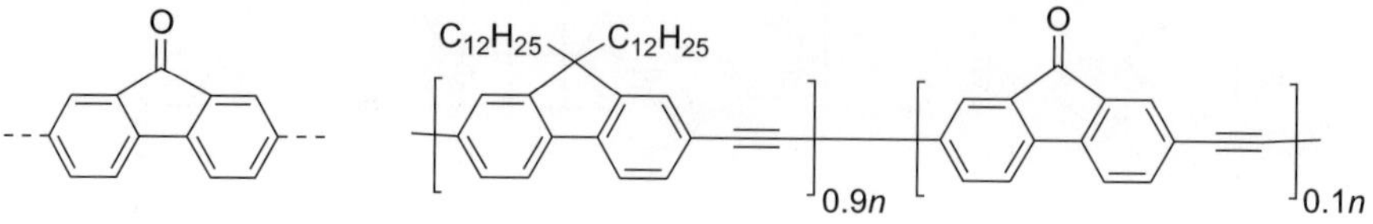

Scheme 47 The solid-state emission from the fluorenone-containing copolymer **102** matches the long-wavelength emission from polyfluorenes attributed to an emissive defect **69**

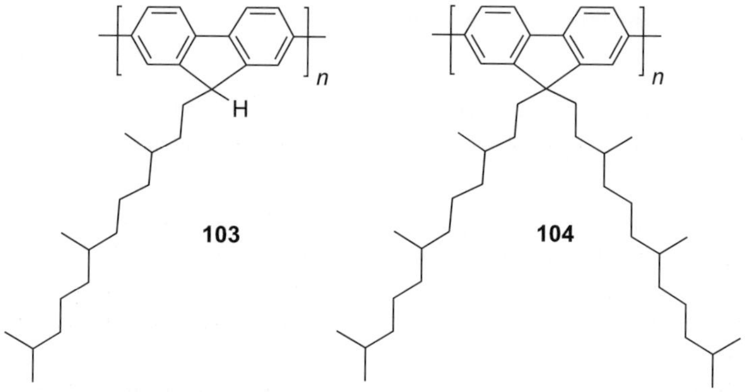

Scheme 48 Mono- and dialkylated polyfluorenes used to test ketone-defect hypothesis

fluorenones during annealing in air. This suggested that the appearance of the yellow emission band might be at least partially due to oxidation of fluorene units in the polymer to fluorenone [201].

Scherf and coworkers [204] prepared the polymer **103** (Scheme 48) with only one alkyl substituent at the 9-position by the Yamamoto method, and found that even pristine material contained fluorenone units as shown by a carbonyl stretching band at 1721 cm^{-1} in the IR absorption spectrum, and that the solid-state PL and EL spectra of **103** were dominated by a low-energy emission band centred at 533 nm due to fluorenone emission. A band at exactly the same position was seen from the corresponding dialkylfluorene polymer **104** after photooxidation in air or 30-min continuous operation of an LED. They accordingly suggested that the long wavelength emission band in PDAFs is due to energy transfer to ketone defect sites. The greater contribution of this band to the EL than the PL spectrum they attributed to the fluorenone units acting as electron-trapping sites, thus favouring recombination at the defect sites. Further support for this view came from time-delayed PL measurements of polymer **93** in solution by Lupton et al. [205].

To account for the formation of the defect sites in pristine **103**, they proposed that, during the polymerisation, some of the alkylfluorene units were reduced by the nickel(0) to monoalkylfluorenyl anions **105**, which

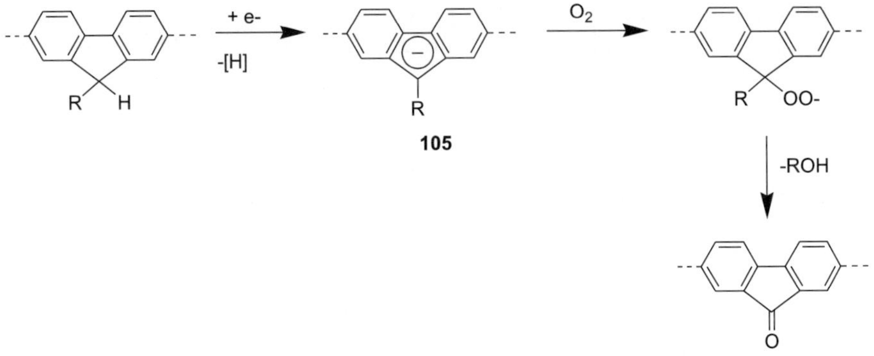

Scheme 49 Proposed mechanism for formation of ketone defects **69**

were then oxidised by atmospheric oxygen to the ketones **69** during the workup (Scheme 49). The oxidation of the 2-bromofluorenyl anion by atmospheric oxygen has been used to prepare 2-bromofluorenone in high yield, thus demonstrating the susceptibility of fluorenyl anions to attack by oxidation [206].

The defect hypothesis was challenged by Zeng et al., who reported that a pronounced emission band at 520 nm appeared upon annealing **85** under a nitrogen atmosphere, which they suggested ruled out oxidation [207]. However, in this experiment they did not use totally anaerobic (glove box) conditions. List and coworkers have found that heating a film of **104** at 200 °C in a dynamic vacuum ($< 10^{-4}$ mbar) produced no long wavelength emission band, even when the sample was simultaneously illuminated, while heating in air produced a strong band at 530 nm [208]. This demonstrates conclusively that the emission band at 530 nm from PFs is due to the formation of ketone defects by oxidation during synthesis and/or handling. This does not mean that the demonstrated interchain interactions produced during annealing of PDAFs play no part in the appearance of the long-wavelength emission, as increased interchain interactions (aggregation) would enhance exciton migration to the defect sites.

Considerable effort has gone into developing stable blue emission from PDAFs. Fractionation of **89** to remove low molecular mass material has been reported to reduce the long-wavelength emission [209]. This is at first sight a somewhat surprising result, as the probability of a polymer chain containing a defect increases with increasing chain length. However, it is possible that the oligomers in the low molecular mass fraction assist in interchain charge and energy migration, and so removing them would reduce exciton migration to defect sites. Meijer and coworkers have shown that if a dialkylfluorene monomer is treated with base to remove residual monoalkylfluorenes before polymerisation then the resulting polymers display much stabler blue emission [210]. Since the materials were pure by all standard analytical methods before this extra purification, the amount of the monoalkylated impurities must have been less than 1 mol %.

Stable blue emission has been obtained by Neher and coworkers from blends of the copolymer **106** with hole-transporting molecules, e.g. **107** (Scheme 50) [211]. This was attributed to the dopants acting as hole traps, thus reducing the amount of charge trapping at the defect sites, and thus the emission therefrom.

A group at IBM prepared (Scheme 51) polymers **108** end-capped with vinyl groups by addition of bromostyrene as an end-capping reagent to a Yamamoto polycondensation [212]. These polymers can be thermally cross-linked at 150–200 °C to produce insoluble materials which show stable blue emission [212]. This is attributed to a reduction in chain mobility, due to the cross-linking, which hinders exciton migration to the defect sites.

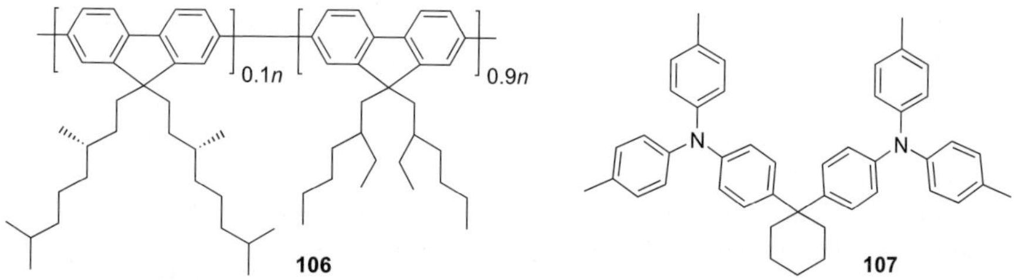

Scheme 50 Polyfluorene and typical hole-transporting material blended with it to obtain stable blue emission

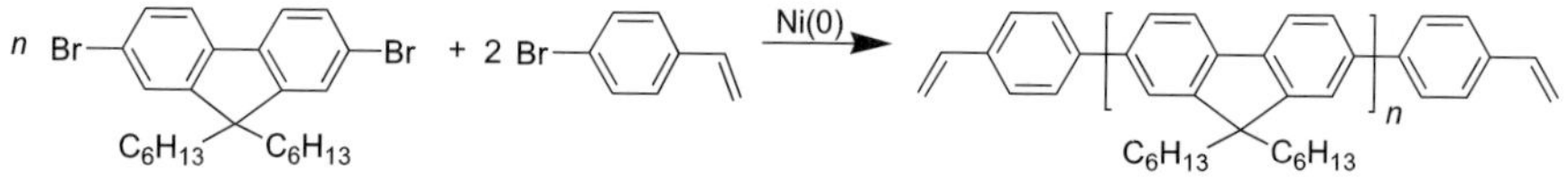

Scheme 51 Preparation of a polyfluorene end-capped with a cross-linkable group

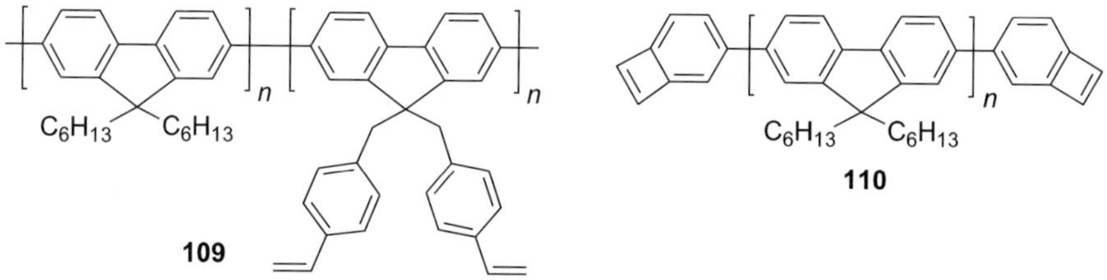

Scheme 52 Other cross-linkable polyfluorenes

By contrast, cross-linking of polymers **109** with styryl groups as side chains (Scheme 52) [212] does not produce efficient suppression of the long-wavelength emission. This suggests that reducing chain-end mobility is important in reducing interchain interactions. No suppression in defect emission is observed in thermally cross-linked polymers **110** end-capped with cyclobutene groups [213]. It is not clear why this is so, but it may be that more oxidation occurs at the higher conversion temperature (250 °C) used in this process or that the cross-linking is less efficient.

Other approaches that have been found to improve the stability of blue emission from PDAFs are to attach bulky substituents such as cyclohexyl groups [214] or polyhedral siloxanes [215–217] to the ends of the alkyl chains. These presumably work by reducing exciton migration to defect sites.

4.1.5
Colour Control by Copolymerisation

Miller and coworkers have prepared copolymers **111** (Scheme 53) of dihexylfluorene with anthracene by the Yamamoto method. These show stable blue emission (λ_{max} = 455 nm) even after prolonged annealing [182, 218]. As the ratio of anthracene (15%) to fluorene (85%) units was too low to produce significant steric repulsion between the polymer chains, they attributed the absence of long-wavelength emission to trapping of the exciton at the anthracene sites and subsequent emission therefrom [218, 219]. Similar copolymers **112** with dioctylfluorene are reported to show PL and EL maxima at respectively 446 nm and 435 nm [220]. The reason for this difference is not apparent.

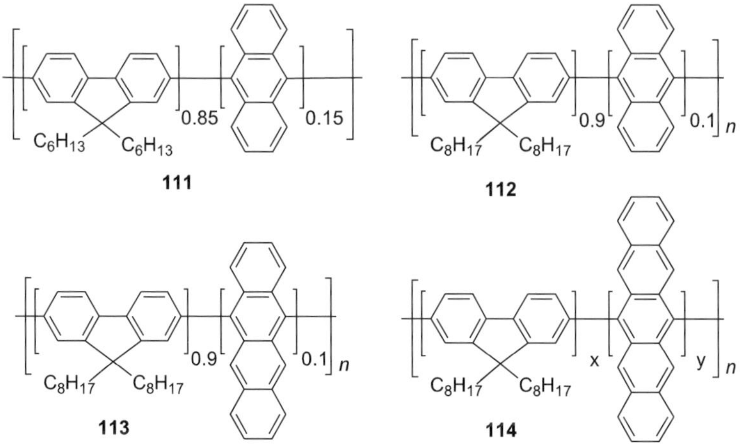

Scheme 53 Copolymers of fluorene with acenes

Copolymers **113** and **114** containing tetracene and pentacene units show emission from the acene units to give green (λ_{max} = 520 nm) and red (λ_{max} = 623 nm) PL and EL, respectively [220]. However, due to rapid degradation of the acenes by oxidation their emission rapidly turns blue and the emission efficiency drops markedly. The pentacene copolymer **114** shows a lower turn-on voltage and higher efficiency with 1 mol % as opposed to 10 mol % of pentacene.

Similar exciton trapping has also been observed in copolymers containing perylene **115** (an equal mixture of 3,9- and 3,10-linked units) or cyanostilbene **116** units (Scheme 54) made by Yamamoto copolymerisation [219, 221]. Thus, the perylene copolymer **115** (m = 15%) showed green emission (λ_{max} = 540 nm), while by altering the amount of stilbene the emission from **116** could be tuned between λ_{max} = 466 nm (m = 5%) and λ_{max} = 510 nm (m = 50%).

Such exciton trapping has been exploited by Müllen and coworkers, who prepared copolymers of dioctylfluorene- and perylene-based dyes (1– 5 mol %) by Yamamoto copolymerisation (Scheme 55) [222–224]. These materials were designed so that by efficient Förster energy transfer from the fluorene to the dye units, efficient emission across the whole visible spectrum could be obtained. Perylene dyes were chosen as the chromophores due to their high solid-state PL quantum yields, and their excellent thermal and photochemical stability.

Four different dyes were used, with emission maxima in the green (λ_{max} = 525 nm), yellow (λ_{max} = 540 nm), orange (λ_{max} = 584 nm), and red (λ_{max} = 626 nm), respectively. In the copolymers with the green-emitting **117** and yellow-emitting **118** dyes, the conjugation is maintained between the fluorenes and the chromophore. It should be noted that these dyes are inseparable 1 : 1 mixtures of 3,9- and 3,10-dibromo compounds, but there is no reason to believe that this has any adverse effect upon their optical or electronic properties. By contrast, in the copolymers with the orange-emitting **119** and red-emitting **120** dyes, the imide groups interrupt the conjugation.

Since the efficiency of energy transfer between the fluorene and the dye is dependent upon the overlap of the fluorene emission spectrum and the absorption spectrum of the dye, it was anticipated that energy transfer to the

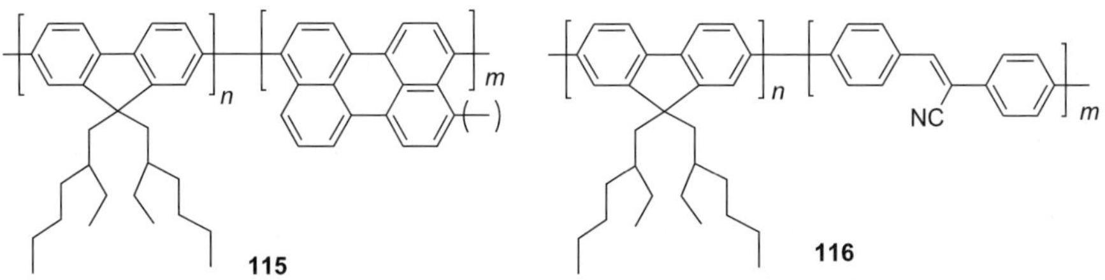

Scheme 54 Copolymers of fluorene with green- and red-emitting chromophores

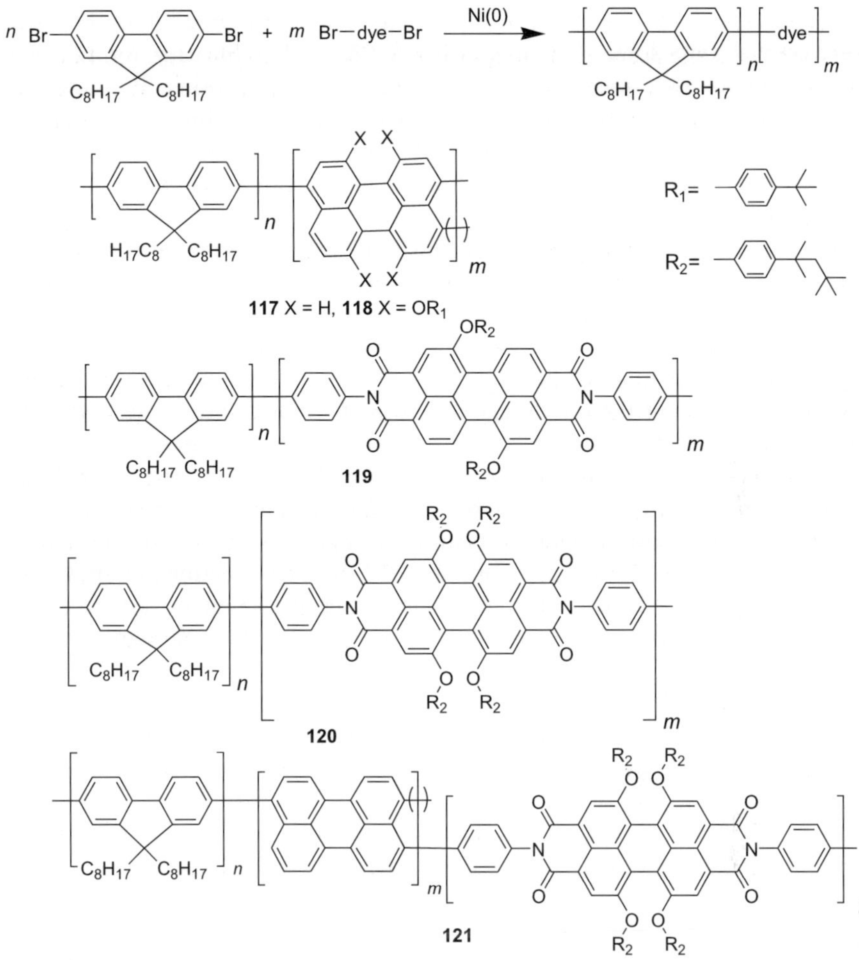

Scheme 55 Synthesis of copolymers of fluorene- and perylene-based dyes

orange and red dyes would be less efficient than to the green and yellow ones. Calculations based upon the spectral overlap indicated that the rate constant for energy transfer to the red dye was an order of magnitude smaller than for transfer to the green dye. Accordingly, a copolymer **121** was made containing both green (3 mol %) and red (3 mol %) emitting chromophores in order to determine if sequential energy transfer might be more efficient. The molecular weights of the copolymers were a little lower than for the homopolymer **89** (32 000–75 000 versus 85 000 g/mol), and their polydispersities were higher (3.6–4.9 versus 2.8).

The PL spectra of these materials in solution showed emission from both components roughly proportional to their relative mole ratios, but in films energy transfer occurred so that the PL spectrum was dominated by emission from the dyes. The PL quantum efficiencies of the copolymers **117–120** were only slightly lower (33–51%) than for **89** (55%), but that for **121** was very low (7%). This must be due to aggregation or some other fluorescence-quenching process. In the EL spectra only emission from the dyes was seen, which is due to a combination of energy transfer from the fluorene and the dyes acting as charge traps. The EL spectra closely resembled the PL emission from the dyes except for the copolymer **119** with the orange-emitting dye, where the EL spectrum was broader and red shifted compared with the PL spectrum. External efficiencies of 0.2–0.6% were obtained from these devices, which are comparable with other polyfluorene-based devices. The emission colours are stable, unlike devices using a blend of dye and **89**, where phase separation results in a loss of half the luminescence intensity within a few minutes, accompanied by a change in emission colour as the emission from the host polymer **89** reappears as the energy transfer becomes less efficient. As all the copolymers contain at least 95 mol % of dioctylfluorene units their blends with each other and with **89** should be stable, thus enabling stable emission of practically any colour desired, including white, to be obtainable.

Efficient energy transfer was also obtained from a copolymer **122** in which the perylene was attached as an end-capping group (Scheme 56) [224]. By using a ratio of fluorene to monobromo dye of 18 : 1 in a Yamamoto poly-

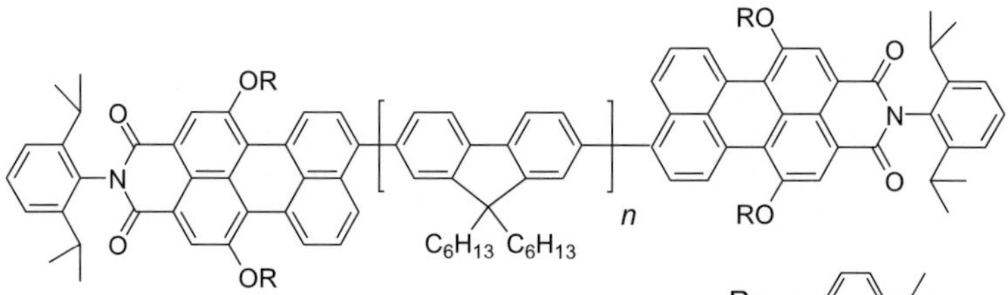

Scheme 56 Polyfluorene end-capped with a perylene dye

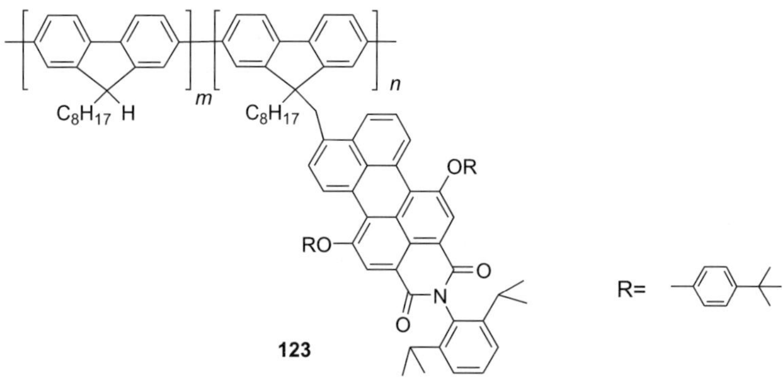

Scheme 57 Polyfluorene with dyes as side chains

condensation, a polymer with $M_n = 21\,000$ and a polydispersity of 2.1 was obtained, corresponding to a degree of polymerisation of about 40. As with the random copolymers **117–121**, no energy transfer was seen in solution, but in the solid state the emission was almost entirely from the dye moieties ($\lambda_{max} = 613$ nm). A problem with using the dye as an end-capping group is that due to a competing debromination reaction, complete end-capping in Yamamoto reactions is usually not obtained. In this case it was estimated that only 40 mol % of the polymer chains contained two dye units.

Due to the low mole ratio of dye units present, the above copolymers, with perylene dyes in the main chain or as end groups, show energy transfer only in the solid state. If the dyes are attached on the side chain, then copolymers containing much higher mole ratios of chromophore are accessible. The copolymer **123** (Scheme 57) in which 33% of the fluorene units have dyes attached ($m:n = 2:1$) showed energy transfer in solution as well as in a thin film [224]. The emission colour differed slightly between the two states, with the emission maximum appearing at $\lambda_{max} = 561$ nm in solution with a shoulder at 599 nm, and at $\lambda_{max} = 599$ nm in the solid state. This is probably due to interaction of the chromophores in the solid phase.

4.1.6
Dendronised Polyfluorenes – Towards Stable Blue Emission

In order to suppress long interchain interactions without losing solubility, bulky substituents can be attached. The IBM group prepared PDAFs end-capped with Fréchet-type dendrons of generations one (**124**, Scheme 58) to four [225]. By using a fluorene to end-capper ratio of 9 : 1 they obtained polymers with about 80 fluorene units. They found that the polymers with third- or fourth-generation dendrimer end groups showed no long-wavelength emission, even after annealing at 200 °C.

They have also prepared polymers with Fréchet-type dendrons of generations one to three as side chains by Yamamoto polycondensation (Scheme 59) [226]. A molecular weight of 51 000 was obtained for the homopolymer **125** with a first-generation dendrimer attached, but only oligomers ($M_n < 10\,000$ g/mol) were obtained from polymerisation of the fluorenes

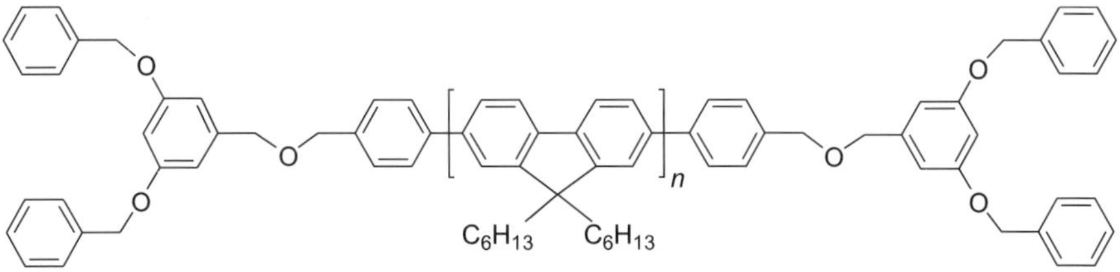

124

Scheme 58 Polyfluorene end-capped with a Fréchet-type dendron

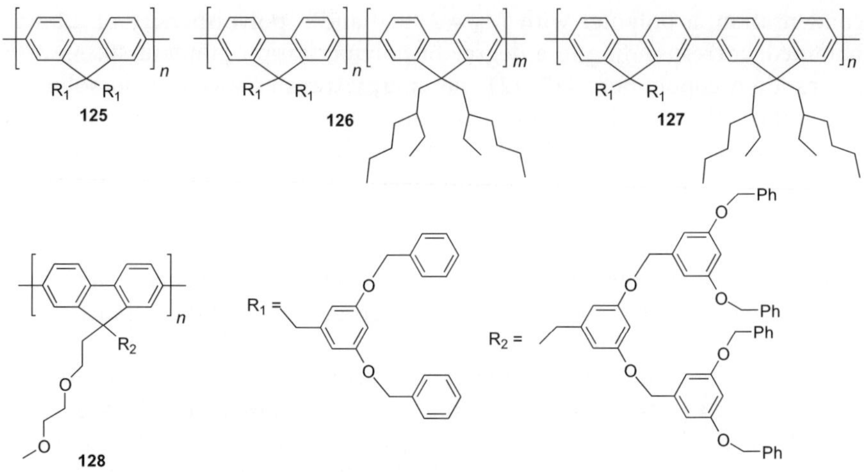

Scheme 59 Polyfluorenes with Fréchet dendrons as side chains

with higher-generation dendrimers attached. Random copolymers with ethylhexylfluorene, e.g. **126**, were made by the same method and showed much higher molecular weights (M_n = 26 000–68 000 g/mol), but the relative amount of dendrimer incorporated decreased with increasing generation from $m:n$ = 1 : 1.5 for the first-generation dendrimer to 1 : 0.07 for the third-generation dendrimer. This shows that, as might be expected, the bulkier monomers are much less reactive under Yamamoto conditions. Alternating copolymers, e.g. **127**, were made by Suzuki coupling. The molecular weights were lower (M_n = 4000–16 000 g/mol) with a degree of polymerisation for the monomers with first- and third-generation dendrimers of only three (six fluorene units). The second-generation dendrimer monomer was more reactive, showing a degree of polymerisation of eight. The PL and EL spectra of the polymers with second- or third-generation dendrimer side chains showed no long-wavelength emission, but the lifetime of the LEDs was short, as only very thin films could be deposited by spin coating. The highest PL efficiencies were obtained from the polymers and copolymers with second-generation dendrimer substituents. Pure blue PL (λ_{max} = 423, 447 nm) has also been reported from the polymer **128** with a single second-generation dendrimer side chain [227].

Müllen and coworkers have prepared a polymer **129** with a first-generation polyphenylene dendron on the side chain (Scheme 60) [228]. Unlike Fréchet dendrimers, such dendrimers are rigid and shape-persistent. Interestingly, the bulky side chains produced no blue shift in the absorption or emission of the polymer compared with PDAFs such as **89**, indicating that there is no increased steric repulsion between adjacent fluorene units. LEDs using this polymer display device characteristics (turn-on voltage, emission efficiency,

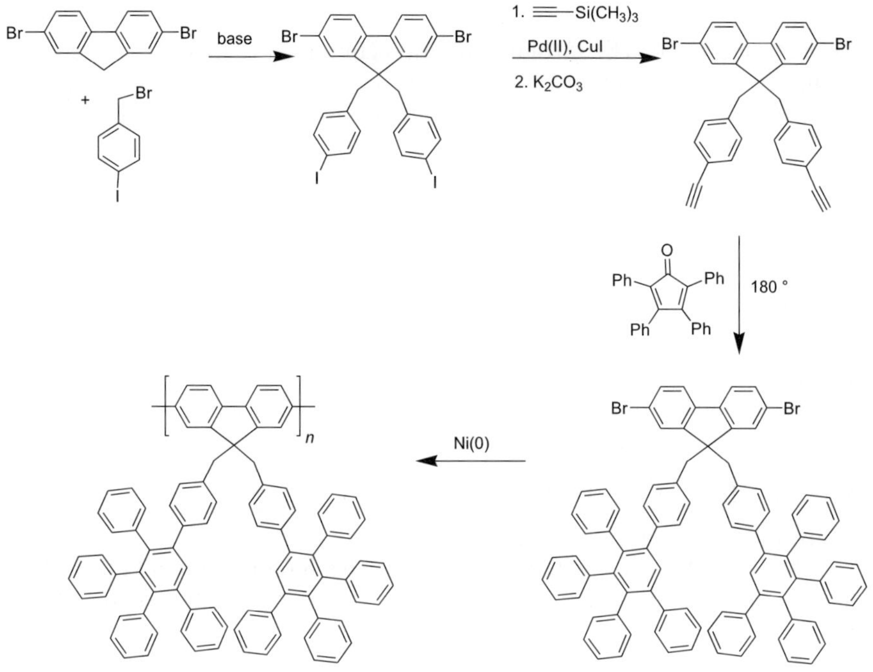

Scheme 60 Synthesis of a polyfluorene with a polyphenylene dendron on the side chain

and brightness) comparable to those for the PDAFs such as **89**, but with no long-wavelength emission being seen at driving voltages below 12 V.

Though the emission stability of **129** is much better than for PDAFs, long-wavelength emission is seen with time, possibly due to the susceptibility of the benzyl linkages to oxidation. Accordingly, a new class of polyfluorenes with aryl substituents at the 9-position has been developed (Scheme 61) [228]. Aryl substituents are much less susceptible to oxidation to fluorenone than alkyl groups and their bulk reduces interchain interactions, and hence exciton diffusion to any defect sites that may be formed.

As aryl groups, unlike alkyl groups, generally cannot be directly substituted onto the 9-position of fluorenes, the monomers were prepared by addition of aryl lithium reagents to the biphenyl-2-carboxylic acid methyl ester **130**, followed by ring closure of the resulting carbinols in hot acetic acid, using a small amount of concentrated sulfuric acid. This produced a dark-blue or lilac colour due to the formation of the cation, followed by the appearance of a white precipitate of the desired 9,9-diarylfluorenes **131**. These were then polymerised using nickel(0).

The diphenylfluorene polymer **132** (Scheme 62) was virtually totally insoluble. Characterisation by MALDI-TOF of a soluble fraction, obtained by prolonged Soxhlet extraction with toluene, suggested that it consisted mainly of oligomers up to the hexamer ($n = 6$), though chains of up to 15 fluorene units were detected [229]. The insoluble fraction contained chains with degrees of polymerisation of up to 25, which suggests that some chain addition occurred even after the material was no longer soluble in the reaction medium.

By contrast, the materials **133** with di-*tert*-butylphenyl and **134** with first-generation dendron side chains were processable from toluene [228]. The degrees of polymerisation for these polymers were rather low, with the dendronised polymer **134** having M_n of only 10 000 g/mol ($n = 10$), probably due to the bulkiness of the substituents. Polymer **134** was also found to be exceptionally thermally stable, with no significant mass loss being observed in thermal gravimetric analysis (TGA) until 570 °C (cf. 463 °C for **89** [224]). Substituted first-generation and unsubstituted second-generation dendrimers have been attached to fluorene by a modification of the above route in which an ethynyl-substituted benzene is attached to the fluorene 9-position and the dendrimers then built up from it [230]. The degree of polymerisation of **134** can be increased ($M_n = 32\,000$ g/mol, $n = 32$) by using a dichloro instead of a dibromo monomer in the Yamamoto polymerisation [231].

Stable blue PL emission has been obtained from films of both **133** and **134**, with no sign of an emission band at 530 nm even after annealing for 24 h in

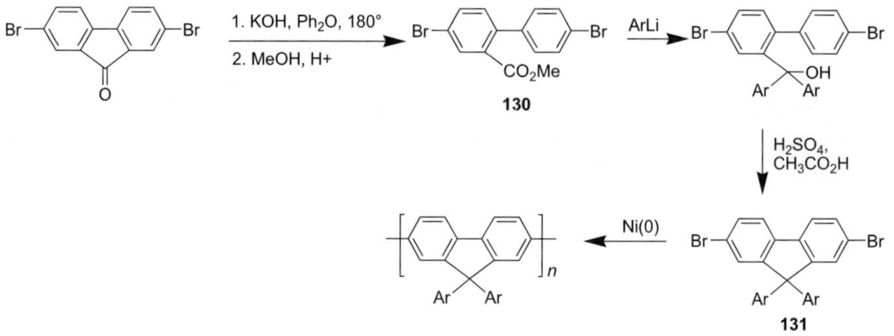

Scheme 61 Synthetic route to poly(9,9-diarylfluorene)s

132 R = H
135 R = OC$_8$H$_{17}$

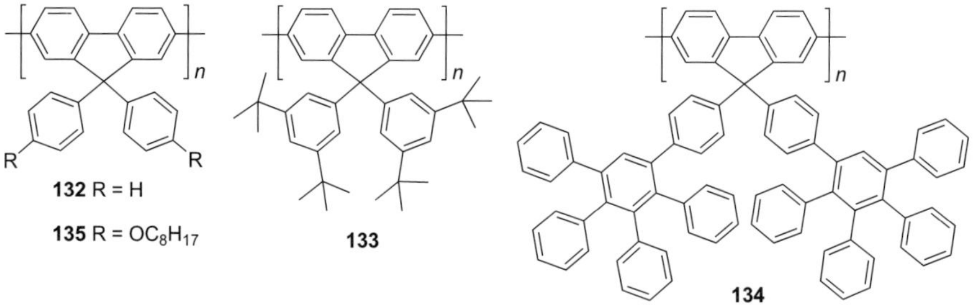

Scheme 62 Poly(9,9-diarylfluorene)s

air at 100 °C. This shows that aryl substituents are much stabler towards oxidation than alkyl groups, as annealing of **89** under these conditions produces a strong defect emission band. Drop-cast films of **133** showed a slight red shift in the PL spectrum upon annealing, which may be attributed to changes in the chain packing, as no such change was seen from spin-cast films, which illustrates the importance of processing conditions in determining the properties of films of conjugated materials. Stable blue emission has also been obtained from the polymer **135** bearing 4-octyloxyphenyl substituents [232]. This was prepared by addition of phenol to dibromofluorenone, followed by alkylation and polymerisation (cf. Scheme 70, below).

Both **133** and **134** show blue EL with no significant defect emission even after several minutes operation [223, 233]. When devices using **134** were run continuously at 8 V for 30 min, a slight increase in long-wavelength emission was observed [233]. This partially disappeared upon reversing the bias, which was attributed to a shift in the recombination zone within the device. By subtracting the original spectrum from that of the aged device it was found that the extra emission corresponded to the defect emission band at 530 nm, which suggests that some oxidation of **134** was occurring, possibly due to incomplete encapsulation of the device. When the device was run in pulsed mode, however, stable blue blue EL with a luminance of 191 cd/m^2 was obtained with no sign of degradation even over prolonged periods [233].

Comparative photophysical studies of **90** and **129** suggest that in dendronised polyfluorenes the bulkiness of the substituents hinders exciton migration due to increased chain separation [234]. Delayed PL studies [235] of **90** and **134** found that both polymers display a defect emission band at 530 nm, and a band at 480 nm previously attributed to interchain aggregates [205]. This is an interesting result as interchain aggregates had previously been assumed to arise from π-stacking of the polymer chains. Calculations of

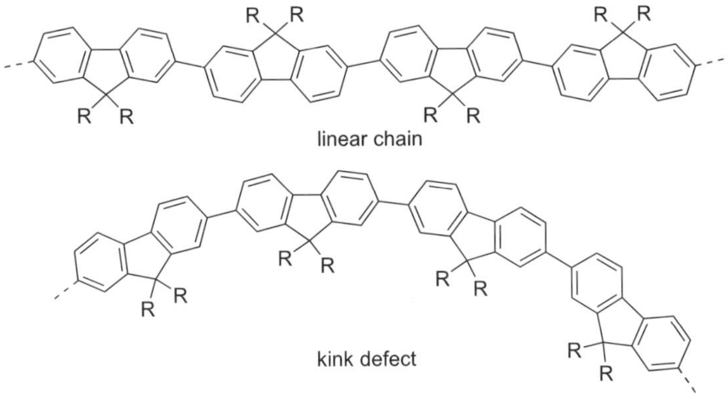

Scheme 63 Possible conformations of polyfluorene chains

oligomers of **89** and **90** show that by adopting a planar conformation the dioctylfluorene units in **89** can π-stack with a separation of 5 Å, while the fluorenes with the ethylhexyl substituents in **90** cannot approach within 7 Å of each other, which precludes π-stacking [236]. The separation of dendronised fluorene units can be expected to be even greater than for the units in **90**. Thus interchain interactions in these polymers must arise not from cofacial π-stacking, but at point defects on the chains. These defects can arise because fluorene is not linear, but is bent, with an angle between the bonds to the 2- and 7- positions of about 20°. Normally in polyfluorene chains the fluorene units align with successive units having their substituents pointing in opposite directions so as to give overall a linear chain. However, 'kink' defects can form where the chain bends due to successive fluorene units aligning with the 9-substituents on the same side (Scheme 63). A bent conformation has been detected for an end-capped fluorene trimer in a polymer matrix [237]. In theory the chain can even loop back on itself. At such points two chains might be able to interact without interference from the side chains.

4.1.7
Polyfluorenes with Improved Charge Injection

A second problem with PDAFs is that of charge injection. The energy levels of the HOMO and LUMO orbitals of **89** have been determined to be 5.8 eV and 2.12 eV by cyclic voltammetry [238]. This means that there are large barriers to charge injection from electrodes such as calcium (work function 2.9 eV) and ITO (4.7–5.0 eV). The value for the HOMO-LUMO gap of 3.7 eV determined by electrochemistry is much larger than the optical band gap of 2.95 eV calculated from the onset of absorption, which suggests that one or both of the electrochemically derived orbital energy levels are inaccurate. As the efficiency of devices using calcium electrodes seems to be limited by hole injection [239], most probably it is the LUMO value which is at fault (too high).

A number of approaches have been adopted to overcome the problem of obtaining efficient charge injection. Satisfactory electron injection seems to be obtained by use of calcium as cathode, and a composite LiF/Ca/Al cathode is reported to give better electron injection than a simple calcium cathode [240], but there usually remains a large barrier to hole injection due to the difference between the HOMO and ITO energy levels. As a result, hole-injecting layers are used in PDAF devices in order to obtain good efficiency (Scheme 64) [241]. Poly(3,4-ethylenedioxythiophene) (**136**) doped with poly(styrene sulfonate) (PEDOT:PSS) [178] is most commonly used, but recently poly(4,4'-dimethoxybithiophene) (**137**) has been reported to be a superior hole-injection material [239]. Scott and coworkers at IBM found that the efficiency of devices using emissive layers of cross-linked polymer **108** could be increased by factors of 400 and 800 respectively by use of electron- and hole-injecting layers, though the latter device also showed some

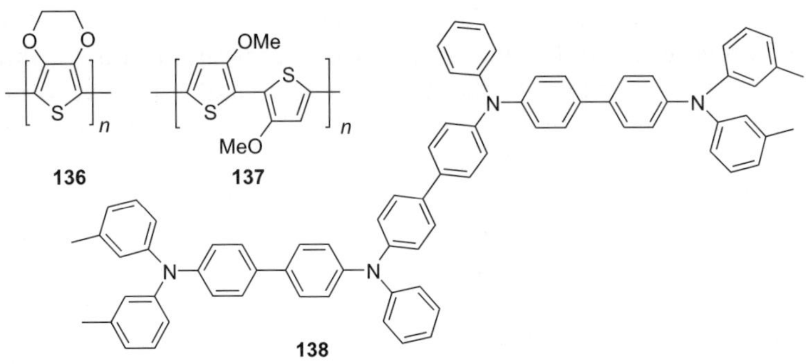

Scheme 64 Materials used to improve hole injection or transport in polyfluorenes

red-orange emission from the triarylamine hole-transporting layer [242]. A triple-layer device with both hole- and electron-injecting layers was even more efficient, with efficiencies of over 1% being obtained, although again some weak emission from the hole-transporting layer was observed.

A second method is to blend charge-transporting materials into the emissive polymer, though this raises the problems of phase separation. Neher and coworkers have examined blends of the copolymer **106** (Scheme 50) with hole-transporting small molecules [211]. Not only did this increase the efficiency and luminance of the devices, but as mentioned above it also suppressed the green emission from the ketone defects. The highest efficiency (0.87 cd/A) and luminance (800 cd/m^2) (cf. 0.04 cd/A and 70 cd/m^2 for undoped **106**) were obtained by incorporation of 3 wt % of the oligomeric triarylamine **138** (Scheme 64). No phase separation leading to device degradation was observed by them during the characterisation of their devices, but the long-term stability of such blends is uncertain.

Another approach is to incorporate charge-transporting groups into the polymer, either as end groups, as units in the polymer chain or as side chains. The IBM group [243] have compared the efficiency, using both calcium and aluminium cathodes, of triblock polymers (Scheme 65) containing an anthracene–dialkylfluorene copolymer as the emissive block with hole-transporting poly(vinyl triphenylamine) (**139**) or electron-transporting poly(vinyl oxadiazole) (**140**) blocks at each end with the simple copolymer **111** (Scheme 53).

In single-layer devices with calcium cathodes the EL efficiency of the hole-transporting triblock **139** was nearly double that of **111** (0.35% versus 0.02%), but that of the electron-transporting triblock **140** was slightly lower (0.014%). This shows that with calcium electrodes, hole injection is the limiting factor for efficiency. With aluminium electrodes, electron injection by contrast became the limiting factor so that the highest efficiency (0.01%) was obtained with **140**, although the efficiency of **139** (0.008%) was still much higher than

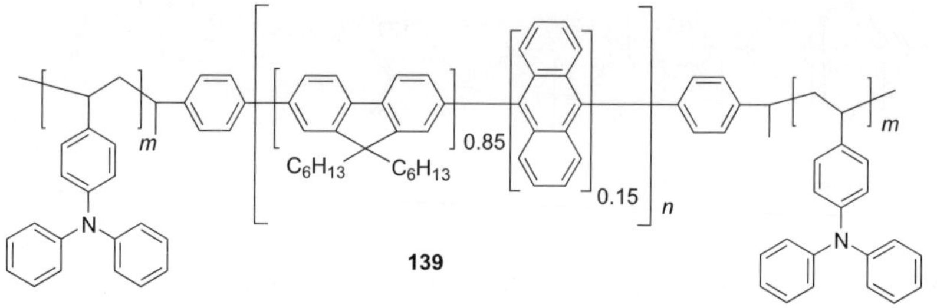

Scheme 65 Triblock copolymers containing both emissive and charge-transporting blocks

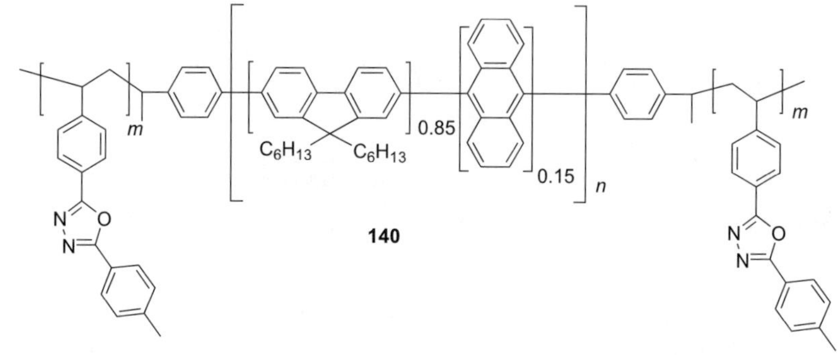

of **111** (0.001%). The triblock **139** showed a strong orange-red-emission band due to emission from the triphenylamine units in the EL, but not in the PL spectrum. The highest efficiency (0.54%) was obtained from a double-layer device of **139** with a hole-transporting layer and a calcium cathode. Again some emission from the triphenylamine units was seen, though it was much weaker than in the single-layer device.

Scherf, Neher and coworkers prepared liquid-crystalline polyfluorenes, **141** and **142** (Scheme 66), by adding bromo-substituted triarylamines as end-capping reagents to a Yamamoto polymerisation [244]. By varying the amount of end-capper they were able to control the molecular weight and polydispersity of the polymer, so that for **141** 2 mol % of end-capper gave a polymer with $M_n = 102\,000$ and $M_w/M_n = 1.4$, while 9 mol % of end-capper produced a polymer with $M_n = 12\,000$ and $M_w/M_n = 2.6$. NMR analysis showed that the end-capping was incomplete so that 2 mol % of end-capper produced 1.8% of triarylamine end groups and 9 mol % resulted in 8.3% incorporation of the end groups. Films of these polymers deposited on an aligned polyimide layer showed polarised emission with better efficiencies (up to 0.75 cd/A) than for the homopolymer **90**. Even better efficiencies (up to 2.7 cd/A) have been obtained from devices using an emissive layer of **141** (2 mol % end-capper) and a series of three cross-linked hole-transporting layers of varying work function [245].

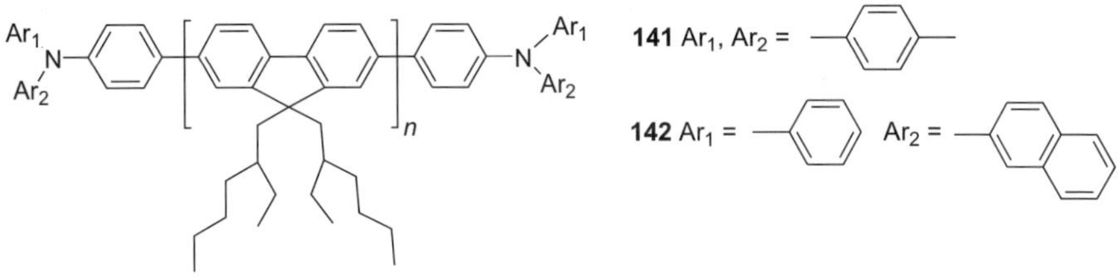

Scheme 66 Polyfluorenes end-capped with hole-transporting groups

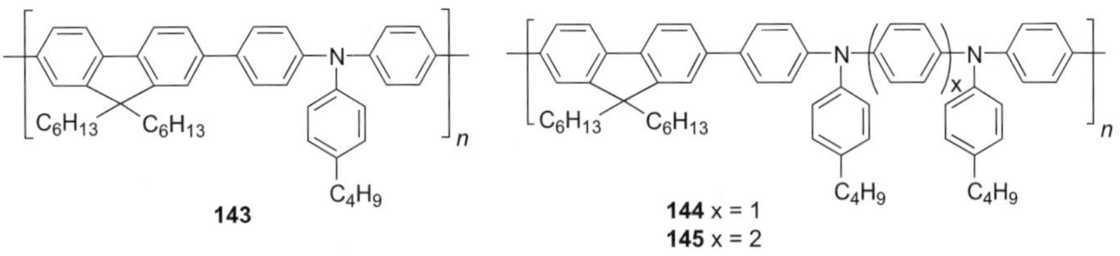

Scheme 67 Copolymers of fluorene with hole-transporting triarylamines

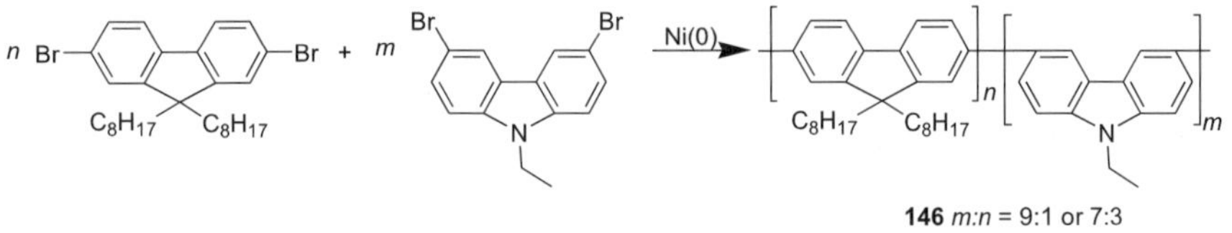

Scheme 68 Synthesis of a fluorene–carbazole copolymer

The Dow group have prepared alternating copolymers of dialkylfluorene with hole-transporting triphenylamines (Scheme 67), e.g. **143**, by Suzuki coupling of fluorene bisboronic acids with dibromotriarslamines [174–177, 241]. Blue emission has been reported from some of these polymers, e.g. **144** [246] (λ_{max} = 486 nm) and **145** [174] (λ_{max} = 481 nm). It would appear, however, that these materials are primarily intended to be used as hole-transporting rather than emissive materials.

Xia and Advincula have prepared copolymers **146** containing hole-transporting carbazole units by Yamamoto copolymerisation (Scheme 68) [247]. Cyclic voltammetry showed that the HOMO energy level increased from 5.8 eV to 5.6 eV with 10 mol % carbazole and to 5.5 eV with 30 mol % carbazole. They also found that films of these copolymers showed stabler blue PL than the homopolymer **89** with the green ketone emission band appearing only slowly upon annealing at 200 °C.

The IBM group have investigated the incorporation of charge-transporting groups into the copolymer **110** [182]. Introduction of hole-transporting triphenylamine units produced a slight red shift (λ_{max} = 462 nm) in the EL spectrum, while electron-accepting diphenylsulfone units caused a small blue shift (λ_{max} = 445–449 nm) in the emission, in both cases with a decrease

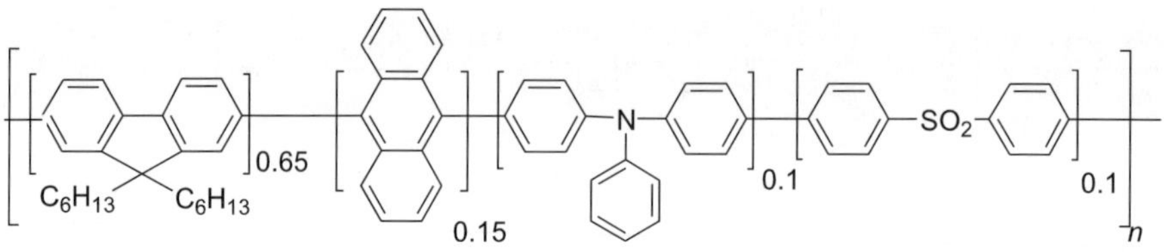

Scheme 69 Copolymer of fluorene with both hole- and electron-transporting units

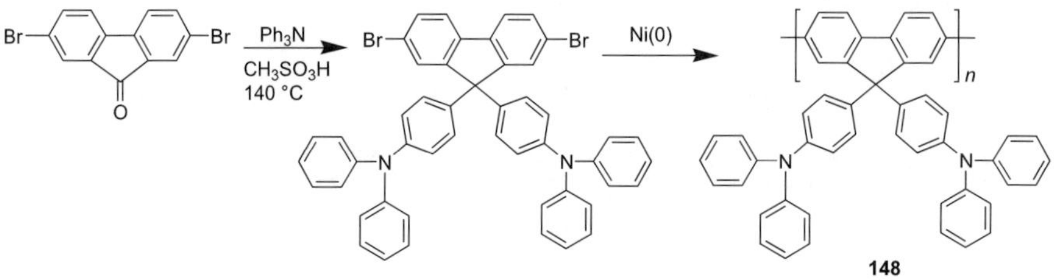

Scheme 70 Synthesis of polyfluorene with triphenylamine substituents

in the EL efficiency. By contrast, the copolymer **147** (Scheme 69) containing 10 mol % of both groups showed stable blue EL (λ_{max} = 460 nm) with nearly twice the efficiency of **110**.

Müllen and coworkers have recently prepared the polymer **148** in which hole-transporting triphenylamine units are attached at the 9-position (Scheme 70) [248]. The triphenylamine substituents are introduced by Friedel–Crafts alkylation of triphenylamine with 9-fluorenyl cations produced from dibromofluorenone under the reaction conditions. The degree of polymerisation is not very high (n = 14) due to the limited solubilising power of the substituents. The HOMO of **148** has been determined by cyclic voltammetry to be at 5.5 eV, so the barrier to hole injection from ITO is much smaller than for PDAFs. Polymer **148** shows stable blue EL (λ_{max} = 428 nm), but the overall EL efficiency is lower than for PDAFs despite the better hole-accepting properties, probably because of the lower PL efficiency of **148** (22% versus 55% for **89** [178]). The use of a PVK hole-injecting layer does not improve the efficiency as it does with PDAFs, indicating that hole injection into **148** is superior.

Müllen and coworkers have also prepared the blue-emitting (λ_{max} = 450 nm) polyketal **149**, films of which can be converted by exposure to dichloroacetic acid vapour to the orange-emitting (λ_{max} = 580 nm) polyfluorenone **150** (Scheme 71) [249]. The carbonyl groups enhance the electron-accepting properties of **150**, and this polymer shows useful electron-transporting properties, though the acidic residues from the conversion are a potential source of problems for electronic applications [250].

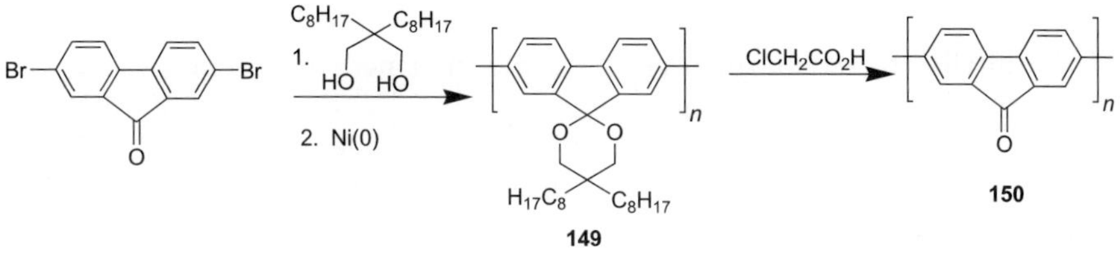

Scheme 71 Precursor route to polyfluorenone

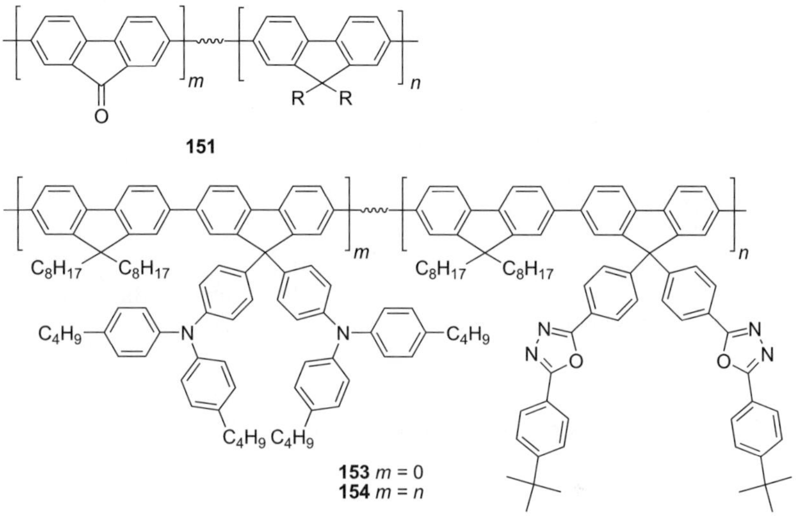

Scheme 72 Fluorene copolymers with electron-accepting units

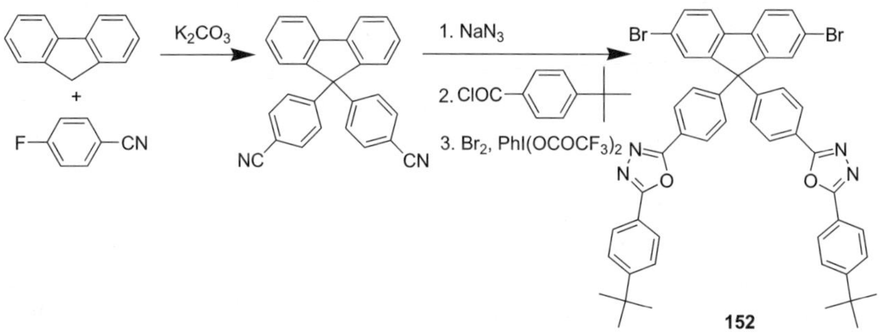

153 $m = 0$
154 $m = n$

Scheme 73 Synthesis of a fluorene with oxadiazole substituents

Copolymers **151** (Scheme 72) containing fluorenone units have been used as green EL materials ($\lambda_{max} = 535$ nm) with the optimal efficiency of 0.19% being obtained for the copolymer with 3 mol % of fluorenone [251].

Recently, Shu, Jen and coworkers [252] prepared a fluorene **152** with two electron-accepting oxadiazole substituents by nucleophilic substitution of 4-fluorobenzonitrile with the fluorenyl anion, and subsequent conversion of the nitriles to oxadiazoles via tetrazole intermediates (Scheme 73). The alternating copolymer with dioctylfluorene **594** (Scheme 72) showed better colour stability and higher efficiency in a single-layer device (0.52%) than poly(dioctylfluorene) (**560b**) (0.2%) due to its improved charge-accepting ability. The copolymer **595** ($m : n = 1 : 1$) with both hole- and electron-accepting substituents displayed even better efficiency (1.21%) [253].

4.1.8
Alternating Copolymers of Fluorenes and Other Arylenes

Copolymers with alternating fluorene and phenylene units (Scheme 74) can be readily prepared by Suzuki cross-coupling. The biphenylene copolymers **155** show violet-blue EL ($\lambda_{max} = 416$ nm) [254, 255], while the phenylene copolymers **156** are blue emitting ($\lambda_{max} = 420$ nm) [254–260]. Substitution of the phenylene ring with alkyl groups as in **157** produces a marked blue shift in the PL emission ($\lambda_{max} = 404$–409 nm) [207, 259], presumably due to increased torsion of the phenylene units. By contrast, the alkoxy substituents as in **158** produce almost no effect on the PL and EL maxima ($\lambda_{max} = 420$–425 nm) [207, 261, 262]. The reported EL spectrum of **158** shows no long wavelength emission band, and annealing of a film produces much less green emission than for PDAFs [207], suggesting that this material is stabler.

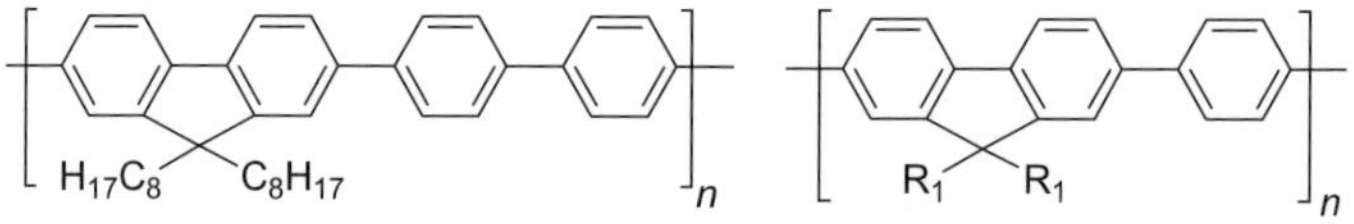

155

156 $R_1 = C_6H_{13}$ or C_8H_{17}, $R_2 = H$
157 $R_1, R_2 = C_6H_{13}$
158 $R_1 = C_6H_{13}$, $R_2 = OC_{10}H_{21}$

Scheme 74 Alternating copolymers of fluorene and other arylenes

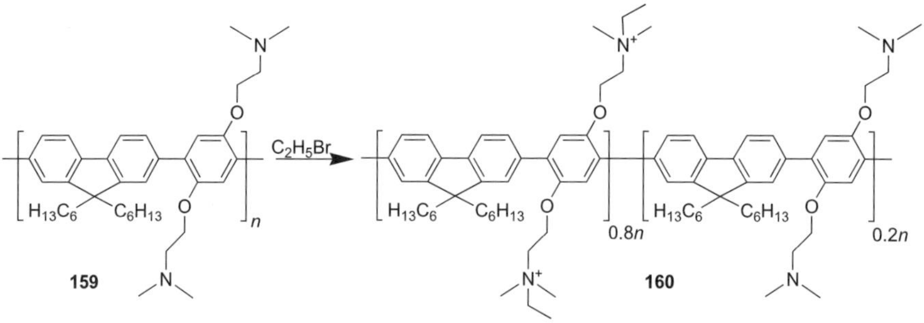

159 **160**

Scheme 75 Preparation of a water-soluble fluorene copolymer

Quaternisation of the amino side chains on the copolymer **159** (Scheme 75) with ethyl bromide causes a complete inversion in the solubility, with the 80% quaternised polymer **160** being soluble in DMSO, methanol, and water, but insoluble in chloroform and THF [263]. The emission shows some solvent dependence, with the maximum appearing at $\lambda_{max} = 409$ nm in water and methanol, and at $\lambda_{max} = 419$ nm in DMSO. A further effect of the quaternisation is that the PL quantum efficiency drops. The efficiency is much higher in methanol than in water (86% versus 25%), which was attributed to aggregation in aqueous solution.

A similar water-soluble copolymer **161** (Scheme 76) has been made by Bazan and coworkers [264] and used in a novel biosensor application [265, 266], in which energy transfer between **161** and a fluorescent dye attached to a strand of probe DNA occurs only in the presence of a DNA sequence complementary to that of the probe.

The carboxylate substituents on the phenylene in **162** (Scheme 77) produce a red shift in the PL ($\lambda_{max} = 443$ nm) but a drop in PL efficiency (20% versus 78–81% for **156**) [259]. Jen and coworkers have prepared the copolymers **163** ($\lambda_{max} = 447$ nm) and **164** ($\lambda_{max} = 477$ nm) with electron-accepting dicyanobenzene units by Suzuki polycondensations [267]. The alternating copolymer **163** shows only marginally higher EL efficiency than the PDAF homopolymer **85** (0.057% versus 0.044%), but the other copolymer **164** with only 25% of dicyanobenzene units ($m = n$) shows an EL efficiency over 10 times higher (0.50%) in double-layer devices with a PEDOT hole-transporting layer and calcium cathodes [267].

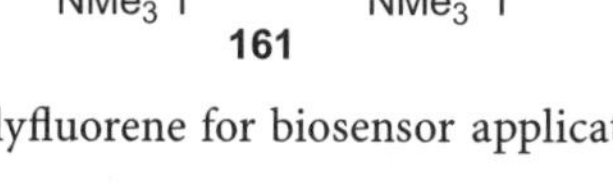

Scheme 76 Water-soluble polyfluorene for biosensor applications

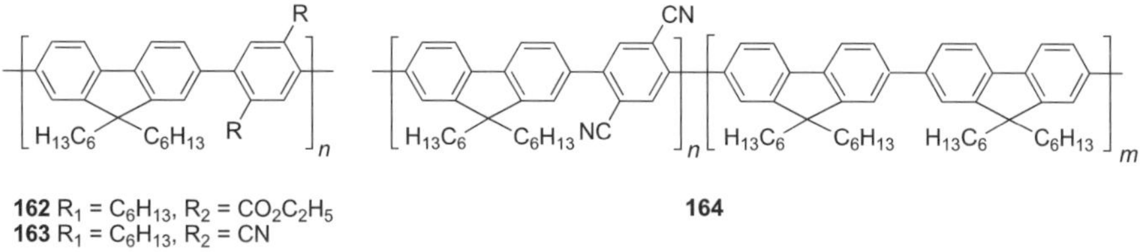

162 $R_1 = C_6H_{13}$, $R_2 = CO_2C_2H_5$
163 $R_1 = C_6H_{13}$, $R_2 = CN$

164

Scheme 77 Alternating fluorene–arylene copolymers with electron-accepting substituents

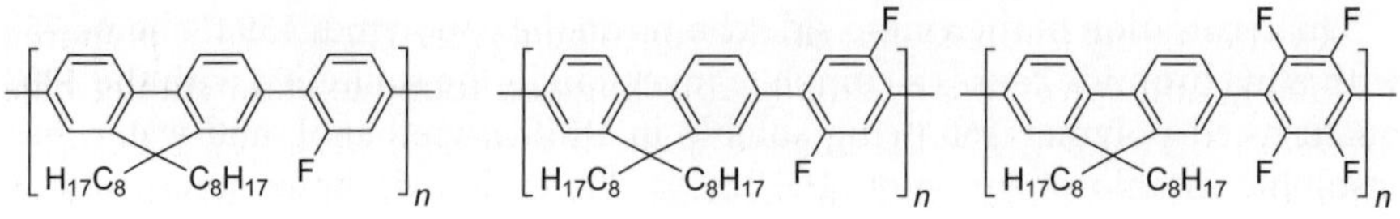

Scheme 78 Fluorophenyl–fluorene copolymers

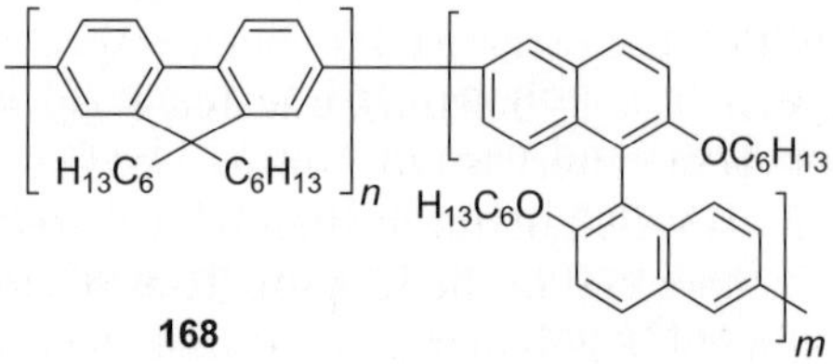

Scheme 79 Copolymer of fluorene and binaphthyl

A Japanese group have prepared the copolymers **165–167** (Scheme 78) [260]. The bulky fluorine substituents produce a blue shift in the PL emission due to increased steric repulsion, so that the PL maxima in solution shift from $\lambda_{max} = 400$ nm for **165** to $\lambda_{max} = 398$ nm and 382 nm for **166** and **167**. This blue shift is accompanied by a drop in the PL efficiencies (measured in solution) from 34% for **165** and **166** to 29% for **167** (cf. 78–81% for **156**).

Jen and coworkers have prepared a copolymer **168** ($n : m = 3 : 1$) of dihexylfluorene and binaphthyl by Suzuki copolymerisation (Scheme 79), which showed two EL maxima at $\lambda_{max} = 420$ and 446 nm in two-layer devices, whose relative intensity depended upon the device structure [268, 269]. External efficiencies of up to 0.79% were obtained.

An industrial group have reported obtaining efficient red, green, and blue emission from polyfluorene-based alternating copolymers prepared by Suzuki polycondensation, but without publishing their structures [270, 271].

4.1.9
Polymers Containing Spirobifluorene Units

Spirobifluorenes have been investigated by Salbeck [272] and found to be promising materials for use in blue LEDs, and so some effort has been made to incorporate these units into polymers (Scheme 80). Blue ($\lambda_{max} \sim 425$ nm) EL emission has been obtained from poly(spirobifluorene oligophenylene) copolymers **169** ($m = 0$–2) prepared by Suzuki cross-coupling [273]. These materials have somewhat limited solubility and so have not been fully characterised. By contrast, the polymer **170** bearing solubilising alkoxy groups is soluble in organic solvents [274]. No green PL emission was observed from a film of **170** that had been annealed in air at 200 °C for 3 h.

An alternating fluorene–spirobifluorene copolymer **171** has been prepared by Huang and coworkers by Suzuki coupling and found to show stabler blue emission than PDAFs with long-wavelength emission only being detected after the polymer film was heated to 150 °C [207, 275]. Blue emission with efficiencies of 0.12% and 0.54% was obtained from single- and double-layer (with a copper phthalocyanine hole-transporting layer) devices [275].

Meerholz and coworkers [276] have used Suzuki cross-coupling of a spirobifluorene bisboronate with various dibromo-comonomers to prepare cross-linkable copolymers **172–174** (Scheme 81) which respectively produced blue (λ_{max} = 457 nm), green (λ_{max} = 507 nm), and red (λ_{max} = 650 nm) EL. Illumination of films of these materials with ultraviolet light in the presence of a photoacid induced cationic polymerisation of the oxetane rings to cross-link the materials. Comparison of the EL properties of cross-linked **172** with a similar non-cross-linkable polymer showed that the cross-linking had no significant effect upon the luminescence of the materials. They used these materials to make a pixelated red, green, blue display in which each of the materials was patterned by irradiation through a mask.

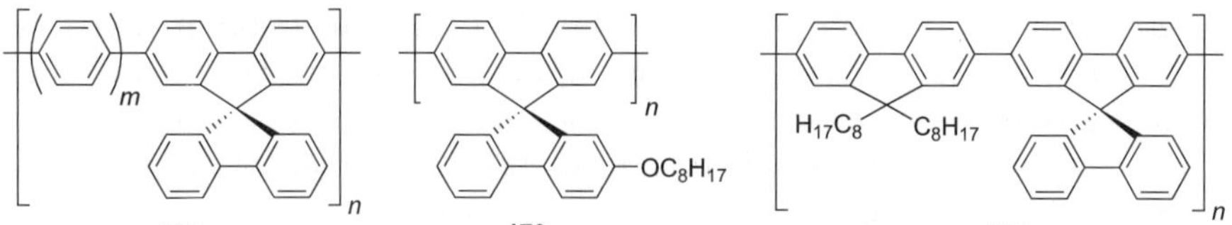

Scheme 80 Polymers containing spirobifluorene units

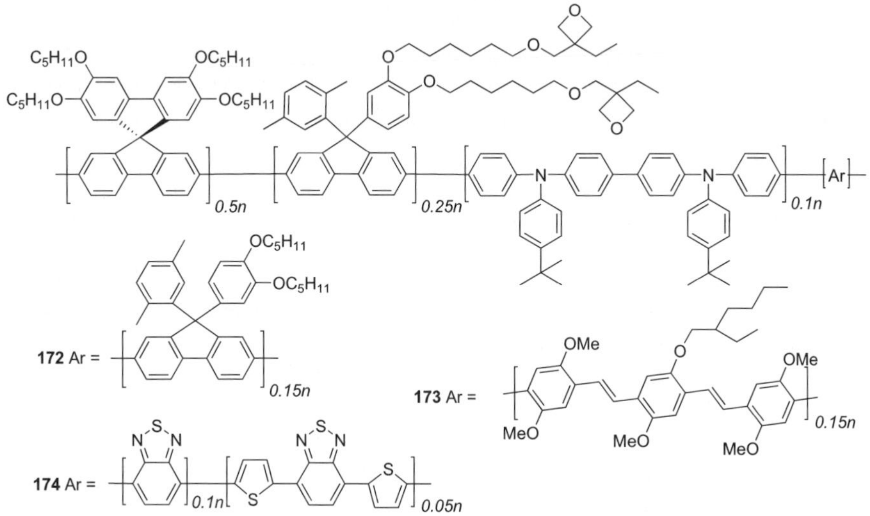

Scheme 81 Cross-linkable copolymers containing spirobifluorenes

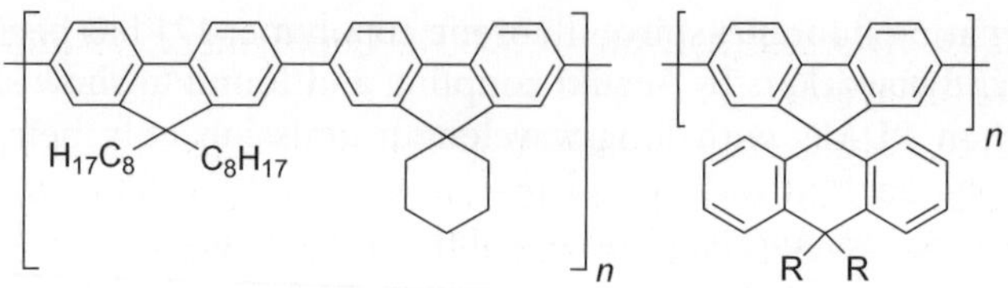

Scheme 82 Polymers containing spiro-functionalised fluorenes

Spiro-functionalisation with groups at C9 has also been examined as a means to stabilise blue emission from polyfluorenes. While polymers with spirocycloalkyl groups, e.g. **175** (Scheme 82), display green emission upon annealing [277], the polymer **176** with aryl groups at C9 produces stable blue PL and EL emission even after annealing [278].

4.2
Poly(indenofluorene)s

Poly(tetraalkylindenofluorene)s (PIFs, Scheme 83), which are intermediate in structure between PDAFs and LPPP, have been prepared by Müllen and co-workers by Yamamoto coupling [279]. In solution they showed strong blue PL with maxima around 430 nm, which made them attractive candidates for use in blue LEDs. By extrapolating the absorption and emission maxima of oligomers the effective conjugation lengths were determined to be $n = 5$–6 (15–18 benzene rings) for emission and $n = 6$–7 for absorption. Unfortunately, as with PDAFs, obtaining stable blue emission in the solid state has presented difficulties. The tetraoctylpolymer **177** is a green emitter in the solid state, with the PL and EL spectra being dominated by a broad emission band at 560 nm [280]. By contrast, films of the polymer **178** with ethylhexyl substituents show blue PL (λ_{max} = 429, 450 nm). The EL from **178** is initially blue, but rapidly red shifts, with a simultaneous loss of emission intensity, so that the devices have a half-life (time for the intensity to drop to half of its initial value) of less than 1 h. By contrast, the green emission from **177** is

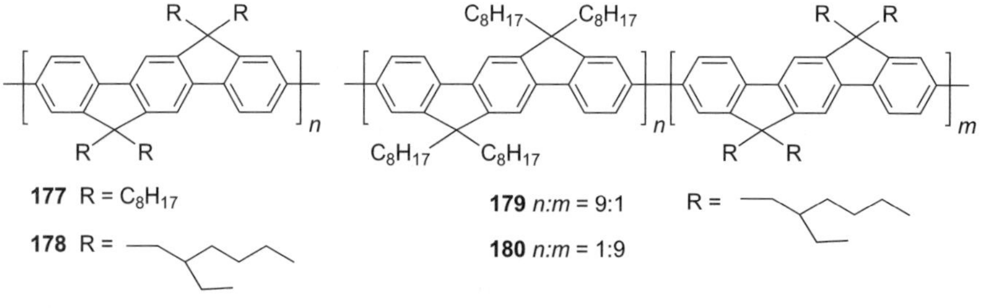

Scheme 83 Alkylindenofluorene-based polymers

stable, with an estimated half-life of 5000 h. Copolymers **179** and **180** show intermediate behaviour.

The source of this green emission might be either excimers or defects. The amount of green emission in the solid-state spectra has been shown to correlate well with the presence of long ordered structures due to π-stacking in the film morphology revealed by atomic force microscopy studies [280]. These results are consistent with the green emission arising from aggregates. Certainly the greater solid-state PL efficiency for **178** (36%) and the copolymers **179** (40%) and **180** (50%) compared with **177** (24%) is consistent with the bulkier branched side chains reducing interchain interactions and so reducing the possibility of non-radiative decay. An alternative explanation would be that the emission arises from ketone defects **181** (Scheme 84), which would be expected to show green emission intermediate between the defects **69** (530 nm) and **70** (565 nm) proposed for PDAFs and LPPPs, respectively. The above EL results would then be explicable in terms of the relative ability of excitons to diffuse to defect sites in the polymer films. To resolve this, the ketone **182** has been prepared as a model compound for such defects. The emission properties of **182** match those of defects detected by time-resolved PL measurements of **177**–**178** [281].

The phase behaviour of PIFs **177** and **178** resembles that of the corresponding PDAFs **89** and **90**, but with higher phase-transition temperatures. Thus, the octyl polymer **177** shows two nematic phases with transition temperatures at 250 and 290 °C (reverse transitions at 270 and 140 °C), while the ethylhexyl polymer **178** has only a single nematic phase with a transition at 290 °C (220 °C for the reverse transition) [279].

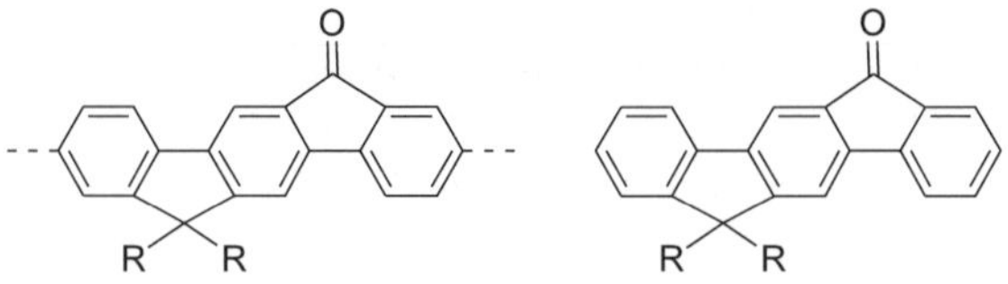

Scheme 84 Proposed emissive defect in polyindenofluorenes

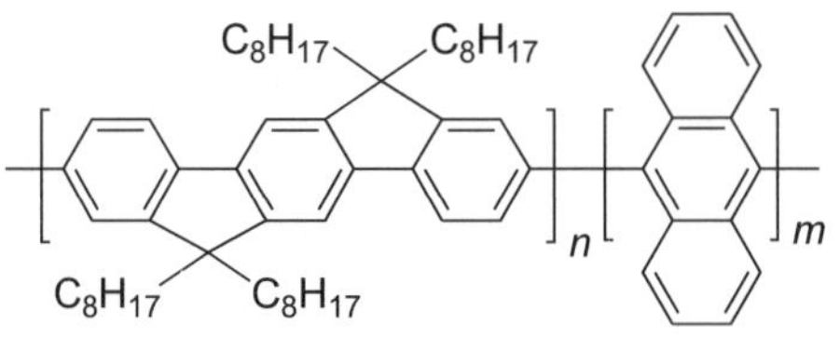

Scheme 85 Indenofluorene–anthracene copolymer which produces stable blue emission

As with PDAFs, copolymers with anthracene **183** (Scheme 85, $n:m = 85:15$) show stable blue EL ($\lambda_{max} = 445$ nm), probably due to exciton confinement [282].

Since aryl substituents have been shown to greatly enhance the stability of blue emission from PDAFs, the effect of such substituents in PIFs has also been investigated. The synthesis of the tetraarylmonomers (Scheme 86) follows a route similar to that used for the diarylfluorenes **131** (see Sect. 4.1, above). Suzuki coupling of a dibromoterephthalate [157] with commercially available 4-trimethylsilylbenzeneboronic acid gives the terphenyl **184** in high (92%) yield. Treatment of this with four equivalents of an aryl lithium followed by electrophilic displacement of the silyl group with bromine and then ring closure produces tetraarylindenofluorene monomers **185** in good (70%) yield. These were then polymerised with nickel(0) [230].

Due to the limited solubilising power of the substituents, only oligomers ($n = 2$–6) of the *tert*-butylphenyl polymer **186** (Scheme 87) were obtained. The octylphenyl groups in **187** provided much better solubility, so that this polymer was obtained with $M_n = 66\,400$ g/mol, $M_w/M_n = 3.86$ (measured against a polyphenylene standard), corresponding to a degree of polymerisation of about $n = 66$ (~ 200 phenylene rings).

The PL emission from **187** is blue both in solution ($\lambda_{max} = 428$ nm) and in thin films ($\lambda_{max} = 434$ nm), with no sign of the green-emission band seen for **177**. Some unoptimised LEDs have been constructed using **187**, which have rather high operating voltages (16 V) due to problems with preparing good-quality films. The EL spectrum has its maximum at 434 nm in the blue but with a long tail extending into the red region of the spectrum, so the emission is blue-white in colour. The emission is much stabler than from **177**–**179** above with only a slight increase in the long-wavelength emission being seen after several minutes operation of a device at 16 V. This suggests that improved devices with lower operating voltages might show stable blue EL

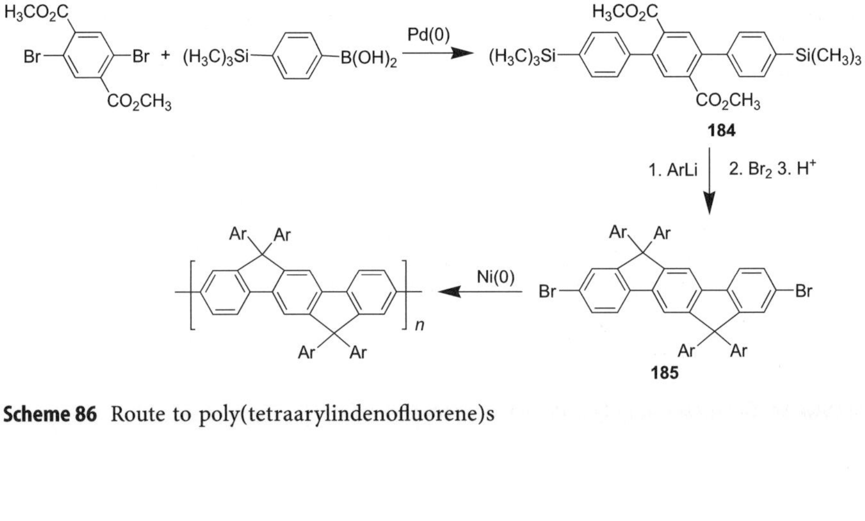

Scheme 86 Route to poly(tetraarylindenofluorene)s

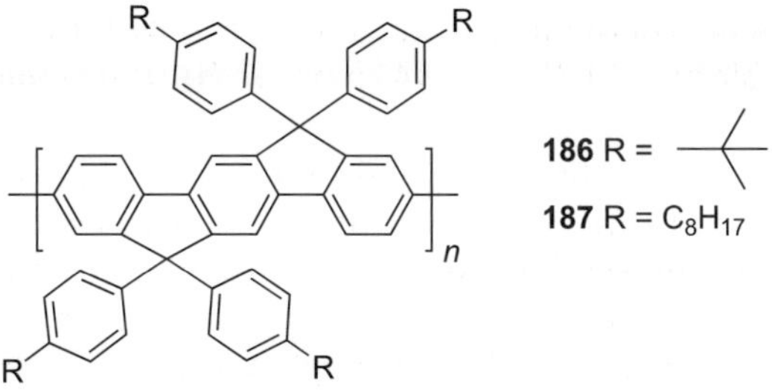

Scheme 87 Poly[tetra(4-alkylphenyl)indenofluorene]s

186 R = —C(CH₃)₃ *(tert-butyl group shown)*
187 R = C₈H₁₇

emission. Blending 0.3 wt % of the ketone **182** into **187** has been found to totally suppress the blue emission [281], thus demonstrating that even very low levels of emissive defects can profoundly affect the emission properties of conjugated polymers.

4.3
Poly(ladder-type pentaphenylene)s

A modification of the synthesis of the tetrarylindenofluorenes **186** shown in Scheme 86 has been used to prepare ladder-type pentaphenylenes (Scheme 88) [283]. The dibromoterephthalate was coupled with two equivalents of a fluorene boronate instead of a benzene boronic acid to give a pentaphenylene **188**. Reaction with an aryl lithium, followed by ring closure with boron trifluoride, gave the ladder-type pentamers **189**, which were brominated to prepare dibromomonomers that were then polymerised using nickel(0).

The polymer **190** produces blue PL and EL with an emission maximum at 445 nm and a very small Stokes' shift, and so closes the gap in emission colour between PIFs and LPPPs. The optical properties are very similar to those of

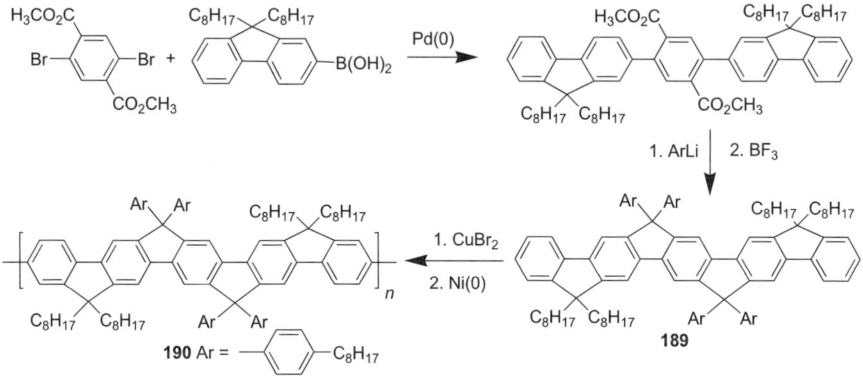

Scheme 88 Synthesis of poly(ladder-type pentaphenylene)s

LPPP **3**, but the blue emission is much more stable. Polymer **190** also has a remarkably large persistence length of 25 nm [284], indicating a highly rigid structure.

4.4
Other 'Stepladder'-type Polyphenylenes

PFs, PIFs, and poly(ladder-type pentaphenylene)s have methine bridges, and so are structurally related to LPPP **3**. Stepladder polymers have also been prepared with other types of bridges.

4.4.1
Poly(tetrahydropyrene)s

Poly(tetrahydropyrene)s **65** (Scheme 28), like polyfluorenes, have a bridged biphenyl unit, but with two ethane bridges the angle between the two phenylene rings is about 20°. Poly(2,7-dioctyl-4,5,9,10-tetrahydropyrene) (**191**) has been synthesised by Müllen and coworkers (Scheme 89) [285] by Yamamoto coupling of the dibromomonomer **192**. The monomer was prepared by a photocyclisation of a 2,2'-bis(1''-alkenyl)biphenyl **193**, followed by bromination. The molecular mass of the polymer (M_n = 20 000 g/mol) corresponds to about 46 monomer units (92 phenylene rings).

The emission from **191** is blue in solution (λ_{max} = 425 nm) and in thin films (λ_{max} = 457 nm) [285] and **191** has been used to make a blue-green-emitting LED (λ_{max} = 457 nm) with an efficiency of 0.10–0.15% [7, 286]. The large red shift between the solution and solid-state emission may be due to aggregation. Interestingly, the effective conjugation length for absorption in **191** is about 20 benzene rings [120], which is longer than for LPPP (**3**) (11 rings), but comparable with the values for PDAFs (24 rings) and PIFs (18–21 rings).

The monomer **192** is a mixture of *cis*- and *trans*-diastereomers, which can be separated, thus making stereoregular polymers accessible. Such polymers might show different properties, due to different stacking of the polymer chains.

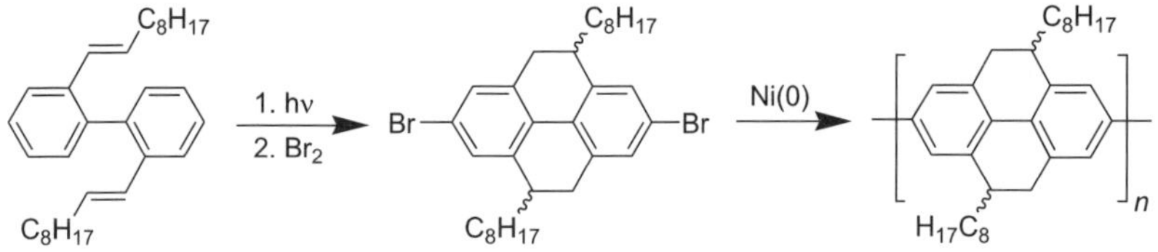

Scheme 89 Synthesis of poly(2,7-dioctyl-4,5,9,10-tetrahydropyrene)

4.4.2
Polycarbazoles

Carbazole is isoelectronic with fluorene, but the nitrogen bridge significantly alters the chemical behaviour. Whereas electrophilic substitution of fluorene occurs preferentially at the 2- and 7-positions, the strong electronic effect of the nitrogen directs substitution to the 3- and 6-positions of carbazole. As a result 3,6-carbazole polymers are much more synthetically accessible than 2,7-carbazole polymers. The chemistry and applications of carbazole-based polymers have been the subject of two major recent reviews to which the reader is referred [287, 288].

Poly(N-alkyl-3,6-carbazole)s **194** are readily obtained by condensation of N-alkyl-3,6-dibromocarbazoles with electrogenerated nickel(0) (Scheme 90) [289–291]. These polymers show blue (λ_{max} = 470 nm) PL [291] and EL [292].

Poly(N-alkyl-2,7-carbazole)s are harder to prepare than the 3,6-polymers **194** as the 2,7-halocarbazole monomers cannot be obtained by halogenation of carbazole. Morin and Leclerc have developed a short efficient route to 2,7-dichlorocarbazole **195** (Scheme 91) [293]. Suzuki condensation of 4-chlorobenzeneboronic acid with 2-bromo-5-chloronitrobenzene occurs preferentially with the more reactive bromide to give 4,4′dichloro-2-nitrobiphenyl, which is then reductively ring closed with triethyl phosphite to give **195** in 55% overall yield. They also prepared 2,7-iodocarbazole **196** by a longer and much less efficient route in which the ring closure was achieved by heating a 2-azidobiphenyl. Müllen and coworkers have synthesised 2,7-dibromocarbazole **197** in 51% overall yield by nitration of 4,4′-dibromobiphenyl followed by ring closure with triethyl phosphate [294].

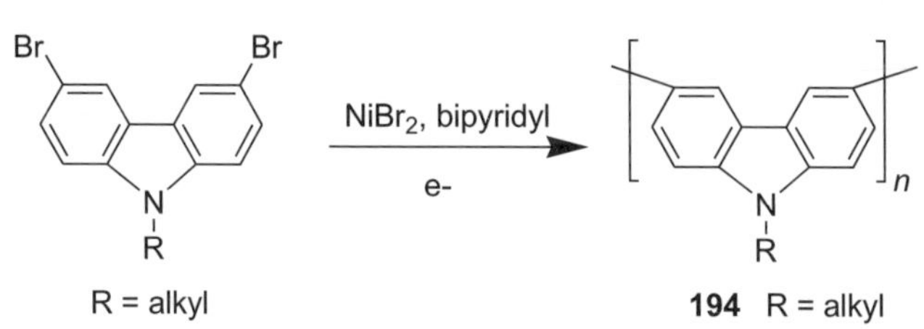

Scheme 90 Poly(N-alkyl-3,6-carbazole)s by electropolymerisation

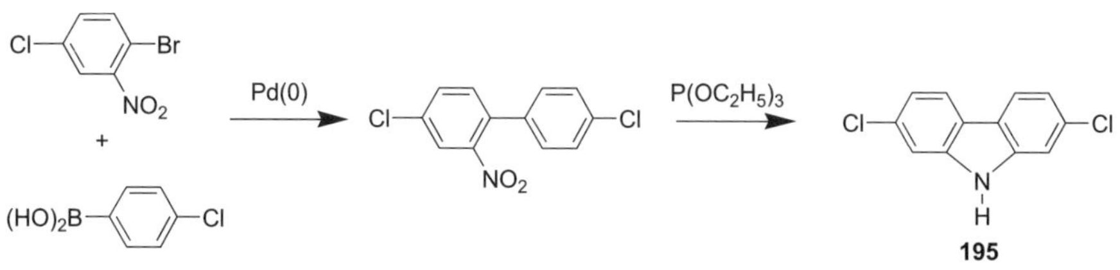

Scheme 91 Synthesis of 2,7-dichlorocarbazole

Poly(*N*-alkyl-2,7-carbazole)s (**198**) have then been made by alkylation of **195–197**, followed by polymerisation with Ni(0) (Scheme 92) [293, 295] or magnesium [296]. A ditosylate monomer **199** has also been used [297]. Due to the lower number of solubilising groups the polymers **198** are less soluble than the corresponding PDAFs, so that to obtain high molecular weight materials large branched chain solubilising groups are needed [296, 298].

Polymers **198** produce blue PL in both solution (λ_{max} = 417 nm) and as a thin film (λ_{max} = 437, 453 nm) [293, 295]. Unlike the structurally very similar PDAFs (Sect. 4.1) these polymers show no long-wavelength emission in the solid state. This is further evidence that the green-emission band seen in PDAFs does not arise from an aggregate, as chains of **198** should be able to π-stack at least as well as those of poly(dioctylfluorene) (**89**). The copolymer **200** with dioctylfluorene was made by Suzuki coupling of a diiodocarbazole and a dioctylfluorene-2,7-bisboronate (Scheme 93) [293]. This copolymer also shows stable blue PL in the solid state (λ_{max} = 450 nm). Suzuki coupling has also been used to prepare other copolymers containing dialkoxyphenylene (**201**), binaphthyl (**202**), and alkylpyridine (**203**) units, which produce stable blue emission [295, 299].

Blue-emitting LEDs have been prepared using **198** with octyl and ethylhexyl substituents, but with only modest efficiencies [300]. Electrochemical

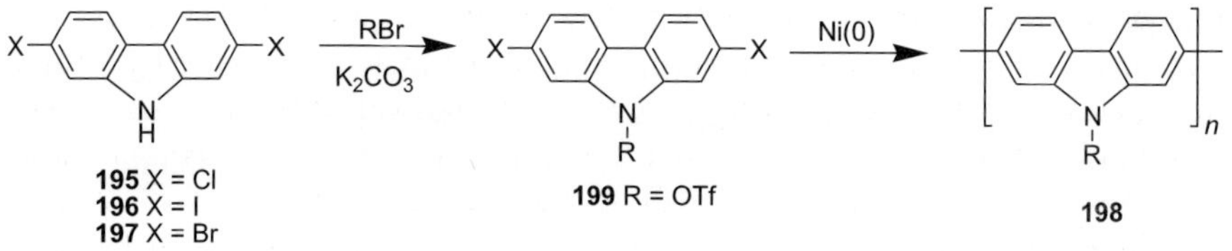

Scheme 92 Synthesis of poly(*N*-alkyl-2,7-carbazole)s

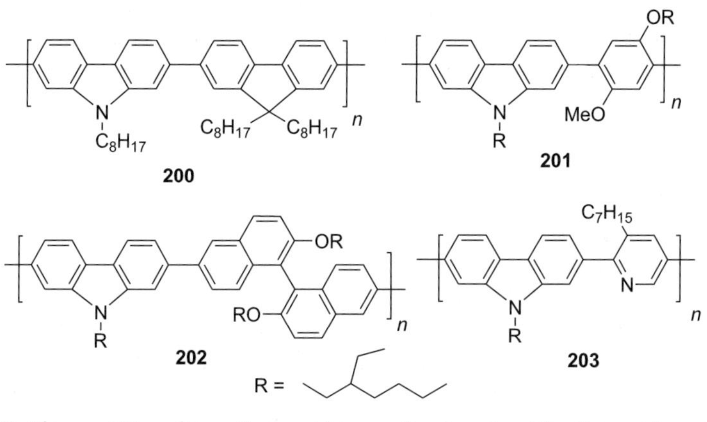

Scheme 93 Blue-emitting alternating copolymer of carbazole with other arylenes

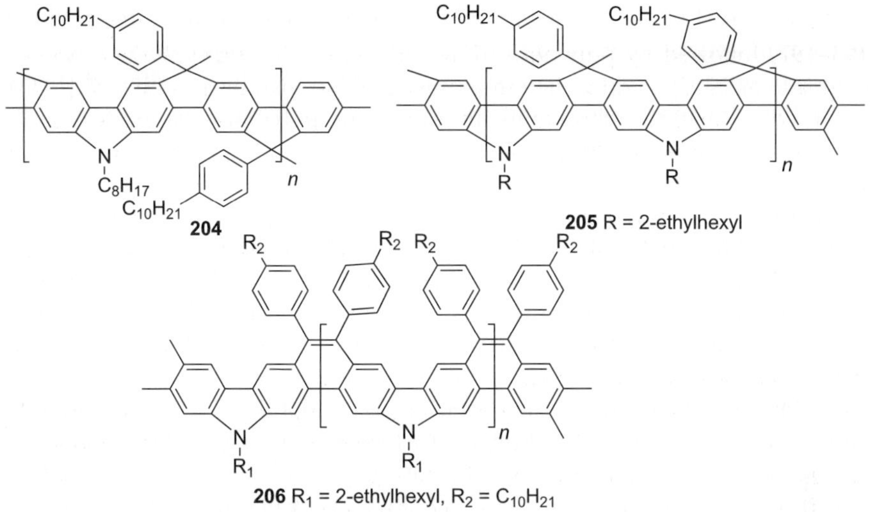

Polyphenylene-type Emissive Materials 65

Scheme 94 Carbazole-based ladder polymers

204
205 R = 2-ethylhexyl
206 R_1 = 2-ethylhexyl, R_2 = $C_{10}H_{21}$

measurements suggest that these materials are unstable under electrolytic conditions, probably due to the high electron densities at the unsubstituted carbazole 3- and 6-positions promoting coupling reactions, which implies that devices using them are likely to have short lifetimes [296]. This problem might be overcome by incorporating carbazoles into ladder structures in which the carbazoles are bridged via the 3- and 6-positions (Scheme 94). Scherf and coworkers have prepared the ladder-type polymer **204** by a modification of the synthesis of Me-LPPP using a carbazole bisboronate in place of a benzene bisboronate [301]. This polymer displays blue-green PL with a maximum at 470 nm, which is very slightly red shifted compared to Me-LPPP. Müllen and coworkers have prepared the ladder-type polycarbazoles **205** and **206**, which are also blue-green emitters in dilute solution, with emission maxima at 471 nm and 467 nm, respectively [302]. At higher concentrations polymer **205** shows broad, red-shifted emission due to aggregation.

5
Polyphenylene-based Block Copolymers

Polyphenylene-based block copolymers come in two varieties: rod–coil copolymers with a rigid rod-like PPP segment attached to a (usually non-luminescent) flexible coil or coils, and rod–rod copolymers in which the PPP rod(s) are attached to another conjugated block. To date most polyphenylene-based block copolymers reported have been rod–coil copolymers. Here the coils may contain charge-transporting units, for example the triblock copolymers **138** and **139**, or may act as solubilising or phase-forming units. Phase separation between the rod and coil units in rod–coil copolymers has been shown to produce ordered nanostructures, which may have profound effects upon the film morphology and the electronic properties of these materials [303].

5.1
Rod–Coil Block Copolymers

There exist two approaches to the synthesis of rod–coil block copolymers: first, the 'grafting from' approach where one component is used as a macroinitiator for the preparation of the other, and, secondly, the 'grafting onto' method where the second component is covalently attached to a suitably end-functionalised first component. The first approach was used to prepare block copolymers of polystyrene (PS) with PPP **207** (Scheme 95) by François and coworkers [304–308] and by Advincula and coworkers [309]. In this synthesis, cyclohexadiene was added to 'living' polystyrene to give a precursor polymer **208**, which was then aromatised to **207** using chloranil in refluxing xylene.

The ultraviolet absorption spectrum of **207** reveals that the PPP units are at most 11–12 benzene rings long, which is shorter than the length of the poly(cyclohexadiene) block in the precursor **208**, indicating that PPP blocks contain defects [304–306]. As in the synthesis of PPP via a poly(cyclohexadiene), 1,2- as well as 1,4-linkages are formed during the polymerisation of the cyclohexadiene. Light-scattering experiments showed that the copolymer forms micelle-like structures with aggregates of PPP blocks surrounded by a shell of polystyrene coils [304, 307]. As a result, films of **207** show an unusual honeycomb-like morphology [308, 310]. As a result of the aggregation of the PPP units, **207** exhibits different optical properties

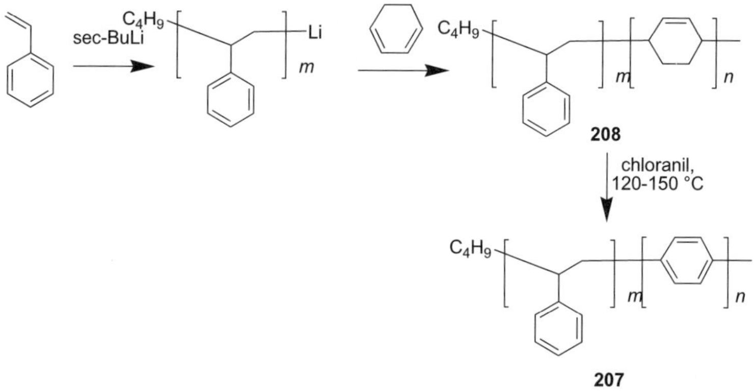

Scheme 95 Synthesis of PPP-PS block copolymers by 'grafting from' method

to PPP (**1**) and the blue EL emission width is much narrower than the PL emission [311, 312].

In the above case, the coil was used as the macroinitiator for the synthesis of the rod. The reverse case has been used by Müllen and coworkers to make the PDAF-PEO copolymer **209** (Scheme 96) [313]. Here addition of bromobenzene to the Suzuki polycondensation of a 2-bromofluorene-7-boronic acid **210** partially end-capped the growing chain at one terminus with a phenyl group, with the remaining chains bearing a terminal hydrogen at this position, while the other terminus completely retained its bromine substituent. The incomplete end-capping is due to the well-known tendency for aryl boronates to fall off under the coupling conditions. Suzuki coupling of this monobromo-PDAF with 4-formylbenzeneboronic acid produced a PDAF **211** end-capped completely at one chain terminus with an aldehyde functionality. This had a molecular weight by GPC and NMR analysis of about 3500 g/mol corresponding to a degree of polymerisation of 8. Reduction of **211** with borohydride, followed by titration with potassium naphthalide, gave the alkoxide **212** which acted as the macroinitiator for the anionic polymerisation of ethylene oxide to give the desired copolymer **209**. NMR analysis suggested that the length of the PEO coil was about 5–6 units, while GPC suggested that it was about 10 units.

Müllen and coworkers have also used the alternative 'grafting onto' approach to make a number of rod–coil copolymers, by making end-capped conjugated rods and attaching non-conjugated coils (Scheme 97). Thus, they produced an end-capped PPP **213** by the same methodology as above [314]. The degree of polymerisation in **194** was also low ($n \sim 10$). Addition of a 'living' polystyrene chain to **213** gave the PPP-PS copolymer **214**, while condensation of **213** with a commercially available amino-terminated poly(ethylene oxide) produced a PPP-PEO copolymer **215**.

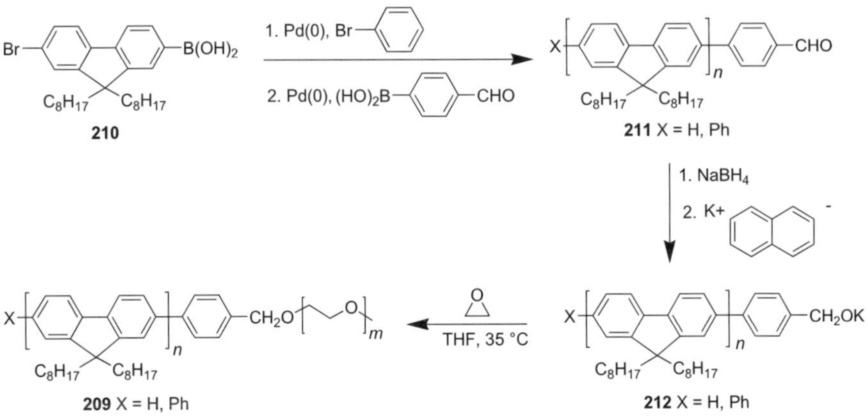

Scheme 96 Synthesis of polyfluorene–poly(ethylene oxide) rod–coil copolymer

The doubly end-capped polyfluorenes **216** and **217**, with phenol or carboxylic acid functional groups at the chain termini, have been made by addition of 13 mol % of suitable end-capping reagents to the Yamamoto polymerisation of a dibromofluorene followed by conversion of the protected functionalities (Scheme 98) [315]. The materials appeared to be totally end-capped by ^{1}H NMR analysis.

Polystyrene chains have been grafted onto the bisacid **217** (M_n = 7140 g/mol, M_w/M_n = 236, n = 18) by activation with N,N'-carbonyldiimidazole, followed by reaction with a hydroxy-terminated polystyrene (M_n = 9300 g/mol, M_w/M_n = 1.07, m = 87) to give the coil–rod–coil triblock copolymer **218** (M_n = 26 400 g/mol, M_w/M_n = 2.39) (Scheme 99) [315].

The addition of an acid-terminated PEO chain (M_n = 5860 g/mol, M_w/M_n = 1.06, m = 112) to the bisphenol **216** (M_n = 11 300 g/mol, M_w/M_n = 1.79, n = 29) under the same conditions was found to give a mixture of the diblock **219** (M_n = 17 230 g/mol, M_w/M_n = 2.10) and triblock **220** (M_n = 23 170 g/mol, M_w/M_n = 2.37) copolymers, which were separable owing to their differential solubility in methanol (Scheme 100) [315].

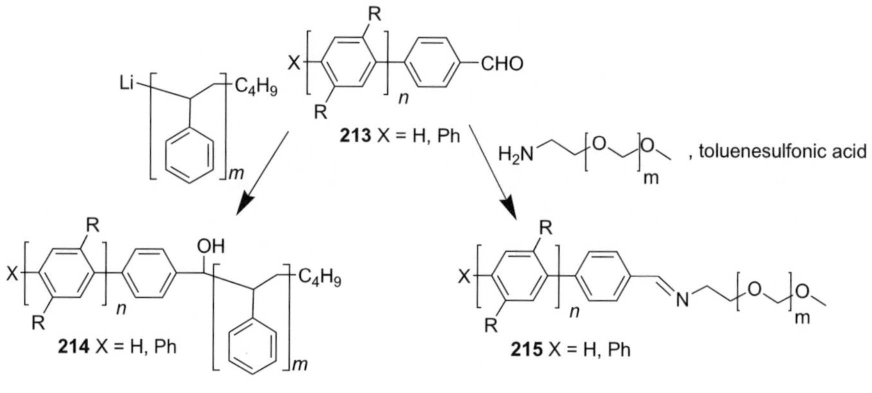

Scheme 97 PPP-based rod–coil copolymers by 'grafting onto' method

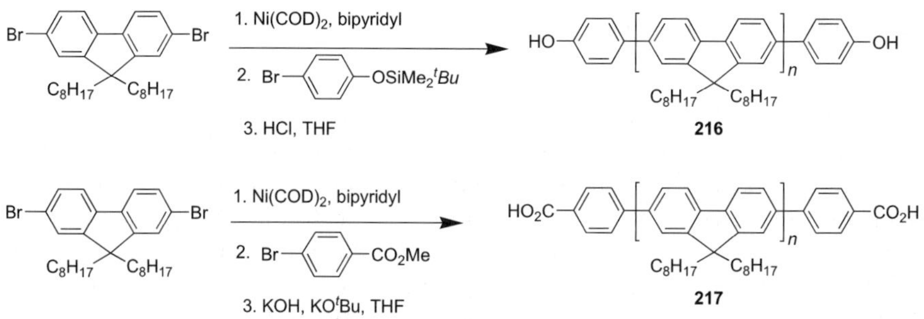

Scheme 98 Synthesis of doubly end-capped polyfluorenes as units for rod–coil copolymers

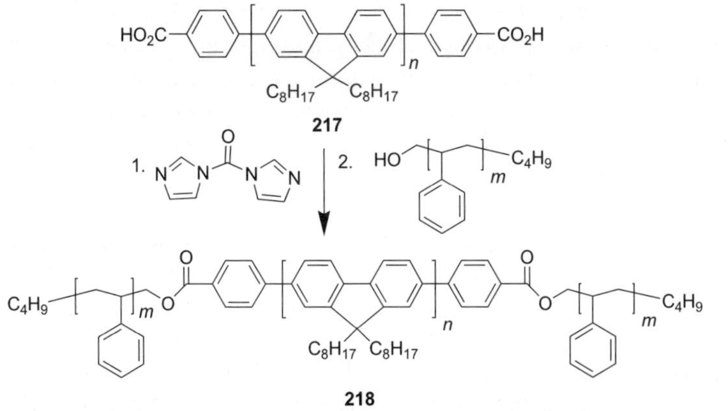

Scheme 99 Synthesis of polyfluorene–polystyrene coil–rod–coil triblock copolymers

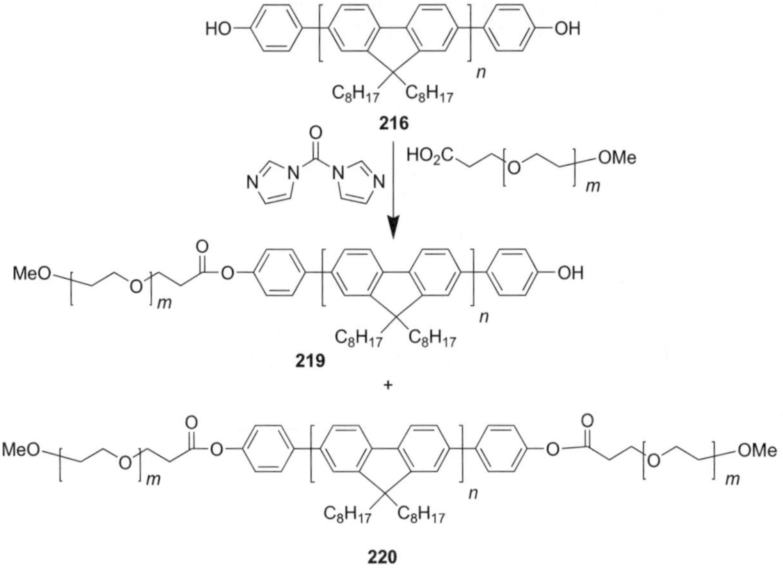

Scheme 100 Synthesis of polyfluorene–poly(ethylene oxide) di- and triblock copolymers

Scherf and coworkers have prepared di- and triblock copolymers of polyfluorene and polyaniline (Scheme 101) by attaching suitable end groups in both Suzuki and Yamamoto polymerisations. Addition of 10 mol % of 4-nitrobromobenzene as an end-capping reagent to an AA-BB Suzuki polycondensation of a fluorene bisboronate and a dibromofluorene (Scheme 41), followed by reduction with hydrogen, gave a mixture of **221** and **223** (M_n = 7700 g/mol, M_w/M_n = 2.9, $n \sim 19$). MALDI-TOF analysis of the nitro-end-capped material showed that some of the mono-end-capped chains had

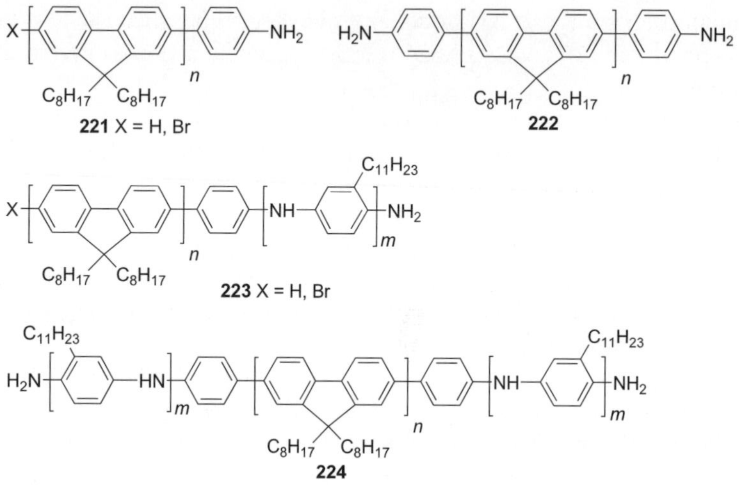

Scheme 101 Polyfluorene–polyaniline block copolymers and precursors

a bromine while others had a hydrogen at the other terminus but a quantitative analysis was not possible. These compounds were oxidatively coupled with 2-undecanylaniline to give a mixture of the PF-PANI diblock **223** and triblock **224** copolymers ($M_n = 14\,400$, $M_w/M_n = 4.3$) [316]. NMR analysis of the end-capped polyfluorenes suggested that the end-capping introduced about 1–1.2 end groups per polymer chain, so that the final product was largely the diblock copolymer **223**. From the UV-VIS spectrum, the PANI blocks appeared to be about 6–7 units long. By contrast, addition of 15 mol % of 4-bromoaniline to a Yamamoto polycondensation of a dibromofluorene (Scheme 42) produced complete end-capping to give only **222** ($M_n = 10\,400$ g/mol, $M_w/M_n = 3.58$, $n \sim 25$) [317]. Oxidative coupling of this with undecanylaniline then gave the triblock copolymer **224** ($M_n = 36\,900$ g/mol, $M_w/M_n = 1.55$). Elemental analysis suggested that the PANI blocks were about 7–8 units long.

The optical properties of the PPP copolymers **214** and **215** do not differ from the homopolymer, but the PDAF copolymers **213** and **218–220** show slight differences in their solid-state PL spectra from the homopolymer **89**, which suggests that there is a higher degree of order in the films [313–315]. Studies of the morphology of some of these rod–coil copolymers have shown that they form extended one-dimensional fibrillar structures due to π-stacking of the rods [236, 318]. The ratio of the block sizes in PDAF-PEO copolymers has been shown to strongly influence the packing behaviour, with fibrillar structures being observed only for copolymers with an average volume ratio of PEO of 0.3 or below [319]. However, no study of the effects of the coils on the luminescence properties of these polymers has been reported yet.

Though their synthesis has not been published, it is probable that the triblock copolymers **138** and **139** [243] were prepared by a similar grafting onto procedure.

There also exist syntheses of rod-coil copolymers which do not fit neatly into either of the above categories. Miller and coworkers have prepared coil-rod-coil triblock copolymers in a one-pot method by using end-capping agents in the Yamamoto polymerisation of dibromodihexylfluorene, which also act as initiators for the formation of the coils [320]. This method can be thought of as a combination of the two approaches. Thus, nickel(0) polymerisation of dibromodihexylfluorene and living radical polymerisation of styrene in the presence of the end-capper/initiator **225** proceeded together in the same pot to produce **226** (M_n = 36 000 g/mol, M_w/M_n = 1.52), which by NMR analysis incorporated the styrene and fluorene monomers in a mole ratio of 7 : 3 (Scheme 102).

Similarly, nickel(0) polymerisation of the fluorene and ring-opening polymerisation of caprolactone in the presence of **227** proceeded to give **228** (Scheme 103). By increasing the mole ratio of caprolactone to fluorene monomer in the reaction mixture from 8 : 1 to 25 : 1, the molecular mass of the polymer could be increased from M_n = 17 000 g/mol to 36 000 g/mol, with an accompanying drop in the polydispersity from M_w/M_n = 2.39 to 1.65. NMR analysis showed that the relative amount of the fluorene in the polymer was only about 65–75% of the mole fraction of the fluorene monomer in the reaction mixture.

The copolymer with PMMA **229** (Scheme 104), however, had to be made in a step-wise 'grafting onto' procedure, as the initiator for the atom transfer radical polymerisation (ATRP) of methyl methacrylate was not stable under Yamamoto reaction conditions.

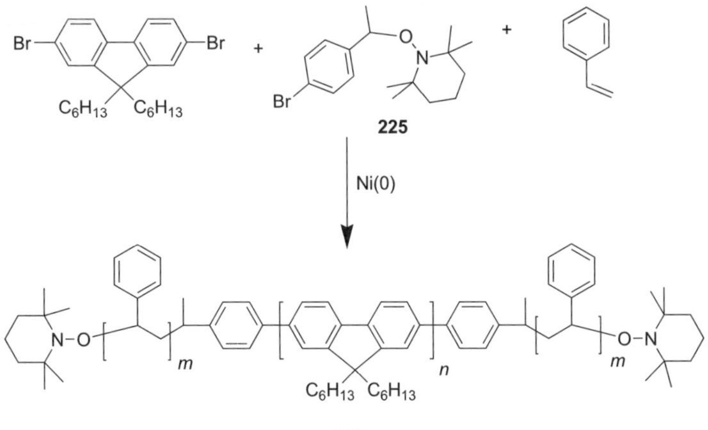

Scheme 102 One-pot synthesis of polyfluorene–polystyrene triblock copolymers

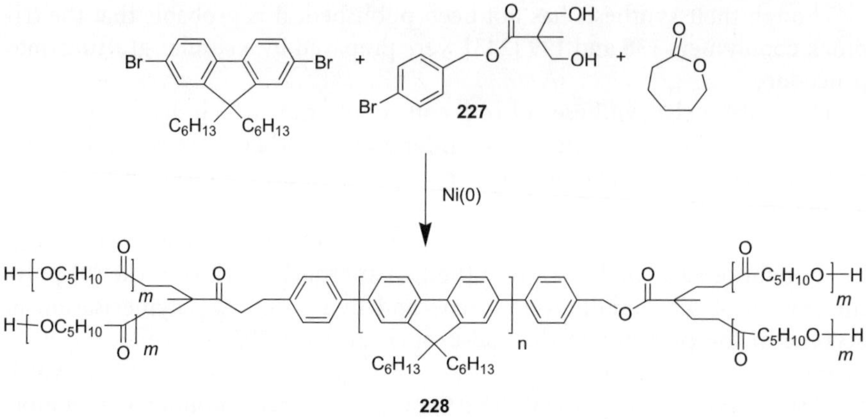

Scheme 103 One-pot copolymerisation of dioctylfluorene and caprolactone

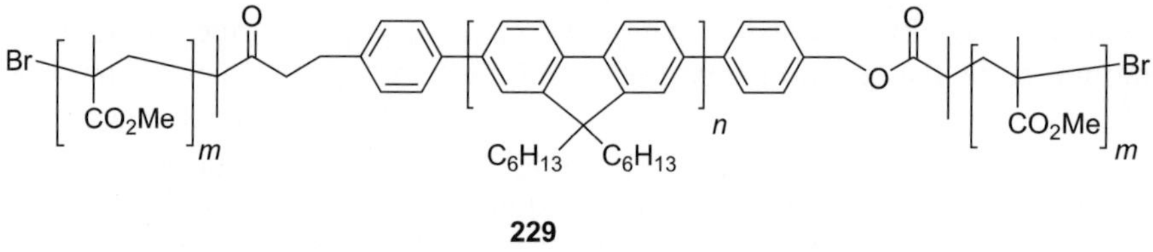

Scheme 104 Polyfluorene–PMMA triblock copolymer made by two-step process

5.2
Rod–Rod Block Copolymers

The only example of rod–rod copolymers with polyphenylene-based rods are the polymers **230** and **231**, in which two polyfluorene blocks are separated by a short arylene vinylene unit (Scheme 105). These were prepared by Suzuki polycondensation of a fluorene monoboronic acid **232** in the presence of the bromo-substituted oligo(arylene vinylene) units **233** and **234** [321]. From the molecular masses of the polymers, the total number of fluorene units in both **230** and **231** is about 18.

The PL in solution comes from the fluorene blocks (λ_{max} = 425 nm), but in the solid state energy transfer to the arylene vinylene units produces emission only from them ($\lambda_{max} \sim$ 525 nm and $\sim$ 490 nm for **230** and **231**, respectively). As the polymers are surely contaminated with the fluorene homopolymer, inter- as well intrachain energy transfer must be occurring. These polymers are promising candidates for polymer lasers [322].

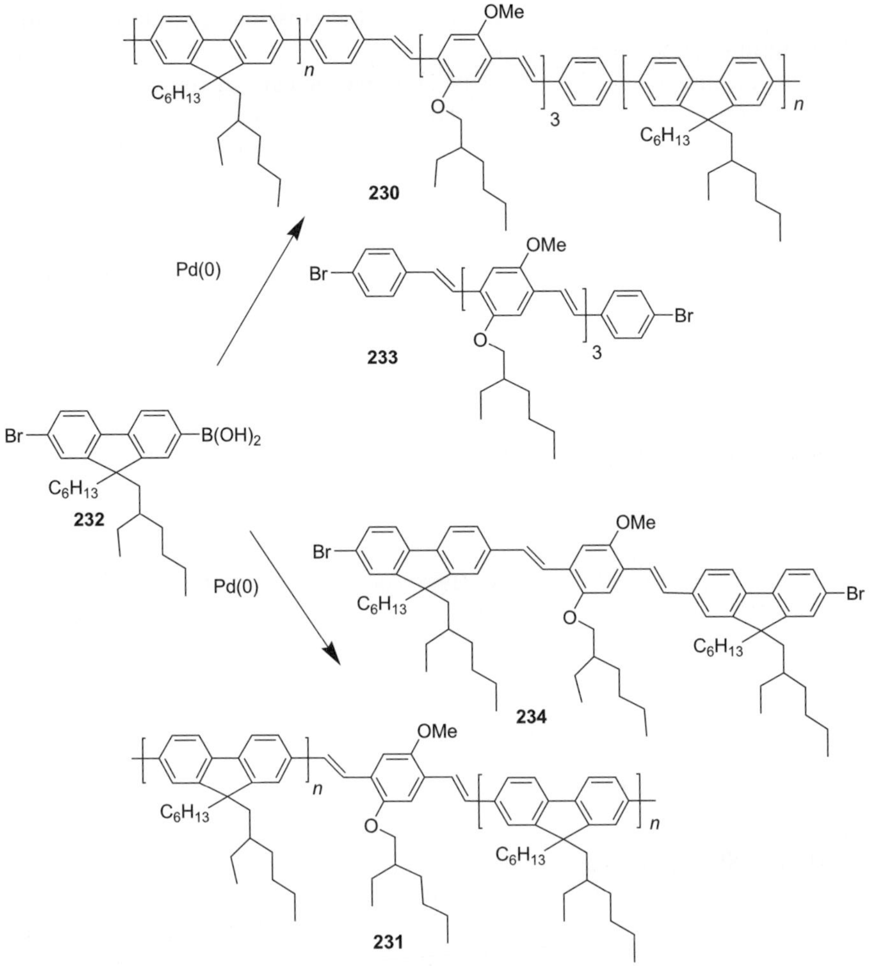

Scheme 105 Fluorene–(arylene vinylene) block copolymers

6
Conclusion

As shown in this review, the properties of polyphenylene-based materials, so as to maximise their potential as active materials in LEDs or polymer lasers, can be controlled by deliberate synthetic design. By incorporating bridges between some or all of the phenylene units to make ladder or stepladder polymers the effective conjugation length of the polymer may be controlled, while the interactions between the chains and the injection of charges may be regulated by careful selection of substituents. By these means it is possible to minimise interchain interactions, which lead to loss of luminescence efficiency and/or undesirable red shifts in the emission spectrum. The goal of obtaining stable blue emission now appears to be attainable if steps are taken to minimise the formation of emissive defects in the polymers and the diffusion of excitons to them. The emission colour can also be tuned efficiently over the whole visible spectrum by incorporation of suitable chromophores. With ongoing interdisciplinary research efforts, the fabrication of polyphenylene-based high-efficiency full-colour LED-based displays with long lifetimes may soon be possible. The prospects for polyphenylene-based lasers also appear good, though much work remains to be done in this field.

Acknowledgements We gratefully acknowledge the financial support given to our research into emissive polyphenylenes by the Bundesministerium für Bildung und Forschung. We also thank Prof. Ullrich Scherf for communicating to us some of his unpublished results and manuscripts.

References

1. Kraft A, Grimsdale AC, Holmes AB (1998) Angew Chem Int Ed 37:403
2. Mitschke U, Bäuerle P (2000) J Mater Chem 10:1471
3. McGehee MD, Heeger AJ (2000) Adv Mater 12:1655
4. Kim DY, Cho HN, Kim CY (2000) Prog Polym Sci 25:1089
5. Grell M, Bradley DDC (1999) Adv Mater 11:895
6. Kaeriyama K (1998) Plast Eng 48:33
7. Scherf U (1999) Top Curr Chem 201:163
8. Schlüter AD (2001) J Polym Sci A Polym Chem 39:1533
9. Yamamoto T (1992) Prog Polym Sci 17:1153
10. Stille JK (1986) Angew Chem Int Ed Engl 25:508
11. Tamao K, Kodama S, Kumada M, Minato A, Suzuki K (1982) Tetrahedron 38:3347
12. Kovacic P, Jones MB (1987) Chem Rev 87:357
13. Miyashita K, Kaneko M (1994) Macromol Rapid Commun 15:511
14. Miyashita K, Kaneko M (1995) Synth Met 68:161
15. Remmers M, Müller B, Martin K, Räder H-J, Köhler W (1999) Macromolecules 32:1073
16. Lee CH, Kang GW, Jeon JW, Song WJ, Seoul C (2000) Thin Solid Films 363:306
17. Song W-J, Seoul C, Kang G-W, Lee C (2000) Synth Met 114:355
18. Kang G-W, Lee C, Song W-J, Seoul C (2001) Proc SPIE 4105:362
19. Graupner W, Grem G, Meghdadi F, Paar C, Leising G, Scherf U, Müllen K, Fischer W, Stelzer F (1994) Mol Cryst Liq Cryst 256:549
20. Grem G, Martin V, Meghdadi F, Paar C, Stampfl J, Sturm J, Tasch S, Leising G (1995) Synth Met 71:2193
21. Leising G, Köpping-Grem G, Meghdadi F, Niko A, Tasch S, Fischer W, Pu L, Wagaman MW, Grubbs RH, Althouel L, Froyer G, Scherf U, Huber J (1995) Proc SPIE 2528:307
22. Meghdadi F, Leising G, Fischer W, Stelzer F (1996) Synth Met 76:113
23. Leising G, Ekström O, Graupner W, Meghdadi F, Moser M, Kranzelbinder G, Jost T, Tasch S, Winkler B, Athouel L, Froyer G, Scherf U, Müllen K, Lanzani G, Nisoli M, DeSilvestri S (1996) Proc SPIE 2852:189

24. Leising G, Tasch S, Meghdadi F, Athouel L, Froyer G, Scherf U (1996) Synth Met 81:185
25. Meghdadi F, Tasch S, Winkler B, Fischer W, Stelzer F, Leising G (1997) Synth Met 85:1441
26. Koch N, Pogantsch A, List EWJ, Leising G, Blyth RIR, Ramsey MG, Netzer FP (1999) Appl Phys Lett 74:2909
27. Yanagi H, Okamoto S (1997) Appl Phys Lett 71:2563
28. Era M, Tsutsui T, Saito S (1995) Appl Phys Lett 67:2436
29. Andrew A, Matt G, Brabec CJ, Sitter H, Badt D, Seyringer H, Sariciftci NS (2000) Adv Mater 12:629
30. Niko A, Tasch S, Meghdadi F, Brandstätter C, Leising G (1997) J Appl Phys 82:4177
31. Tasch S, Brandstätter C, Meghdadi F, Leising G, Froyer G, Athauel L (1997) Adv Mater 9:33
32. Leising G, Meghdadi F, Tasch S, Brandstätter C, Graupner W, Kranzelbinder G (1997) Synth Met 85:1213
33. Leising G, Tasch S, Brandstätter C, Graupner W, Hampel S, List EWJ, Meghdadi F, Zenz C, Schlichting P, Rohr U, Geerts Y, Scherf U, Müllen K (1997) Synth Met 91:41
34. Niko A, Tasch S, Meghdadi F, Brandstätter C, Leising G (1998) Opt Mater 9:188
35. Leising G, List EWJ, Zenz C, Tasch S, Brandstätter C, Graupner W, Markart P, Meghdadi F, Kranzelbinder G, Niko A, Resel R, Zojer E, Schlichting P, Rohr U, Geerts Y, Scherf U, Müllen K, Smith R, Gin D (1998) Proc SPIE 3476:76
36. Gin DL, Conticello VP (1996) Trends Polym Sci 4:217
37. Ballard DGH, Courtis A, Shirley IM, Taylor SC (1983) J Chem Soc Chem Commun: 954
38. Ballard DGH, Courtis A, Shirley IM, Taylor SC (1988) Macromolecules 21:294
39. Grem G, Leditzky G, Ullrich B, Leising G (1992) Synth Met 51:383
40. Grem G, Leising G (1993) Synth Met 55–57:4105
41. Tasch S, Brandstätter C, Graupner W, Hampel S, Hochfilzer C, List EWJ, Meghdadi F, Leising G, Schlichting P, Rohr U, Geerts Y, Scherf U, Müllen K (1997) Mater Res Soc Symp Proc 471:325
42. McKean DR, Stille JK (1987) Macromolecules 20:1787
43. Kim HK, Suck J-K, Zyung T (1996) Polym Mater Sci Eng 75:255
44. Kim HK, Kim K-D, Zyung T (1997) Mol Cryst Liq Cryst 295:27
45. Gin DL, Conticello VP, Grubbs RH (1992) J Am Chem Soc 114:3167
46. Gin DL, Conticello VP, Grubbs RH (1994) J Am Chem Soc 116:10507
47. Gin DL, Conticello VP, Grubbs RH (1994) J Am Chem Soc 116:10934
48. Gin DL, Avlyanov JK, MacDiarmid AG (1994) Synth Met 66:169
49. Chaturvedi V, Tanaka S, Kaeriyama K (1992) J Chem Soc Chem Commun: 1658
50. Chaturvedi V, Tanaka S, Kaeriyama K (1993) Macromolecules 26:2607
51. Komaba S, Amano A, Osaka T (1997) J Electroanal Chem 430:97
52. Argyrakis P, Kobryanskii MV, Sluch MI, Vitukhnovsky AG (1997) Synth Met 91:159
53. Tanigaki N, Yase K, Kaito A (1996) Thin Solid Films 273:263
54. Yamamoto T, Hayashi Y, Yamamoto A (1978) Bull Chem Soc Jpn 51:2091
55. Hamaguchi M, Sawada H, Kyokane J, Yoshino K (1996) Chem Lett: 527
56. Schlüter AD, Wegner G (1993) Acta Polym 44:59
57. Vahlenkamp T, Wegner G (1994) Macromol Chem Phys 195:1933
58. McCarthy TF, Witteler H, Pakula T, Wegner G (1995) Macromolecules 28:8350
59. Remmers M, Schulze M, Wegner G (1996) Macromol Rapid Commun 17:239
60. Vanhee S, Rulkens R, Lehmann U, Rosenauer C, Schulze M, Köhler W, Wegner G (1996) Macromolecules 29:5136

61. Kowitz C, Wegner G (1997) Tetrahedron 53:15 553
62. Frahn J, Karakaya B, Schäfer A, Schlüter A-D (1997) Tetrahedron 53:15 459
63. Goodson FE, Wallow TI, Novak BM (1998) Macromolecules 31:2047
64. Schlüter S, Frahn J, Karakaya B, Schlüter A-D (2000) Macromol Chem Phys 201:139
65. Percec V, Pugh C, Cramer E, Okita S, Weiss R (1992) Makromol Chem, Macromol Symp 54–55:113
66. Percec V, Okita S, Weiss R (1992) Macromolecules 25:1816
67. Percec V (1993) US Patent: 5 241 044
68. Percec V, Bae J-Y, Zhao M, Hill DH (1995) Macromolecules 28:6726
69. Percec V, Zhao M, Bae J-Y, Hill DH (1996) Macromolecules 29:3727
70. Ueda M, Abe T, Awao H (1992) Macromolecules 25:5125
71. Huang J, Zhang H, Tian W, Hou J, Ma Y, Shen J, Liu S (1997) Synth Met 87:105
72. Tanigaki N, Masuda H, Kaeriyama K (1997) Polymer 38:1221
73. Wang Y, Quirk RP (1995) Macromolecules 28:3495
74. Klavetter FL, Gustafsson GG, Heeger AJ (1993) Polym Mater Sci Eng 69:153
75. Jing W-X, Kraft A, Moratti SC, Grüner J, Cacialli F, Hamer PJ, Holmes AB, Friend RH (1994) Synth Met 67:161
76. Hamaguchi M, Yoshino K (1995) Jpn J Appl Phys 34:L587
77. Yang Y, Pei Q, Heeger AJ (1996) J Appl Phys 79:934
78. Yang Y, Pei Q, Heeger AJ (1996) Synth Met 78:263
79. Cimrová V, Remmers M, Neher D, Wegner G (1996) Adv Mater 8:146
80. Cimrová V, Schmidt W, Rulkens R, Schulze M, Meyer W, Neher D (1996) Adv Mater 8:585
81. Wegner G, Neher D, Remmers M, Cimrová V, Schulze M (1996) Mater Res Soc Symp Proc 413:23
82. Chen S-A, Chao C-I (1996) Synth Met 79:93
83. Edwards A, Blumstengel S, Sokolik I, Yun H, Okamoto Y, Dorsinville R (1997) Synth Met 84:639
84. Grüner J, Remmers M, Neher D (1997) Adv Mater 9:964
85. Chao C-I, Chen S-A (1998) Appl Phys Lett 73:426
86. Annan KO, Scherf U, Müllen K (1999) Synth Met 99:9
87. Shin S-W, Park J-S, Park J-W, Kim HK (1999) Synth Met 102:1060
88. Fu Y (1997) Polym Prepr 38(1):410
89. Cimrová V, Výprachtický D, Pecka J, Kotva R (2000) Proc SPIE 3939:164
90. Kim S, Jackiw J, Robinson E, Schanze KS, Reynolds JR, Baur J, Rubner MF, Boils D (1998) Macromolecules 31:964
91. Baur JW, Kim S, Balanda PB, Reynolds JR, Rubner MF (1998) Adv Mater 10:1452
92. Wittemann M, Rehahn M (1998) Chem Commun: 623
93. Thunemann AF, Ruppelt D, Schnablegger H, Blaul J (2000) Macromolecules 33:2124
94. Hamaguchi M, Yoshino K (1997) Polym Adv Technol 8:399
95. Akagi K, Oguma J, Shirakawa H (1998) J Photopolym Sci Technol 11:249
96. Chen SH, Conger BM, Mastrangelo JC, Kende AS, Kim DU (1998) Macromolecules 31:8051
97. Park JH, Lee CH, Akagi K, Shirakawa H, Park YW (2001) Synth Met 119:633
98. Fiesel R, Huber J, Apel V, Enkelmann V, Hentsche R, Scherf U, Cabrera K (1997) Macromol Chem Phys 198:2623
99. Fiesel R, Huber J, Scherf U (1998) Enantiomer 3:383
100. Fiesel R, Scherf U (1998) Acta Polym 49:445
101. Katsis D, Chen H-MP, Chen SH, Tsutsui T (2000) Proc SPIE 4107:77

102. Taylor PN, O'Connel MJ, McNeill LA, Hall MJ, Aplin RT, Anderson HL (2000) Angew Chem Int Ed 39:3456
103. Cacialli F, Wilson JS, Michels JJ, Daniel C, Silva C, Friend RH, Severin N, Samori P, Rabe JP, O'Connell MJ, Taylor PN, Anderson HL (2002) Nat Mater 1:160
104. Baskar S, Lai Y-H, Valiyaveetil S (2001) Macromolecules 34:6255
105. Park KC, Dodd LR, Levon K, Kwei TK (1996) Macromolecules 29:7149
106. Rhee TH, Choi T, Chang EY, Suh DH (2001) Macromol Chem Phys 202:906
107. Reddinger JL, Reynolds JR (1997) Macromolecules 30:479
108. Mikroyannidis JA (2001) Macromol Chem Phys 202:2367
109. Bao Z, Chan WK, Yu L (1995) J Am Chem Soc 117:12 426
110. Bao Z, Yu L (1995) Proc SPIE 2528:210
111. Raney MB, Reynolds JR (1999) Polym Prepr 40(2):1207
112. Kaeriyama K, Tsukahata Y, Negoro S, Tanigaki N, Masuda H (1997) Synth Met 84:263
113. Jiang B, Tilley TD (2000) Polym Prepr 41(1):829
114. Birgerson J, Fahlman M, Bröms P, Salaneck WR (1996) Synth Met 80:125
115. Birgerson J, Kaeriyama K, Barta P, Bröms P, Fahlman M, Granlund T, Salaneck WR (1996) Adv Mater 8:983
116. Salaneck WR (1997) Philos Trans R Soc Lond A 355:789
117. Edwards A, Blumstengel S, Sokolik I, Dorsinville R, Yun H, Kwei TK, Okamoto Y (1997) Appl Phys Lett 70:298
118. Huang J-S, Zhang H-F, An H-Y, Tian W-J, Hou J-Y, Chen B-J, Liu S-Y, Shen J-C (1996) Chin Phys Lett 13:944
119. Scherf U (1999) J Mater Chem 9:1853
120. Grimme J, Kreyenschmidt M, Uckert F, Müllen K, Scherf U (1995) Adv Mater 7:292
121. Scherf U, Müllen K (1991) Makromol Chem Rapid Commun 12:489
122. Scherf U, Müllen K (1992) Macromolecules 25:3546
123. Scherf U, Müllen K (1995) Adv Polym Sci 123:1
124. Scherf U, Müllen K (1997) ACS Symp Ser 672:358
125. Fiesel R, Huber J, Scherf U (1996) Angew Chem Int Ed Engl 35:2111
126. Grimme J, Scherf U (1996) Macromol Chem Phys 197:2297
127. Schindler F, Jacob J, Grimsdale A, Scherf U, Müllen K, Lupton JM, Feldmann J (2005) Angew Chem Int Ed 44:1520
128. Stampfl J, Graupner W, Leising G, Scherf U (1995) J Luminesc 63:117
129. Hüber J, Müllen K, Saalbeck J, Schenk H, Scherf U, Stehlin T, Stern R (1994) Acta Polym 45:244
130. Leising G, Grem G, Leditzky G, Scherf U (1993) Proc SPIE 1910:70
131. Grüner J, Wittmann HF, Hamer PJ, Friend RH, Huber J, Scherf U, Müllen K, Moratti SC, Holmes AB (1994) Synth Met 67:181
132. Mahrt RF, Siegner U, Lemmer U, Hopmeier M, Scherf U, Heun S, Göbel EO, Müllen K, Bässler H (1995) Chem Phys Lett 240:373
133. Köhler A, Grüner J, Friend RH, Müllen K, Scherf U (1995) Chem Phys Lett 243:456
134. Yang X, Hou Y, Wang Z, Chen X, Xu Z, Xu X (2000) Thin Solid Films 363:211
135. Scherf U, Bohnen A, Müllen K (1992) Makromol Chem 193:1127
136. Haugeneder A, Lemmer U, Scherf U (2002) Chem Phys Lett 351:354
137. Tasch S, Niko A, Leising G, Scherf U (1996) Appl Phys Lett 68:1090
138. Tasch S, Niko A, Leising G, Scherf U (1996) Mater Res Soc Symp Proc 413:71
139. Stagira S, Zavelani-Rossi M, Nisoli M, DeSilvestri S, Lanzani G, Zenz C, Mataloni P, Leising G (1998) Appl Phys Lett 73:2860
140. Kallinger C, Hilmer C, Haugeneder A, Perner M, Spirkl W, Lemmer U, Feldmann J, Scherf U, Müllen K, Gombert A, Wittwer V (1998) Adv Mater 10:920

141. Riechel S, Kallinger C, Lemmer U, Feldmann J, Gombert A, Wittwer V, Scherf U (2000) Appl Phys Lett 77:2310
142. Tasch S, Gao J, Wenzl FP, Holzer L, Leising G, Heeger AJ, Scherf U, Müllen K (1999) Electrochem Solid State Lett 2:303
143. Tasch S, List EWJ, Hochfilzer C, Leising G, Schlichting P, Rohr U, Geerts Y, Scherf U, Müllen K (1997) Phys Rev B Condens Matter 56:4479
144. List EWJ, Tasch S, Hochfilzer C, Leising G, Schlichting P, Rohr U, Geerts Y, Scherf U, Müllen K (1998) Opt Mater 9:183
145. Tasch S, List EWJ, Ekström O, Graupner W, Leising G, Schlichting P, Rohr U, Geerts Y, Scherf U, Müllen K (1997) Appl Phys Lett 71:2883
146. Lupton JM (2002) Chem Phys Lett 365:366
147. Romaner L, Heimel G, Weisenhofer H, Scandiucci de Freitas P, Scherf U, Bredas J-L, Zojer E, List EJW (2004) Chem Mater 16:4667
148. Qiu S, Lu P, Liu X, Lu FS, Liu L, Ma Y, Shen J (2003) Macromolecules 36:9823
149. Patil S, Scherf U, personal communication
150. Lupton JM, Pogantsch A, Piok T, List EWJ, Patil S, Scherf U (2002) Phys Rev Lett 89:7401
151. Forster M, Scherf U (2000) Macromol Rapid Commun 21:810
152. Chmil K, Scherf U (1993) Makromol Chem Rapid Commun 14:217
153. Chmil K, Scherf U (1997) Acta Polym 48:208
154. Kirstein S, Cohen G, Davidov D, Scherf U, Klapper M, Chmil K, Müllen K (1995) Synth Met 69:415
155. Goldfinger MB, Swager TM (1994) J Am Chem Soc 116:7895
156. Tour JM, Lambda JJS (1993) J Am Chem Soc 115:4935
157. Lambda JJS, Tour JM (1994) J Am Chem Soc 116:11 723
158. Grem G, Paar C, Stampfl J, Leising G, Huber J, Scherf U (1995) Chem Mater 7:2
159. Grüner J, Hamer PJ, Friend RH, Huber H-J, Scherf U, Holmes AB (1994) Adv Mater 6:748
160. Stern R, Schenk H, Salbeck J, Stehlin T, Scherf U, Müllen K, Leising G (1999) US Patent: 5 856 434
161. Grüner J, Friend RH, Huber J, Scherf U (1996) Chem Phys Lett 251:204
162. Leclerc M (2001) J Polym Sci A Polym Chem 39:2867
163. Neher D (2001) Macromol Rapid Commun 22:1365
164. Scherf U, List EWJ (2002) Adv Mater 14:477
165. Fukuda M, Sawada K, Yoshino K (1989) Jpn J Appl Phys 28:L1433
166. Fukuda M, Sawada K, Yoshino K (1993) J Polym Sci Polym Chem 31:2465
167. Ohmori Y, Uchida K, Muro K, Yoshino K (1991) Jpn J Appl Phys 30:L1941
168. Ohmori Y, Uchida M, Morishima C, Fujii A, Yoshino K (1993) Jpn J Appl Phys 32:L1663
169. Uchida M, Ohmori Y, Morishima C, Yoshino K (1993) Synth Met 57:4168
170. Liu B, Chen Z-K, Yu W-L, Lai Y-H, Huang W (2000) Thin Solid Films 363:332
171. Advincula R, Yia C, Inaoka S (2000) Polym Prepr 41(1):846
172. Woo EP, Inbasekaran M, Shiang W, Roof GR (1997) PCT Int Patent Appl: WO97/05184
173. Inbasekaran M, Wu W, Woo EP (1998) US Patent: 5 777 070
174. Bernius MT, Inbasekaran M, Woo EP, Wu W, Wujkowski L (1999) Proc SPIE 3797:129
175. Bernius M, Inbasekaran M, Woo E, Wu W, Wujkowski L (2000) J Mater Sci Mater Electron 11:111
176. Bernius M, Inbasekaran M, O'Brien J, Wu W (2000) Adv Mater 12:1737
177. Inbasekaran M, Woo E, Bernius M, Wujkowski L (2000) Synth Met 111–112:397

178. Grice AW, Bradley DDC, Bernius MT, Inbasekaran M, Wu WW, Woo EP (1998) Appl Phys Lett 73:629
179. Grell M, Knoll W, Lupo D, Meisel A, Miteva T, Neher D, Nothofer H-G, Scherf U, Yasuda A (1999) Adv Mater 11:671
180. Klaerner G, Miller RD (1998) Macromolecules 31:2007
181. Lee SH, Tsutsui T (2000) Thin Solid Films 363:76
182. Miller RD, Klaerner G, Fuhrer T, Kreyenschmidt M, Kwak J, Lee V, Chen W-D, Scott JC (1999) Nonlinear Opt 20:269
183. Grell M, Bradley DDC, Inbasekaran M, Woo EP (1997) Adv Mater 9:798
184. Teetsov J, Fox MA (1999) J Mater Chem 9:2117
185. Grell M, Bradley DDC, Whitehead KS (2000) J Korean Phys Soc 36:331
186. Whitehead KS, Grell M, Bradley DDC, Jandke M, Strohriegl P (2000) Appl Phys Lett 76:2946
187. Whitehead KS, Grell M, Bradley DDC, Inbasekaran M, Woo EP (2000) Synth Met 111–112:181
188. Whitehead KS, Grell M, Bradley DDC, Jandke M, Strohriegl P (2000) Proc SPIE 3939:172
189. Miteva T, Meisel A, Grell M, Nothofer H-G, Lupo D, Yasuda A, Knoll W, Kloppenburg L, Bunz UHF, Scherf U, Neher D (2000) Synth Met 111–112:173
190. Nothofer H-G, Meisel A, Miteva T, Neher D, Forster M, Oda M, Lieser G, Sainova D, Yasuda A, Lupo D, Knoll W, Scherf U (2000) Macromol Symp 154:139
191. Oda M, Nothofer H-G, Lieser G, Scherf U, Meskers SCJ, Neher D (2000) Adv Mater 12:362
192. Oda M, Meskers SCJ, Nothofer HG, Scherf U, Neher D (2000) Synth Met 111–112:575
193. Nothofer H-G, Oda M, Neher D, Scherf U (2000) Proc SPIE 4107:19
194. Tang H, Fujiki M, Motonaga M, Torimitsu K (2001) Polym Prepr 42(1):440
195. Zhang J, Grimsdale AC, Müllen K, unpublished results
196. Pei Q, Yang Y (1996) J Am Chem Soc 118:7416
197. Stéphan O, Collomb V, Vial J-C, Armand M (2000) Synth Met 113:257
198. Yang Y, Pei Q (1997) J Appl Phys 81:3294
199. Kreyenschmidt M, Klaerner G, Fuhrer T, Ashenhurst J, Karg S, Chen WD, Lee VY, Scott JC, Miller RD (1998) Macromolecules 31:1099
200. Lee J-I, Klärner G, Miller RD (1999) Synth Met 101:126
201. Bliznyuk VN, Carter SA, Scott JC, Klärner G, Miller RD, Miller DC (1999) Macromolecules 32:361
202. Lee J-I, Klaerner G, Miller RD (1999) Chem Mater 11:1083
203. Pschirer NG, Byrd K, Bunz UHF (2001) Macromolecules 34:8590
204. List EWJ, Guentner R, Scanducci de Freitas P, Scherf U (2002) Adv Mater 14:374
205. Lupton JM, Craig MR, Meijer EW (2002) Appl Phys Lett 80:4489
206. Pei J, Ni J, Zhou X-H, Cao X-Y, Lai Y-H (2002) J Org Chem 67:4924
207. Zeng G, Yu W-L, Chua S-J, Huang W (2002) Macromolecules 35:6907
208. Gaal M, List EJW, Scherf U (2003) Macromolecules 36:4236
209. Weinfurtner K-H, Fujikawa H, Tokito S, Taga Y (2000) Appl Phys Lett 76:2502
210. Craig MR, de Kok MM, Hofstraat JW, Schenning APHJ, Meijer EW (2003) J Mater Chem 13:2861
211. Sainova D, Miteva T, Nothofer H-G, Scherf U, Glowacki I, Ulanski J, Fujikawa H, Neher D (2000) Appl Phys Lett 76:1810
212. Klärner G, Lee J-I, Lee VY, Chan E, Chen J-P, Nelson A, Markiewicz D, Siemens R, Scott JC, Miller RD (1999) Chem Mater 11:1800
213. Roitman DB, Antoniadis H, Helbing R, Pourmiazaie F, Sheats JR (1998) Proc SPIE 3476:232

214. Suh Y-S, Ko SW, Jang B-J, Shim H-K (2002) Opt Mater 21:109
215. Lee J, Cho H-J, Jung B-J, Cho NS, Shim H-K (2004) Macromolecules 37:8523
216. Chou C-H, Hsu S-L, Dinakaran K, Chiu M-Y, Wei K-H (2005) Macromolecules 38:745
217. Takagi K, Kunii S, Yuki Y (2005) J Polym Sci A Polym Chem 43:2119
218. Klärner G, Davey MH, Chen W-D, Scott JC, Miller RD (1998) Adv Mater 10:993
219. Klärner G, Lee J-I, Davey MH, Miller RD (1999) Adv Mater 11:115
220. Tokito S, Weinfurtner K-H, Fujikawa H, Tsutsui T, Taga Y (2001) Proc SPIE 4105:69
221. Lee J-I, Klaerner G, Davey MH, Miller RD (1999) Synth Met 102:1087
222. Becker S, Marsitzky D, Setayesh S, Müllen K, Friend RH, MacKenzie JD (2000) UK Patent Appl: GBP 0018782.3
223. Becker S, Ego C, Grimsdale AC, List EWJ, Marsitzky D, Pogantsch A, Setayesh S, Leising G, Müllen K (2002) Synth Met 125:73
224. Ego C, Marsitzky D, Becker S, Zhang J, Grimsdale AC, Müllen K, MacKenzie JD, Silva C, Friend RH (2003) J Am Chem Soc 125:437
225. Klaerner G, Miller RD, Hawker CJ (1998) Polym Prepr 39(2):1006
226. Marsitzky D, Vestberg R, Blainey P, Tang BT, Hawker CJ, Carter KR (2001) J Am Chem Soc 123:6965
227. Tang H-Z, Fujiki M, Zhang Z-B, Torimitsu K, Motonaga M (2001) Chem Commun: 2426
228. Setayesh S, Grimsdale AC, Weil T, Enkelmann V, Müllen K, Meghdadi F, List EJW, Leising G (2001) J Am Chem Soc 123:946
229. Trimpin S, Grimsdale AC, Räder HJ, Müllen K (2002) Anal Chem 74:3777
230. Jacob J, Oldridge L, Zhang J, Gaal M, List EJW, Grimsdale AC, Müllen K (2004) Curr Appl Phys 3:339
231. Oldridge L, Grimsdale AC, Müllen K, unpublished results
232. Lee J-H, Hwang D-H (2003) Chem Commun: 2836
233. Pogantsch AF, Wenzl FP, List EWJ, Leising G, Grimsdale AC, Müllen K (2002) Adv Mater 14:1061
234. List EWJ, Pogantsch A, Wenzl FP, Kim C-H, Shinar J, Loi MA, Bongiovanni G, Mura A, Setayesh S, Grimsdale AC, Nothofer HG, Müllen K, Scherf U, Leising G (2001) Mater Res Soc Symp Proc 665:C5.47.1
235. Lupton JM, Schouwink P, Keivanidis PE, Grimsdale AC, Müllen K (2003) Adv Funct Mater 13:154
236. Leclère P, Hennebicq E, Calderone A, Brocorens P, Grimsdale AC, Müllen K, Brédas JL, Lazzaroni R (2003) Prog Polym Sci 28:55
237. Jäckel F, De Feyter S, Hofkens J, Köhn F, De Schryver FC, Ego C, Grimsdale A, Müllen K (2002) Chem Phys Lett 362:534
238. Janietz S, Bradley DDC, Grell M, Giebeler C, Inbasekaran M, Woo EP (1998) Appl Phys Lett 73:2453
239. Gross M, Muller DC, Nothofer H-G, Scherf U, Neher D, Meerholz K (2000) Nature 405:861
240. Brown TM, Friend RH, Millard IS, Lacey DJ, Burroughes JH, Cacialli F (2001) Appl Phys Lett 79:174
241. Bernius M, Inbasekaran M, Woo E, Wu W, Wujkowski L (2000) Thin Solid Films 363:55
242. Chen JP, Klaerner G, Lee J-I, Markiewicz D, Lee VY, Miller RD, Scott JC (1999) Synth Met 107:129
243. Chen JP, Markiewicz D, Lee VY, Klaerner G, Miller RD, Scott JC (1999) Synth Met 107:203
244. Miteva T, Meisel A, Knoll W, Nothofer HG, Scherf U, Müller DC, Meerholz K, Yasuda A, Neher D (2001) Adv Mater 13:565

245. Müller DC, Braig T, Nothofer H-G, Arnoldi M, Gross M, Scherf U, Nuyken O, Meerholz K (2000) Chem Phys Chem 1:207
246. Raymond F, Xiao SS, Nguyen MT (2001) Polym Prepr 42(2):587
247. Xia C, Advincula RC (2001) Macromolecules 34:5854
248. Ego C, Grimsdale AC, Uckert F, Yu G, Srdanov G, Müllen K (2002) Adv Mater 14:809
249. Uckert F, Setayesh S, Müllen K (1999) Macromolecules 32:4519
250. Uckert F, Tak Y-H, Müllen K, Bässler H (2000) Adv Mater 12:905
251. Kulkarni AP, Kong X, Jenekhe SA (2004) J Phys Chem B 108:8689
252. Wu F-I, Reddy S, Shu C-F, Liu MS, Jen AK-Y (2003) Chem Mater 15:269
253. Shu C-F, Dodda R, Wu F-I, Liu MS, Jen AK-Y (2003) Macromolecules 36:6698
254. Donat-Bouillud A, Lévesque I, Tao Y, D'Iorio M, Beaupré S, Blondin P, Ranger M, Bouchard J, Leclerc M (2000) Chem Mater 2000:1931
255. Lévesque I, Donat-Bouillud A, Tao Y, D'Iorio M, Beaupré S, Blondin P, Ranger M, Bouchard J, Leclerc M (2001) Synth Met 122:79
256. Cho HN, Kim DY, Kim JK, Kim CY (1997) Synth Met 91:293
257. Ranger M, Leclerc M (1998) Can J Chem 76:1571
258. Charas A, Barbagallo N, Morgado J, Alcácer L (2001) Synth Met 122:23
259. Liu B, Yu W-L, Lai W-H, Huang W (2001) Chem Mater 13:1984
260. Kameshima H, Nemoto N, Endo T (2001) J Polym Sci A Polym Chem 39:3143
261. Yu W-L, Cao Y, Pei J, Huang W, Heeger AJ (1999) Appl Phys Lett 75:3270
262. Yu W-L, Pei J, Cao Y, Huang W, Heeger AJ (1999) Chem Commun: 1837
263. Liu B, Yu W-L, Lai Y-H, Huang W (2000) Chem Commun: 551
264. Stork M, Gaylord BS, Heeger AJ, Bazan GC (2002) Adv Mater 14:361
265. Gaylord BS, Heeger AJ, Bazan GC (2002) Proc Natl Acad Sci USA 99:10 954
266. Gaylord BS, Heeger AJ, Bazan GC (2003) J Am Chem Soc 125:896
267. Liu MS, Jiang X, Herguth P, Jen AK-Y (2001) Chem Mater 13:3820
268. Jiang X, Liu S, Zheng L, Liu M, Jen AK-Y (2000) Polym Prepr 41(1):873
269. Jiang X, Liu S, Ma H, Jen AK-Y (2000) Appl Phys Lett 76:1813
270. Millard IS (2000) Synth Met 111–112:119
271. O'Connor SJM, Towns CR, O'Dell R, Burroughes JH (2001) Proc SPIE 4105:9
272. Salbeck J (1996) Ber Bunsenges Phys Chem 100:1666
273. Kreuder W, Lupo D, Salbeck J, Schenk H, Stehlin T (1996) Eur Patent Appl: EP 707 020
274. Wu Y, Li J, Fu Y, Bo Z (2004) Org Lett 6:3485
275. Yu W-L, Pei J, Huang W, Heeger AJ (2000) Adv Mater 12:828
276. Müller CD, Falcou A, Reckefuss N, Rojahn M, Wiederhirn V, Rudati P, Frohne H, Nuyken O, Becker H, Meerholz K (2003) Nature 421:829
277. Grisorio R, Mastronilli P, Nobile CF, Romanazzi G, Surana EP, Acierno D, Amendola E (2005) Macromol Chem Phys 206:448
278. Vak D, Chun C, Lee CL, Kim J-J, Kim D-Y (2004) J Mater Chem 14:1342
279. Setayesh S, Marsitzky D, Müllen K (2000) Macromoledules 33:2016
280. Grimsdale AC, Leclère P, Lazzaroni R, MacKenzie JD, Murphy C, Setayesh S, Silva C, Friend RH, Müllen K (2002) Adv Funct Mater 12:729
281. Keivanidis PE, Jacob J, Oldridge L, Sonar P, Carbonnier B, Baluschev S, Grimsdale AC, Müllen K, Wegner G (2005) Chem Phys Chem 6:1650
282. Marsitzky D, Scott JC, Chen J-P, Lee VY, Miller RD, Setayesh S, Müllen K (2001) Adv Mater 13:1096
283. Jacob J, Sax S, Piok T, List EJW, Grimsdale AC, Müllen K (2004) J Am Chem Soc 126:6987
284. Somma E, Loppinet B, Fytas G, Setayesh S, Jacob J, Grimsdale AC, Müllen K (2004) Colloid Polym Sci 282:867

285. Kreyenschmidt M, Uckert F, Müllen K (1995) Macromolecules 28:4577
286. Stern R, Lupo D, Salbeck J, Schenk H, Stehlin T, Müllen K, Scherf U, Huber J (1996) Eur Patent Appl: EP 699 699
287. Grazulevicius JV, Strohriegl P, Pielichowski J, Pielichowski K (2003) Prog Polym Sci 28:1297
288. Morin J-F, Leclerc M, Ades D, Siove A (2005) Macromol Rapid Commun 26:761
289. Ngbilo E, Ades D, Chevrot C, Siove A (1990) Polym Bull 24:17
290. Siove A, Aboulkassim A, Faid K, Ades D (1995) Polym Int 37:171
291. Romero DB, Schaer M, Leclerc M, Adès D, Siove A, Zuppiroli L (1996) Synth Met 80:271
292. Lmimouni K, Legrand C, Chapoton A (1998) Synth Met 97:151
293. Morin J-F, Leclerc M (2001) Macromolecules 34:4680
294. Dierschke F, Grimsdale AC, Müllen K (2003) Synthesis: 2470
295. Morin J-F, Leclerc M (2002) Macromolecules 35:8413
296. Iraqi A, Wataru I (2004) Chem Mater 16:442
297. Zotti G, Schiavon G, Zecchin S, Morin J-F, Leclerc M (2002) Macromolecules 35:2122
298. Li J, Dierschke F, Wu J, Grimsdale AC, Müllen K (2006) J Mater Chem 16:96
299. Morin J-F, Boudreault P-L, Leclerc M (2003) Macromol Rapid Commun 23:1032
300. Morin J-F, Beaupre S, Leclerc M, Levesque I, D'Iorio M (2002) Appl Phys Lett 80:341
301. Patil SA, Scherf U, Kadashchuk A (2003) Adv Funct Mater 13:609
302. Dierschke F, Grimsdale AC, Müllen K (2004) Macromol Chem Phys 205:1147
303. Jenekhe SA, Chen XL (1998) Science 279:1903
304. Zhong XF, François B (1988) Makromol Chem Rapid Commun 9:411
305. Zhong XF, François B (1989) Synth Met 29:E35
306. Zhong XF, François B (1991) Makromol Chem 192:2277
307. François B, Zhong XF (1991) Synth Met 41:955
308. François B, Widawski G, Rawiso M, Cesar B (1995) Synth Met 69:463
309. Mays J, Hong K, Wang Y, Advincula RC (1999) Mater Res Soc Symp Proc 561:189
310. Widawski G, Rawiso M, François B (1994) Nature 369:387
311. Romero DB, Schaer M, Staehli JL, Zuppiroli L, Widawski G, Rawiso M, François B (1995) Solid State Commun 95:185
312. Romero DB, Schaer M, Zuppiroli L, Cesar B, Widawski G, François B (1995) Opt Eng 34:1987
313. Marsitzky D, Klapper M, Müllen K (1999) Macromolecules 32:8685
314. Marsitzky D, Brand T, Geerts Y, Klapper M, Müllen K (1998) Macromol Rapid Commun 19:385
315. Marsitzky D (1999) Funktionalisierung von konjugierten Polymeren auf der Basis von PPP, PhD Thesis, Johannes Gutenberg University, Mainz
316. Schmitt C, Nothofer H-G, Falcou A, Scherf U (2001) Macromol Rapid Commun 22:624
317. Güntner R, Asawapirom U, Forster M, Schmitt C, Stiller B, Tiersch B, Falcou A, Nothofer H-G, Scherf U (2002) Thin Solid Films 417:1
318. Leclère P, Calderone A, Marsitzky D, Francke V, Geerts Y, Müllen K, Brédas J-L, Lazzaroni R (2000) Adv Mater 12:1042
319. Surin M, Marsitzky D, Grimsdale AC, Müllen K, Lazzaroni R, Leclère P (2004) Adv Funct Mater 14:708
320. Klaerner G, Trollsås M, Heise A, Husemann M, Atthoff R, Hawker CJ, Hedrick JL, Miller RD (1999) Macromolecules 32:8227
321. Johansson DM, Grandlund T, Theander M, Inganäs O, Andersson MR (2001) Synth Met 121:1761
322. Theander M, Grandlund T, Johanson DM, Ruseckas A, Sundström V, Andersson MR, Inganäs O (2001) Adv Mater 13:323

Editor: Ullrich Scherf

Spiro Compounds for Organic Electroluminescence and Related Applications

R. Pudzich · T. Fuhrmann-Lieker · J. Salbeck (✉)

Macromolecular Chemistry and Molecular Materials (mmCmm), Institute of Chemistry, and Center of Interdisciplinary Nanostructure Science and Technology (CINSaT), University of Kassel, Heinrich-Plett-Str. 40, 34109 Kassel, Germany
salbeck@uni-kassel.de

1	Introduction	84
2	The Spiro Concept	86
3	General Synthetic Routes to the Spirobifluorene Core	88
3.1	Synthesis of Spirobifluorene	88
3.2	Synthesis of Precursor Compounds	90
4	Functional Demands for Spiro Compounds	93
4.1	Functional Layers in Organic Light-Emitting Diodes	93
4.2	Optical Demands: Color	95
4.3	Amplified Spontaneous Emission and Lasing	96
4.4	Electrochemical Demands: Energy Levels	97
4.5	Charge Transport	98
4.6	Device Efficiency in OLEDs and Solar Cells	99
5	Spiro Compounds Containing Oligoaryls	100
5.1	Synthesis	100
5.2	Thermal and Morphological Properties	105
5.3	Optical Properties	106
5.4	Stimulated Emission	108
5.5	Electrochemical Properties	110
5.6	Charge Transport	110
5.7	Electroluminescent Devices	110
6	Spiro Compounds with Stilbene and Azobenzene Units	112
6.1	Synthesis	112
6.2	Thermal and Morphological Properties	113
6.3	Optical Properties	113
6.4	Electroluminescent Devices	113
7	Spiro Compounds Based on Arylamines	114
7.1	Synthesis	114
7.2	Thermal and Morphological Properties	114
7.3	Optical Properties	115
7.4	Electrochemical Properties	116
7.5	Charge Transport	118
7.6	Electroluminescent Devices	119
7.7	Solar Cells	120

8	Thiophene Compounds	122
8.1	Synthesis	122
8.2	Optical Properties	124
8.3	Electrochemical Properties	124
8.4	Electroluminescent Devices	125
9	Oxadiazole Compounds	125
9.1	Thermal and Morphological Properties	125
9.2	Optical Properties	126
9.3	Electrochemical Properties	126
10	Other N-Containing Heterocycles	127
11	Spiro Compounds with Mixed Chromophores	128
11.1	Synthesis	129
11.1.1	"Left–Right" Unsymmetric Spiro Compounds	129
11.1.2	"Top–Down" Unsymmetric Spiro Compounds	129
11.2	Thermal and Morphological Properties	133
11.3	Optical Properties	133
12	Chirality in Spiro Compounds	135
13	Spiro Polymers	136
13.1	Synthesis	136
13.2	Electroluminescent Devices	137
14	Conclusion	138
	References	138

Abstract A comprehensive review about functional spiro compounds, their synthesis, physical properties and applications in optoelectronic devices is given.

Keywords Amplified spontaneous emission · Charge transport · Field-effect transistors · Molecular glasses · Organic lasers · Organic light-emitting diodes · Solar cells · Spiro compounds

1
Introduction

Advances in modern electronics have caused a great demand for cheap and high-quality displays. In recent years, more and more academic as well as commercial sites have started concentrating on the development of electroluminescence devices based on organic materials. The concept of "electroluminescence" describes the phenomenon of the emission of light from condensed matter by the impact of an external electric field. Electroluminescence of an inorganic material (ZnS powder placed between two electrodes) was first discovered by Destriau in 1936 [1]; the first red-light-emitting diodes based on the inorganic semiconductor GaAsP were made commercially available in the

1960s [2]. Since then, other colors have been made available by the development of further combinations of semiconducting materials. Nevertheless, so far the technique of inorganic-semiconductor-based light-emitting diodes (LEDs) cannot be used for large area lighting applications or the construction of displays due to the fact that inorganic LEDs are based on single-crystal structures, which are limited to several square millimeters in dimension.

The electroluminescence of organic materials was first observed in 1963 by Pope and coworkers [3] and by Helferich and Schneider [4], who demonstrated that a single crystal of anthracene emits light at an applied voltage of 400 V. Even these first experiments opened the possibility of using organic materials as active components in light-emitting devices, but a commercial application was impossible due to the high voltages required, the short lifetimes of organic materials, and the fact that liquid electrodes had to be used for charge-carrier injections. Subsequent research and development concentrated on the further development of inorganic LEDs. The idea of organic electroluminescence was taken up again by Tang and Van Slyke [5] and Adachi et al. [6, 7] 20 years later. Both groups discovered that the use of a multilayered device structure with vacuum-deposited thin films of amorphous low-molecular-weight fluorescence dyes in combination with suitable organic charge transport materials as active components led to high-efficiency devices with low turn-on voltages and increased efficiencies and lifetimes. Another important discovery was the electroluminescence of thin films of conjugated polymers by Friend and coworkers in 1990 [8].

The use of organic materials in the field of electroluminescence offers a multiplicity of advantages compared to their inorganic counterparts. While inorganic semiconducting materials are constricted to suitable combinations of (toxic) elements like GaAs, GaP, AlGaAs, InGaP, GaAsP, etc., which limits the number of possible colors, the tuning of properties of organic materials like emission color, redox behavior, energy levels, and morphology can be done cost-effectively by simple modification of the chemical structure. For the characterization and development of structure–property relationships chemistry offers a variety of analytical techniques. Materials can be purified by classical chemical techniques like recrystallization, chromatography, or sublimation. The materials can, in the case of low-molecular-weight organic compounds, be processed by vacuum evaporation or by solution-based processes, e.g., spin-coating or ink-jet printing. Since film thicknesses are around 100 nm, the use of flexible substrates offers the possibility of constructing flexible displays [9].

An important characteristic that must be optimized in the development of low-molecular-weight organic materials is the ability to form morphologically stable amorphous films. Thermal stress during the operation of a device can lead to phase transitions of the metastable amorphous film into the thermodynamically stable polycrystalline state. It has been found that this effect leads to a fast degradation of the device. Moreover, grain boundaries between polycrystallites act as traps for charge carriers [10, 11]. Additionally, the contact between the electrodes and the organic film is diminished [12]. Both effects lead finally to the destruction of the display.

An important measurand for the stability of the amorphous state of a low-molecular-weight compound or a polymer is the glass transition temperature. If an amorphous film is heated above this temperature, molecular motion increases rapidly, which favors transition into the crystalline state. Therefore, it is of utmost importance to design the molecular structure of low-molecular-weight compounds in such a way that a high glass transition temperature is achieved. The relationship between molecular structure and the stability of the amorphous state was elucidated by Naito and Miura [13, 14]. They found that molecules with a symmetric globular structure, high molecular weight, and small intermolecular cohesion show a high stability of the amorphous state. There have been several concepts to realize such kinds of molecules including the development of dendritic architectures [15] and tetrahedral molecules [16].

2
The Spiro Concept

A very promising concept for the improvement of the morphological stability of low-molecular-weight materials while retaining their functionality is the spiro concept [17]. This concept is based on the idea of connecting two molecular π-systems with equal or different functions (emission, charge transport) via a common sp^3-hybridized atom. The general structure of such molecules is outlined in Fig. 1.

This concept has several benefits. On the one hand, the perpendicular arrangement of the two molecular halves leads to a high steric demand of the resulting rigid structure. This structural feature efficiently diminishes molecular interactions between the π-systems, which leads to a higher solubility of the spiro-linked compounds compared to the non-spiro-linked parent compounds. The unwieldy structure of fluorescent emitters based on the spiro concept also suppresses very effective excimer formation frequently observed in the solid state of many fluorescent dyes, and the emission properties are thereby stabilized. On the other hand, the doubling of the molecular weight in combination with the cross-shaped molecular structure and the

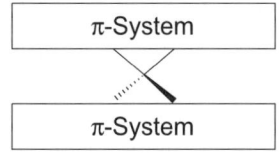

Fig. 1 General structure of spiro-linked low-molecular-weight functional molecules

rigidity of spiro compounds leads to entanglement in the amorphous solid state and prevents crystallization effectively below the glass transition temperature (T_g). The crystallization kinetics above T_g can vary substantially depending on the substitution pattern and is independent of the T_g value. This means that in the typical time scale of a calorimetric experiment, recrystallization can be observed for some spiro compounds (Fig. 2a), whereas other compounds do not crystallize (Fig. 2b).

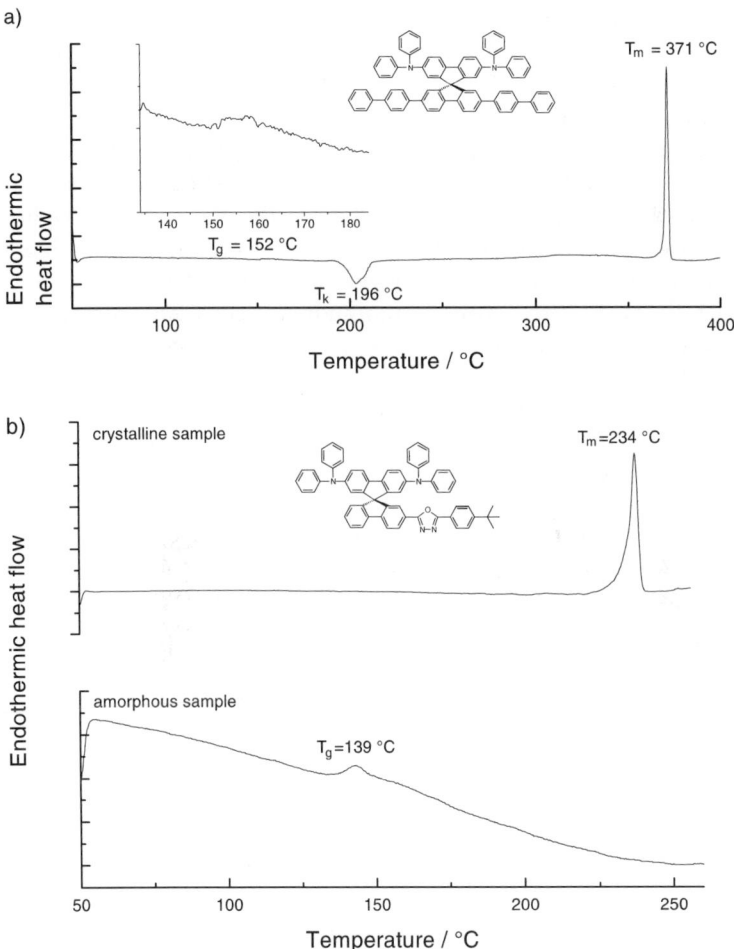

Fig. 2 Differential calorimetric curves for molecular glasses. **a** Spiro-DPSP **115** (second heating curve) and **b** Spiro-DPO **117** (first and second heating curves). The glass transition is indicated by a characteristic step, the melting point by an endothermic peak. In **a** recrystallization occurs above T_g, which can be seen by an exothermic peak. The material in **b** forms a stable amorphous glass. The melting point from the first heating curve of a crystalline sample disappears in the second heating cycle. Only the glass transition is visible

In the following sections we will discuss synthetic pathways to realize such compounds and give an overview of the physical properties of spiro compounds synthesized so far for optoelectronic applications by us and other groups.

3
General Synthetic Routes to the Spirobifluorene Core

From a chemist's point of view, the common feature of the many spiro-linked compounds synthesized so far is the central core of spirobifluorene (1), which is substituted with equal or different substituents in the 2,2' and 7,7' positions, and, in some cases, also in the 3 and 6 or in the 4 and 4' positions, as shown in Fig. 3. There currently exist several pathways to the synthesis of spiro-linked functional materials, whereas the choice of the right pathway depends on the desired substitution pattern. Compounds, which are fully symmetrically substituted in all four para positions, i.e., R1 = R2 = R3 = R4, are generally synthesized from spirobifluorene itself. This strategy is also applied for spiro compounds with a symmetrically sixfold substitution pattern in the 2, 2', 4, 4', 7, and 7' positions. In the case of the horizontally unsymmetrical compounds, in which R1 = R2 ≠ R3 = R4, a more complex strategy has to be applied. Vertically unsymmetrical spiro compounds, in which R1 = R3 ≠ R2 = R4, can be accessed by direct substitution of spirobifluorene provided that the first two groups entering in the 2 and 2' positions deactivate the 7 and 7' positions, so that a subsequent substitution in these positions can be controlled.

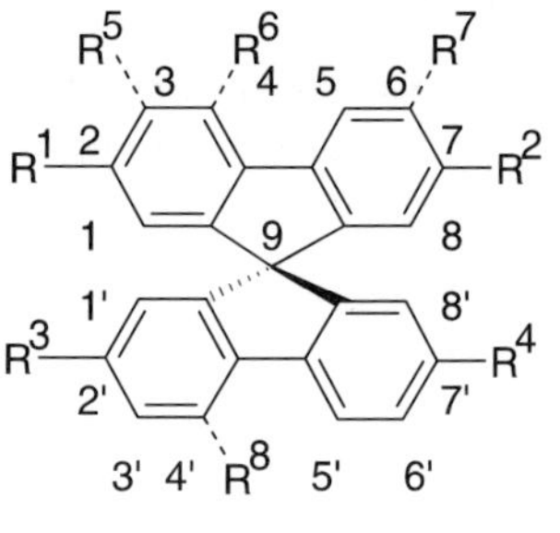

Fig. 3 General structure of substituted 9,9'-spirobifluorene and numbering of ring system

3.1
Synthesis of Spirobifluorene

9,9'-Spirobifluorene itself was first synthesized in 1930 by Clarkson and Gomberg [18]. The original procedure, which is shown in Fig. 4, involves the addition of the Grignard reagent from 2-iodobiphenyl to 9-fluorenone to obtain

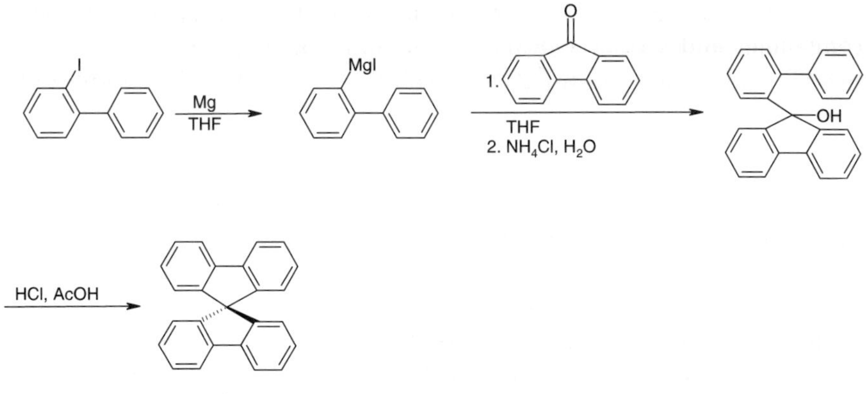

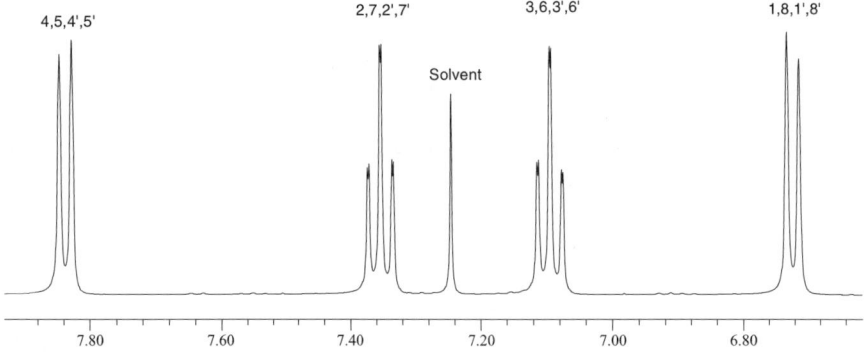

Fig. 4 Synthesis of 9,9′-spirobifluorene

the 9-(biphenyl-2-yl)-9-fluorenol in a yield of 70% after aqueous workup and recrystallization from ethanol.

In the next step, a catalytic amount of hydrochloric acid is added to a boiling solution of the carbinol in acetic acid, which leads to an instantaneous ring-closure reaction, whereupon most of the formed 9,9′-spirobifluorene precipitates on cooling from the solution. For this step, Clarkson and Gomberg reported a yield of 84%. The reaction can also be performed with 2-bromobiphenyl, which is commercially available, although the Grignard reaction between 2-bromobiphenyl and magnesium proceeds slower. Tour and coworkers used *tert*-butyllithium for the metallation of 2-iodobiphenyl and reported a yield of 86% for the formation of the carbinol and 98% for the synthesis of spirobifluorene [19]. Overall yields reported by other groups range from 55 to 94% [20–22].

The NMR spectrum of 9,9′-spirobifluorene is shown in Fig. 5. Due to the characteristic D_{2d} symmetry with a fourfold improper rotation axis (S_4), only

Fig. 5 NMR spectrum of 9,9′-spirobifluorene

four signals can be detected. Another distinct feature are the lowfield shift of the protons in the 4 and 5 positions (7.86 ppm) and the highfield shift of the protons in the 1 and 8 positions (6.75 ppm).

3.2
Synthesis of Precursor Compounds

The synthesis of 2-monosubstituted or 2,2′-disubstituted precursor compounds can be achieved by two different routes: direct electrophilic aromatic substitution or the use of substituted derivatives of 9-fluorenone. In many cases, the first route is complicated due to the fact that the two fluorene moieties in spirobifluorene react chemically independently of each other. The formation of products monosubstituted in the 2-position is therefore often accompanied by the formation of byproducts (Fig. 6). A good example is the nitration of spirobifluorene, which was investigated by Weisburger and coworkers [23]. They discovered that a careful control of reaction times and conditions allowed for the formation of 2-nitro-9,9′-spirobifluorene or 2,2′-dinitro-9,9′-spirobifluorene as the main product. Especially in the case of 2-nitro-9,9′-spirobifluorene, however, the conversions are low because the reaction must be quenched before a suitable amount of the dinitro compound is formed.

In the case of the acylation [24, 25] and formylation [26] of spirobifluorene, the predominant formation of 2-mono or 2,2′-disubstituted products can be controlled by the stoichiometry of the reagents. It must be pointed out that all the substituents dealt with so far lead to a deactivation of the 7 (and 7′) position. For many subsequent reactions, halogen substituents are needed, which do not induce a strong deactivation of the 7 and 7′ positions, and a different procedure has to be applied. The synthesis of 2-monohalogen- and 2,7-dihalogen-substituted spiro compounds can be achieved using substituted fluorenones instead of fluorenone in the synthesis of the spiro core. Following this method, Diederich and coworkers produced 2-bromo-9,9′-spirobifluorene (**8**) in 78% yield; utilizing 2,7-dibromofluorenone, the 2,7-dibromo-9,9′-spirobifluorene (**9**) was synthesized by several groups with overall yields in ranging from 62% to 86% [27, 28]. Unfortunately, this method cannot be adapted for the synthesis of 2,2′-dihalogen-substituted spiro compounds. The preparation of 2,2′-dibromo-9,9′-spirobifluorene (**10**) by direct bromination in the presence of a catalytic amount of ferric chloride has been reported by Sutcliffe [20] and, later, by Pei [28], but these results cannot be confirmed by our group or by Shu's group [29]. All our attempts to synthesize **10** by direct bromination led to a mixture of **10** and 2,7-dibromo-9,9′-spirobifluorene (**9**), which could not be separated simply either by chromatography or by recrystallization. Instead, Shu reports on a synthetic pathway, starting from 2,2′-dinitro-9,9′-spirobifluorene (**5**), which is schematically shown in Fig. 7. A similar pathway for the formation of 2,2′-

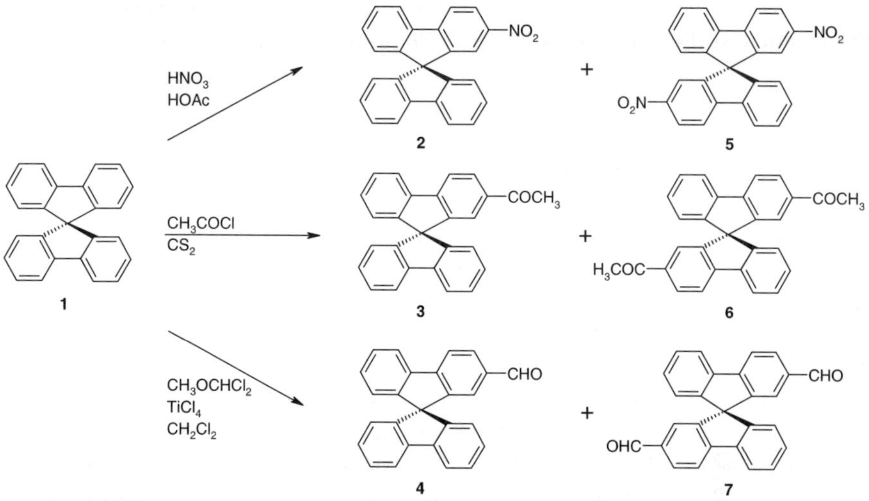

Fig. 6 Mono- and disubstituted derivatives of 9,9′-spirobifluorene, synthesized by electrophilic substitution

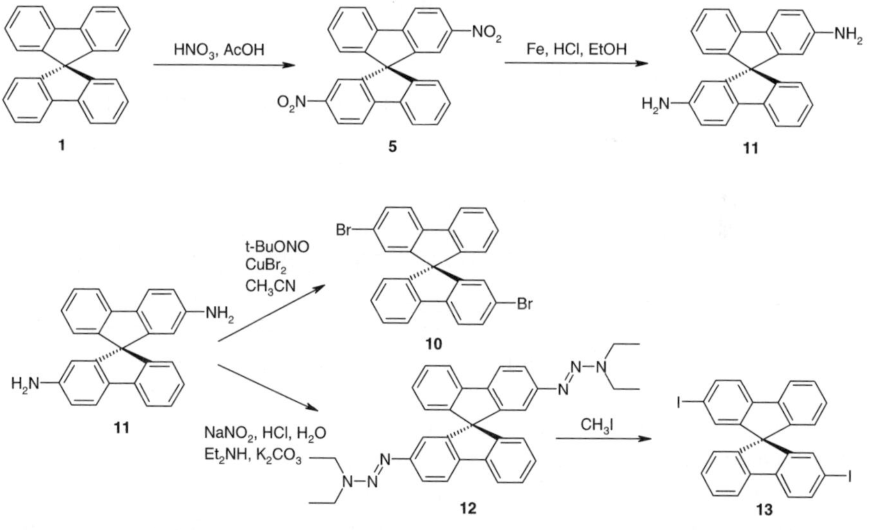

Fig. 7 Synthesis of 2,2′-dibromo-9,9′-spirobifluorene (**10**) and 2,2′-diiodo-9,9′-spirobifluorene (**13**)

diiodo-9,9′-spirobifluorene (**13**) was developed in our group (R. Pudzich and J. Salbeck, unpublished results), utilizing the procedure of Moore et al. [30].

Procedures for the synthesis of tetrahalogenated derivatives of spirobifluorene were reported by several groups. Tour and coworkers utilized a stoichiometric amount of bromine and a catalytic amount of ferric chloride to

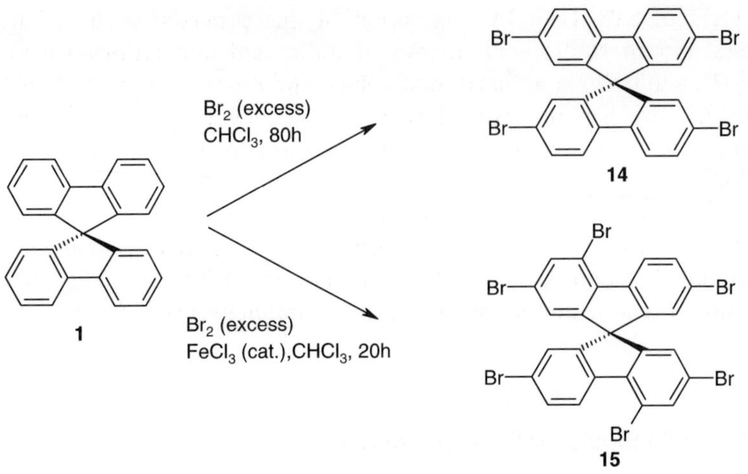

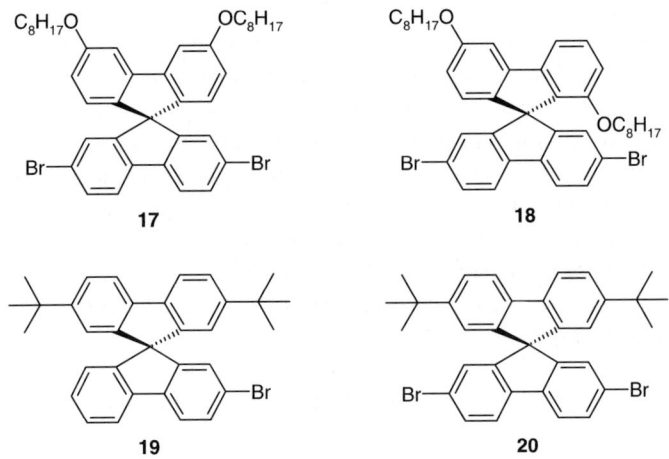

Fig. 8 Synthesis of 2,2′,7,7′-tetrabromospirobifluorene (**14**) and 2,2′,4,4′,7,7′-hexabromospirobifluorene (**15**)

Fig. 9 Halogenated spiro precursors with a mixed substitution pattern

prepare 2,2′,7,7′-tetrabromospirobifluorene (**14**) in a yield of 100% [19, 31]. Although this procedure was later confirmed by Pei [28], all our attempts to obtain pure **14** by the procedure given failed. Instead we found that an excess of bromine without the addition of ferric chloride allowed for selective bromination in the four para positions of spirobifluorene, whereas the addition of ferric chloride combined with an excess of bromine enabled us to prepare the 2,2′,4,4′,7,7′-hexabromospirobifluorene (**15**) (Fig. 8) [32].

2,2′,7,7′-Tetraiodo-9,9′-spirobifluorene (**16**) was prepared by Tour and coworkers with a mixture of concentrated sulphuric acid, iodic acid, and iodine in a yield of 84% [19]. In a variation of this procedure, Lützen and coworkers used periodic acid instead of iodic acid and reported a yield of 51% [33], while Wu et al. used iodchloride and reported a yield of 75% [34]. In our lab, we use the procedure of Merkushev [35], which utilizes the bis(trifluoroacetoxy)iodobenzene-iodine system to prepare **16** in a yield of 95% [32].

Other halogenated spiro compounds with a mixed substitution pattern were synthesized by Lee et al. [36] (compounds **17** and **18**), Kwon et al. [37] (compound **19**), and Spreitzer et al. (compound **20**) [38, 39] using mono- or dibromofluorenone and substituted 2-bromobiphenyls (Fig. 9).

4
Functional Demands for Spiro Compounds

The substituted spirobifluorene precursors discussed in the last section have been used for the design of a broad range of functional materials by appropriate substitution. With respect to their application in organic light emitting diodes (OLEDs), the materials can roughly be divided into three classes: light-emitting fluorescence dyes, hole-transporting compounds, and electron-transporting compounds. Although most of the materials are pure aryl systems, designed for emission in the blue range of the electromagnetic spectrum, some attempts have been undertaken to widen the application area of spiro-linked fluorescence dyes into the green and red part of the electromagnetic spectrum. Spiro-linked amines and heterocycles are primarily used as hole- and electron-transporting materials, respectively.

Before discussing the different classes in detail in the remaining sections of this article, we give a short overview of the physical properties that are important for the respective applications. The multifunctional approach in the design of the materials will be elucidated in the example of the OLED, but similar concepts are also valid in other devices, e.g., in solar cells.

4.1
Functional Layers in Organic Light-Emitting Diodes

First, let us consider the basic structure of organic light-emitting diodes (Fig. 10). Not less than seven layers are used in a state-of-the-art device based on low-molecular-weight materials. The structure is derived from a five-layer concept (three organic layers and two electrodes) by Adachi et al. [6], who extended the original bilayer device of Tang and Van Slyke [5].

The central layer is the emitting layer consisting of a fluorescent or phosphorescent material. Here, the excited states (excitons) are not formed by photoexcitation but by the recombination of an electron-depleted molecule—a radical cation with a "hole" in the highest occupied molecular orbital

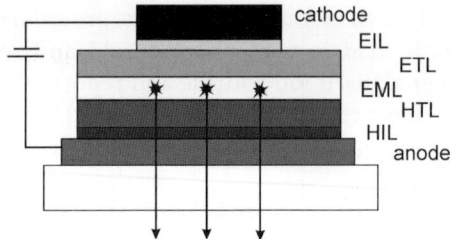

Fig. 10 Structure of an organic light-emitting diode. HIL: hole injection layer (e.g., Cu-Phthalocyanine), HTL: hole transport layer, EML: emission layer, ETL: electron transport layer, EIL: electron injection layer (e.g., LiF). The light generated by the recombination of holes and electrons is coupled out via the transparent anode (e.g., indium-tin-oxide ITO). Typical cathode materials are Mg : Ag, Ca/Ag, Al

(HOMO)—and an excess charge-carrying radical anion having an additional electron in the lowest unoccupied molecular orbital (LUMO), according to the following mechanism:

$$M^+ + M^- \rightarrow M^* + M. \tag{1}$$

The electrons for the radical anions are injected by the cathode, consisting of a metal with a low work function. Usually, calcium, a coevaporated magnesium–silver alloy with a Mg : Ag ratio of 10 : 1, or aluminum can be used. The corresponding work functions are 2.9 eV, 3.7 eV, and 4.3 eV, respectively. The injection of the electrons can be facilitated by an additional layer of lithium fluoride [40]. Several mechanisms have been proposed to explain the electron injection improvement [41, 42]. The most plausible mechanism is the dissociation of LiF toward metallic lithium, which acts as a redox dopant for the electron-transport layer. From the cathode the electrons are then transported through the electron- transport layer on the LUMO level via hopping transport, which is in principle a mutual solid-state redox reaction,

$$E^- + E \rightarrow E + E^-, \tag{2}$$

and then transferred to the emitter molecules M.

On the substrate side, the same process occurs for the holes, but on a different energy level. The holes are injected with a high work function metal or semiconductor like the transparent indium-tin-oxide ITO, which consists of a nonstoichiometric composite of 10–20% SnO_2 and 80–90% In_2O_3. The work function of ITO depends strongly on the surface treatment and lies in the range of 4.4–5.2 eV [43, 44]. As in the case of the cathode, hole injection can be improved by an additional layer of, e.g., copper phthalocyanine [45] or polyethylenedioxythiophene (PEDOT), doped with polystyrenesulfonic acid (PSS) [46]. The holes are injected into the hole transport layer and proceed

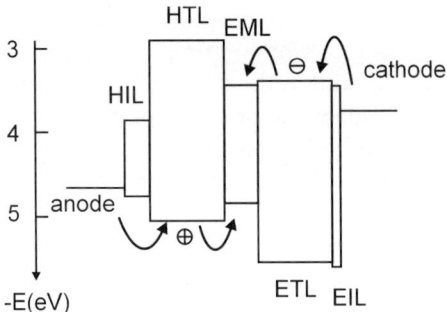

Fig. 11 Energy-level scheme of device in Fig. 10, consisting of electrode work functions and molecular HOMOs (highest occupied molecular orbitals) and LUMOs (lowest unoccupied molecular orbitals)

by hopping on the HOMO level toward the emission layer.

$$H^+ + H \rightarrow H + H^+ \tag{3}$$

The zone of recombination can be made very small, as was shown by Aminaka et al., by doping only a thin layer (5 nm) in the device by a red emission material [47]. By observing the ratio of host and dopant emission, the authors were able to show that the recombination zone of the device was as thin as 10 nm. The emitted light is usually coupled out at the substrate side through the transparent anode. Generally, the electroluminescence spectrum differs little from the photoluminescence spectrum.

Very often, two of the functional layers are replaced by one, using a material with combined emission and electron- or hole-transport capabilities. In addition, the charge-transport layers serve also as blocking layers for the carriers of opposite charge, which helps to confine the zone of recombination to the central area of the device and avoids loss of charges and quenching of the excitons at the electrode interfaces. The control of the recombination zone therefore requires a subtle balance between injection, charge-transport, and charge-blocking kinetics, which is a great challenge for material and device design. The energy level diagram of the OLED, summarizing the charge injection, transport, and recombination processes, is shown in Fig. 11.

4.2
Optical Demands: Color

For the light-emitting applications of molecular glasses, the optical properties in the vitreous solid state are of utmost importance. Due to the close packing in the solid state leading to strong intermolecular interaction and quenching effects, the photophysical and photochemical properties may differ strongly from the behavior in dilute solution.

One of the tasks for display applications is the combination of materials of different colors to produce white-light and mixed-color devices. For that reason, separate strategies aim at the development of blue-, green-, and red-emitting materials with suitable color coordinates.

The color of the luminescence can be defined best in the Commission Internationale de l'Éclairage (CIE) chromaticity diagram in which the color coordinates are defined as follows. The emission spectrum is weighted by the three given color matching functions for the color perception of a standard observer and integrated, giving the tristimulus values x, y and z with the normalization $x + y + z = 1$. In the diagram, usually given two-dimensionally by x and y, the white point is defined by the coordinates $x = y = z = 1/3$.

4.3
Amplified Spontaneous Emission and Lasing

Since the early work on dye lasers in the 1970s it has been known that organic dyes are ideal candidates for stimulated emission and lasing because of their quite large Stokes shift. The Franck–Condon-favored absorption and emission vibronic states form a four-level system that allows population inversion for fluorescence transition by optical pumping. If a Förster energy transfer cascade is inserted, the situation is even better since optical reabsorption is suppressed by the larger separation of absorption and emission wavelengths as described above. With the development of light-emitting semiconducting polymers and molecular glasses, organic solid-state lasers came into focus. Some low-molecular-weight glasses like the spiro-oligophenyls are ideal candidates for solid-state lasing because of their high solid-state quantum yield and a sufficient Stokes shift to avoid reabsorption.

The key prerequisite for optical amplification via stimulated emission is that the emitted photons propagate through the gain medium long enough to initiate further stimulated transitions. This condition can be expressed as

$$g(\lambda) \cdot l > 1, \tag{4}$$

$g(\lambda)$ being the wavelength-dependent overall gain coefficient and l the geometric path length (pump length). The pump-length dependence of the emitted intensity is given by

$$I_{out}(\lambda) \propto \exp(g(\lambda) \cdot l) - 1. \tag{5}$$

Large path lengths are easy to accomplish in the simplest structure achievable with molecular glasses: the thin-film waveguide. If a thin film of a molecular glass—whose refractive index n_2 is usually higher than the substrate $n_3 \approx 1.52$ (glass) and air $n_1 = 1$—is evaporated or spin-coated on a substrate, an unsymmetric slab waveguide is formed. At propagation angles θ_2 (defined relative to the film normal) larger than the angle of total internal reflection, a discrete number of optical modes exist that are confined to the film and characterized by an effective index

$$n_{\text{eff}} = n_2 \cos\theta_2. \tag{6}$$

The effective index represents the dimensionless in-plane component of the propagation vector of the mode (the propagation vectors are in units of $2\pi/\lambda$, λ being the vacuum wavelength). The optical modes can be characterized as transversal electric (TE the electric field is polarized in-plane) and transversal magnetic (TM: the magnetic field is polarized in-plane). For unsymmetric slab waveguides, a minimum thickness (cutoff) exists for each mode to appear [48].

The onset of stimulated emission can be detected by a collapse of the broad emission spectrum to a narrow line. This line narrowing due to optical gain in a waveguide without resonator is commonly referred to as amplified spontaneous emission (ASE) [49], or travelling wave lasing. In the case of additional optical resonators like gratings or microcavities, different modes with much narrower linewidths can be resolved.

The threshold for amplified spontaneous emission can be determined either by plotting the emitted intensity vs. the pump energy density and extrapolating the slope of the curve above threshold to zero emission intensity or by plotting the linewidth of the emission spectrum vs. pump energy density and fitting with a sigmoidal curve [50].

4.4
Electrochemical Demands: Energy Levels

For the electrical injection of charges in an OLED, the energy levels must be aligned carefully in order to allow electrical current to flow without undesired energy barriers. Some hole- and electron-blocking barriers, however, are necessary in order to confine the holes and electrons effectively to the recombination zone. The energy levels are governed by the redox properties of the materials.

At this point it should be noted that a molecular orbital scheme is often used for illustration, but more properly the total energy states of the molecules and their radical cations and anions, which may be subjected to electronic rearrangement, have to be considered. Bearing this in mind, the measured values of redox potentials can be translated into the molecular orbital picture.

Whereas the work function of the electrodes is measured by photoelectron spectroscopy, the organic materials are usually characterized by cyclic voltammetry [51]. The values can be extrapolated to the gas phase by choosing an appropriate reference and neglecting the influence of the polarity of the solvent in which the measurements are taken. In addition, the formation of interfacial potentials due to the formation of dipole layers in the solid-state device is neglected [52]. Making these assumptions, the energy levels obtained by cyclic voltammetry can be compared with the electrode work functions. It is more reliable, though, to compare the redox properties of different materials relative to each other.

Cyclic voltammetry measurements are usually performed in aprotic solvents, e.g., acetonitrile, dichloromethane, or tetrahydrofurane, with a conducting organic salt such as tetrabutylammonium hexafluorophosphate (TBAHFP) [53]. As (pseudo-) reference electrode, Ag/Ag^+ (0.01 mol/l in acetonitrile) or $Ag/AgCl$ is used. For calibration, an internal reference, ferrocene Fc/Fc^+ (+ 0.35 V vs. $Ag/AgCl$) or cobaltocene $CoCp_2/CoCp_2^+$ (− 0.94 V vs. $Ag/AgCl$), is added. The ionization energy of ferrocene is assumed to be 4.8 eV, thus linking the electrochemical potential to the work function scale of the electrodes [54]. The energy of the molecular orbitals is the negative value, i.e.:

$$\varepsilon(MO) = -(4.8\,\text{eV} + E_{1/2}), \tag{7}$$

$E_{1/2}$ being the reversible half-wave potential of the electron transfer reaction with respect to ferrocene. The suggested offset value, however, differs somewhat from group to group.

4.5
Charge Transport

Good hole or electron injection properties do not necessarily mean good transport properties and vice versa. Charge injection is governed by the energetic structure of the molecules, charge transport by the electron-transfer kinetics between a radical ion and a neutral molecule. The charge-transport properties of organic glasses attracted interest very early since the transport materials are good candidates for electrophotography (xerography). Ordered structures favor the overlap of the electron system of neighboring molecules, thus facilitating electron transfer. This means that the charge mobilities usually decrease in the order molecular crystal [55–57] > liquid crystalline glass [58] > amorphous glass. Thus, for amorphous glasses the task is to find materials with sufficiently high mobilities.

Usually, in amorphous molecular materials, charge transport is described by a disorder formalism that assumes a Gaussian distribution of energetic states of the molecules between which the charges jump [59]. The mobility is then given by

$$\mu = \mu_0 \exp\left[-\left(\frac{2\sigma}{3kT}\right)^2\right] \exp\left[C\left(\frac{\sigma^2}{(kT)^2} - \Sigma^2\right) E^{1/2}\right], \tag{8}$$

which expresses the experimentally observed electric field (E) and temperature (T) dependence:

$$\log \mu \propto T^{-2}, \tag{9}$$
$$\log \mu \propto E^{1/2}, \tag{10}$$

where μ_0 is a prefactor mobility (zero field, infinite temperature), C is an empirical constant of 2.9×10^{-4} (cm/V)$^{1/2}$, and σ and Σ express the energetic (diagonal) and positional disorder (off-diagonal), respectively. Other approaches also exist, one of them being based on Marcus theory of charge transfer [60, 61].

Experimentally, the charge mobilities are obtained by time-of-flight measurements or by characterizing field-effect transistor devices made of the materials. In time-of-flight experiments, the mobility μ is directly given by

$$\mu = \frac{d}{Et}, \tag{11}$$

where d is the sample thickness, E the applied electric field, and t the transit time. In transistor measurements, it can be calculated from the current–gate voltage relation [62].

Most of the early experimental work was done on molecular materials imbedded in a polymer matrix at concentrations of 10–80%, so a comparison of the intrinsic properties of the materials is not simple [63]. Another obstacle is the dependence of the charge mobilities on the preparation technology (purity, morphology). However, some data are available for true one-component molecular glasses, a selection of which will be presented in what follows.

4.6
Device Efficiency in OLEDs and Solar Cells

The optical, electrochemical, and electrical properties discussed up to now are properties of distinct materials, but in a multicomponent system the overall efficiency depends on many factors. This means that only the appropriate combination of materials in a device determines whether it is efficient or not. But for the application the device efficiency is the most important parameter. For OLEDs, several efficiency values are distinguished. Since they are related to display applications, the photometric unit system weighted by the photosensitivity of the human eye is used. The *luminance* measures the brightness of a radiating area in cd/m^2; the *luminous efficiency* is given in cd/A and the *power efficiency* in lm/W [64].

For solar cells, the efficiency is given under irradiation with a solar spectrum under air mass 1.5 (AM 1.5) [65] in terms of power conversion efficiency or internal photon-to-electron conversion efficiency (ICPE) [66]. Characteristic parameters that are often used are the open circuit voltage V_{oc}, the short circuit current density I_{sc}, and the fill factor, given by

$$FF = \frac{V_p I_p}{V_{oc} I_{sc}}, \tag{12}$$

where V_p and I_p denote voltage and current density for maximum power output.

5
Spiro Compounds Containing Oligoaryls

5.1
Synthesis

Due to their excellent emission properties in the blue range of the electromagnetic spectrum, oligophenyls are of special interest for use in electroluminescent devices. Unfortunately, with increasing chain length, oligophenyls also exhibit a high tendency toward recrystallization and an extremely low solubility [67–69]. These obstacles can be overcome by linking two oligophenyl

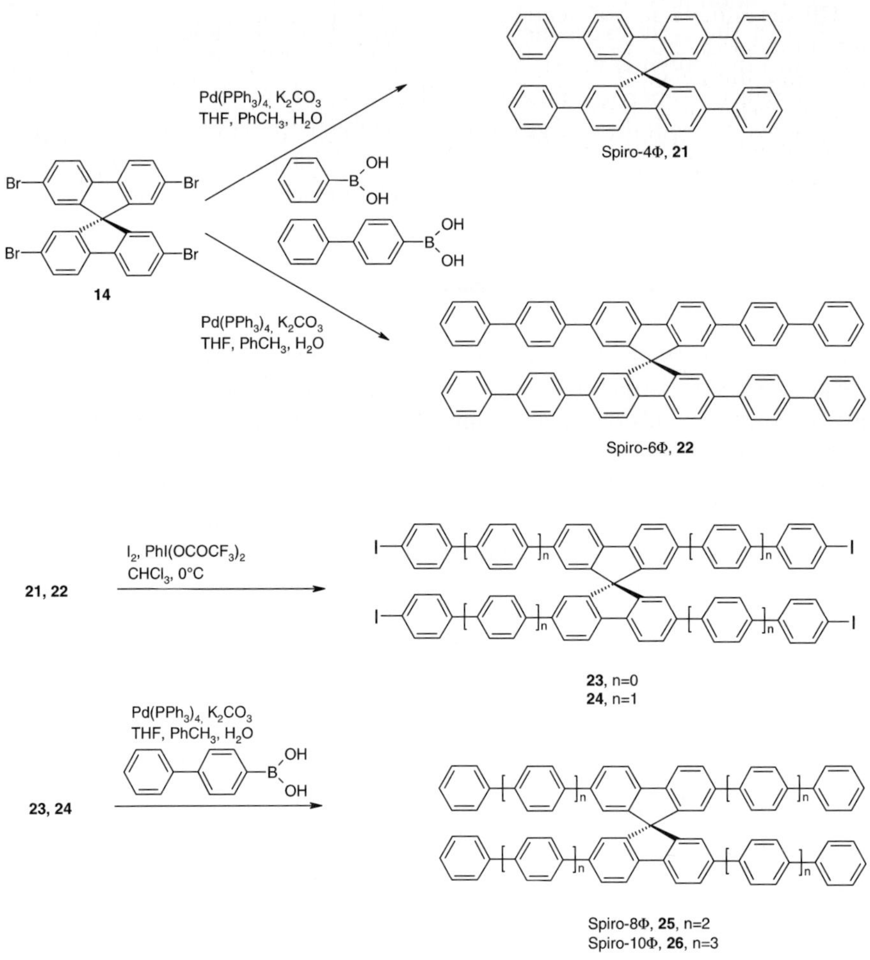

Fig. 12 Synthetic route to spiro-linked oligophenyls

chains by a common central spiro core, as was shown by Salbeck and coworkers, who synthesized soluble spiro-oligophenyls up to a chain length of ten phenyl rings [17, 70, 71]. The synthesis of these compounds involves the subsequent Suzuki coupling [72] of phenyl- or biphenylboronic acid to the central core **14**, as shown in Fig. 12.

To further enhance the morphologic stability and optical properties of these compounds, two strategies were followed. On the one hand, additional substituents in the 4 and 4′ position should enlarge the molecular weight and lead to spiro-oligophenyls with an octahedric, globular structure, increasing not only the glass transition temperature and therefore the stability of the amorphous state but also the solubility of the compounds [17, 70, 71]. These so-called "spiro-octahedric" compounds **27–29**, shown in Fig. 13, were synthesized starting from **15**.

On the other hand, a decrease in rotational freedom of the oligophenyl system was expected to enhance the optical properties, especially the fluorescence quantum yield in neat films and also the morphologic stability. This led to the development of the "starburst-spiro-oligophenyles" **30–32** [73]. Moreover, the terminal fluorene units in these compounds act as "spacers," which prevent interaction between the main chains in the solid state. In the case of

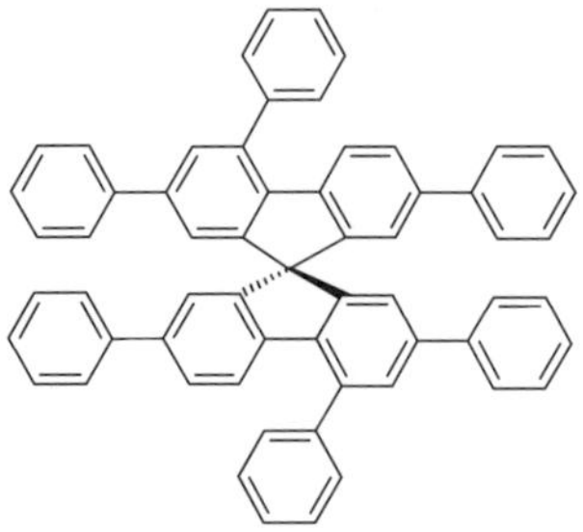

Spiro-octo-1, **27**

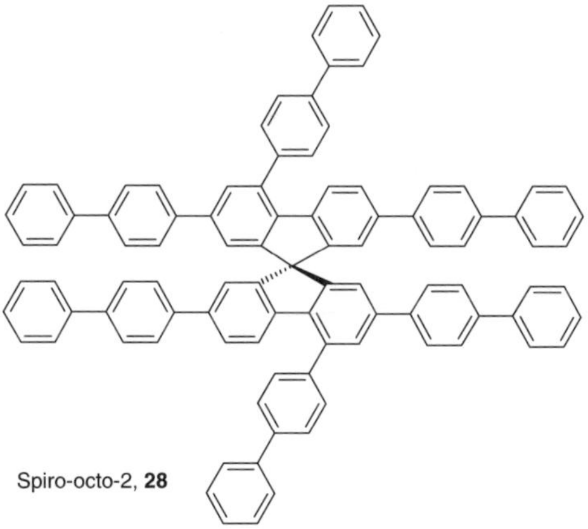

Spiro-octo-2, **28**

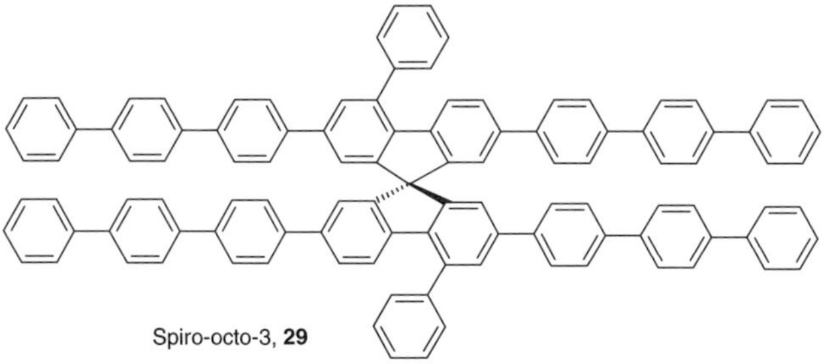

Spiro-octo-3, **29**

Fig. 13 Chemical structure of spiro-octahedric compounds

32, the molecular periphery around the central pentafluorene chains leads to an energetic gradient to the central chain.

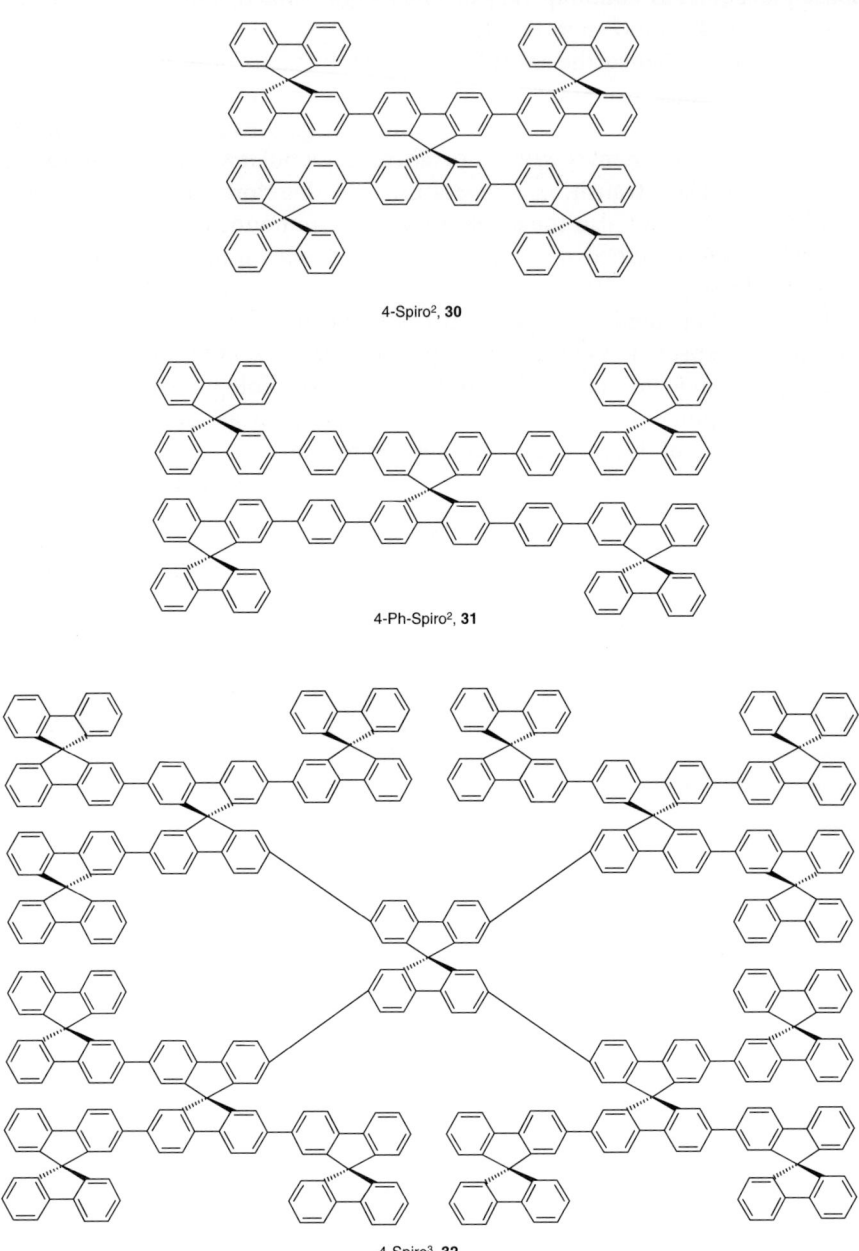

Fig. 14 Chemical structure of starburst spiro-oligophenyles

The compounds **30** and **31** can be synthesized by Suzuki coupling between **14** or **23** and 2-(9,9′-spirobifluoren-2-yl) boronic acid (**33**), which can be obtained from the Grignard reagent of **8** and boronic acid methyl ester [32]. The synthesis of **32** involves a complex strategy, outlined in Fig. 15 [32]. The

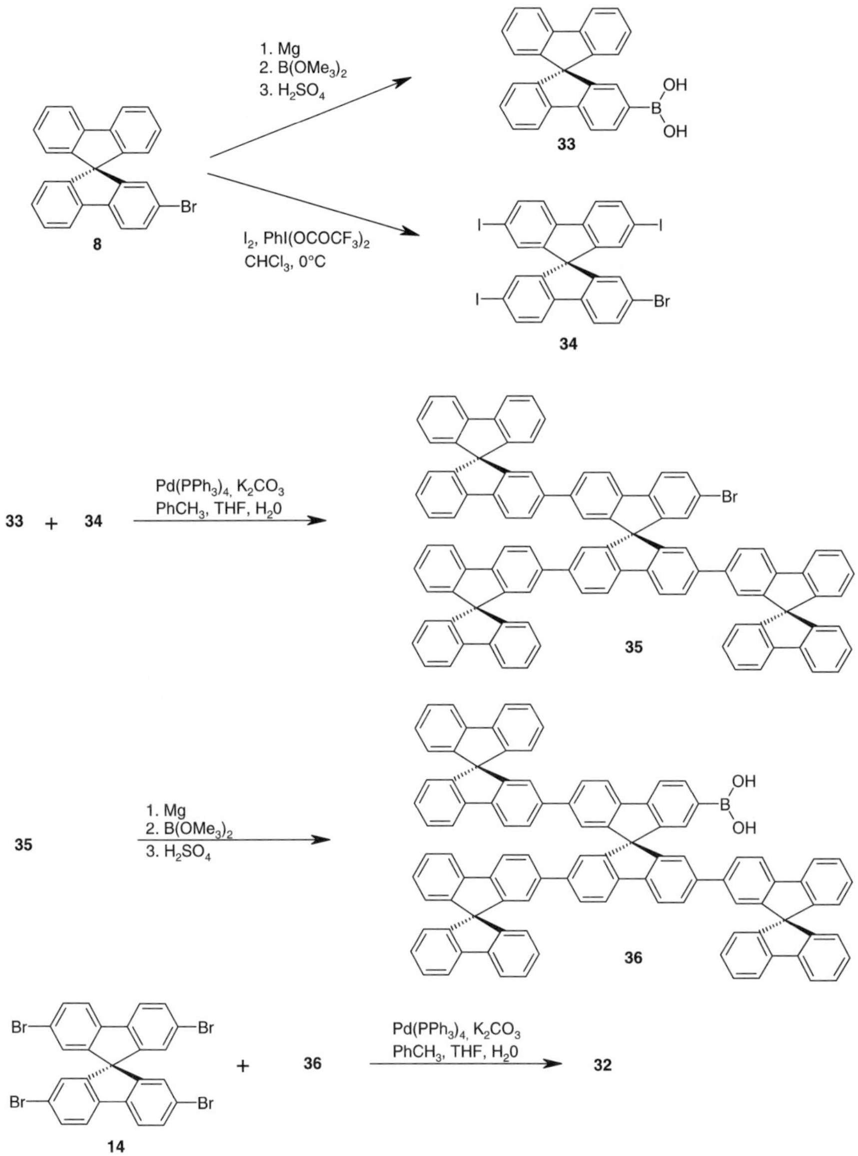

Fig. 15 Synthesis of 4-Spiro^3

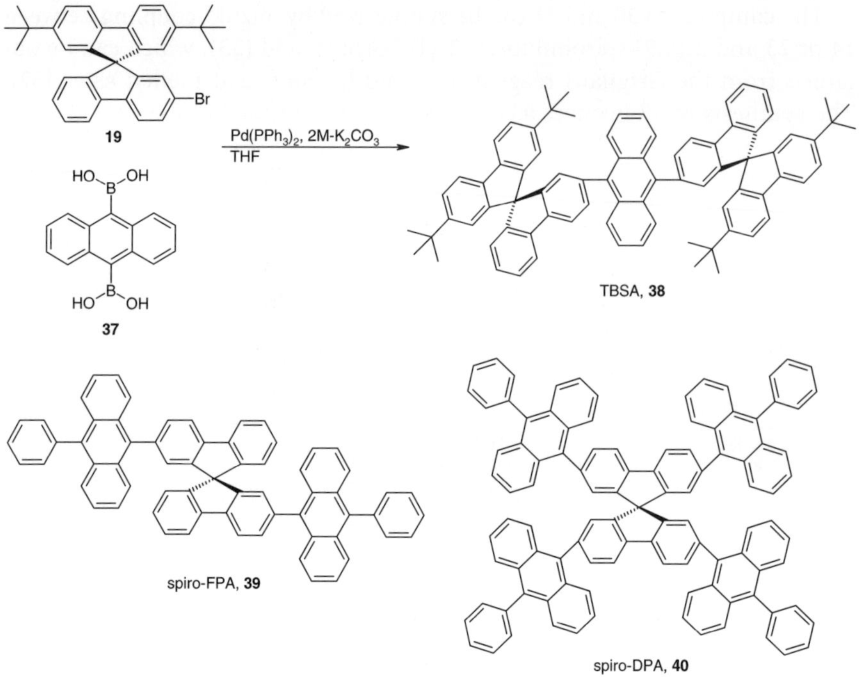

Fig. 16 Synthesis of TBSA **36** and molecular structures of Spiro-FPA **39** and Spiro-DPA **40**

strategy is the utilization of the different reactivities of iodine and bromine substituents in the Suzuki cross-coupling reaction [74].

Spiro-linked aryl systems for use as blue-emitting materials in OLEDs have also been prepared by other groups. Kim et al. synthesized **38**, containing the anthracene moiety as active chromophore by Suzuki coupling of **19** and anthracenediboronic acid (**37**) (Fig. 16) [37]. Anthracene-containing spiro compounds as blue-emitting materials have also been prepared by Shen et al. (compound **39**) [75] and Gerloff et al. (compound **40**) (T. Gerloff and J. Salbeck, unpublished results). Chen and coworkers reported on spiro-linked terfluorenes and spiro-linked ter-, penta-, and heptafluorenes with alkyl side chains, shown in Fig. 17.

Wong and coworkers synthesized the ter(9,9-diarylfluorenes) shown in Fig. 18 [76, 77].

Side-chain-containing spiro-oligophenyls with an increased solubility have also been prepared by Lee and coworkers by Suzuki coupling of **17** or the corresponding diboronic acid [36].

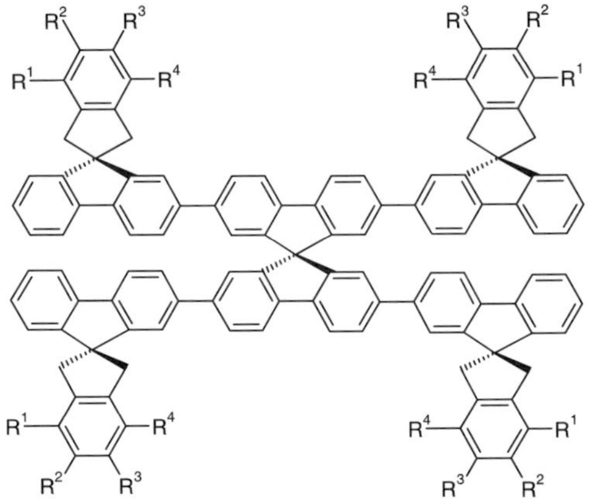

41, R¹=R²=R³=R⁴=H
42, R¹=R⁴=Ph, R²=R³=H
43, R¹=R²=R³=R⁴=Ph

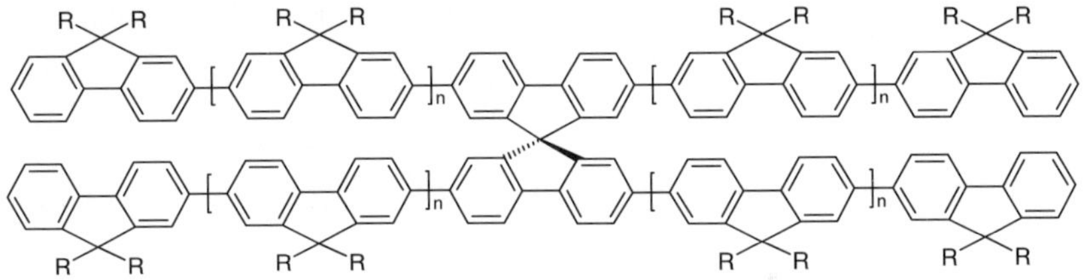

44 - 46, n = 0-2, R = n-C$_8$H$_{17}$
47 - 48, n = 0-1, R = n-C$_3$H$_7$

Fig. 17 Fully spiro-linked terfluorenes and spiro-linked oligofluorenes with alkyl-sidechains

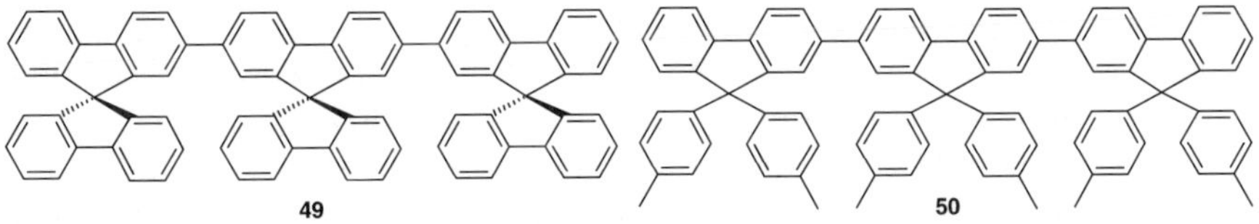

Fig. 18 Structure of ter-(9,9-diarylfluorene)s **49** and **50**

5.2
Thermal and Morphological Properties

In the series of the spiro-oligophenyls, the glass transition temperature increases with increasing chain length in the order 184 °C (Spiro-4Φ **21**) < 212 °C (Spiro-6Φ **22**) < 243 °C (Spiro-8Φ **25**). A further branching leads to even higher T_gs: both Spiro-Octo2 **28** and 4-Spiro² **30** exceed the T_g of Spiro-6Φ, with 236 °C and 273 °C, respectively. The T_g of 4-Spiro³ **32** is as high as 330 °C.

5.3
Optical Properties

The oligophenyls are excellent blue emitters since they exhibit high fluorescence quantum yields and the color can be tuned by extending the oligophenyl chain, but only to a certain extent (Davydov rule [78]). With an increasing number of phenyl rings the extinction coefficient increases and the fluorescence lifetime decreases [79]. Oligophenyls exhibit a large Stokes shift; typical data measured in solution are for p-terphenyl λ_{Abs} = 280 nm, λ_{Em} = 340 nm, for p-quaterphenyl λ_{Abs} = 300 nm, λ_{Em} = 370 nm, for p-quinquephenyl λ_{Abs} = 310 nm, λ_{Em} = 390 nm, and for p-sexiphenyl λ_{Abs} = 320 nm, λ_{Em} = 393 nm [80]. In the ground state of these molecules, the torsional angle between adjacent phenyl rings is 23°. In fluorenes and spiro compounds, the electronic system is modified by fixing some of the torsional angles at zero. Since the equilibrium geometries of the ground state and the excited state differ, the modifications of the oligophenyl chain have a considerable influence on the photophysical properties, namely, on the Stokes shift. In the series of spiro-oligophenyls with 4, 6, 8, and 10 phenyl rings in the chromophore, the absorption maxima are 332 nm, 342 nm, 344 nm, and 344 nm in dichloromethane, respectively. As in the case of the unsubstituted oligophenyls, a limiting value for long chains exists. The first fluorescence maxima increase steadily in the order 359 nm, 385 nm, 395 nm, and 402 nm.

The emission spectra exhibit a clearer vibrational fine structure than the absorption spectra due to the increased planarity and rigidity in the excited state. For Spiro-6Φ **22** a detailed analysis shows that the vibrational splitting of 0.20 eV corresponds to a phenyl breathing mode in the Raman spectrum [81]. In 4-Spiro^2 **30** and oligofluorenes, the torsional constraint is even more pronounced: only rigid biphenyl units are coupled together. For 4-Spiro^2 in comparison with Spiro-6Φ, the absorption maximum is shifted from 346 nm to 353 nm (amorphous films) and the fluorescence maximum from 420 to 429 nm, maintaining the Stokes shift. The corresponding spectra are shown in Fig. 20. The absorption signal at 310 nm in the spectrum

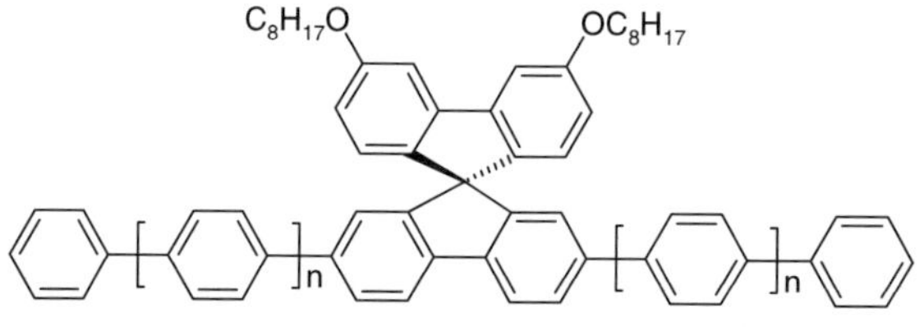

51-53, n = 0-2

Fig. 19 Spiro-oligophenyles containing alkoxy-sidechains

of 4-Spiro² **30** can be attributed to the outer fluorene moieties. The quantum yields for the fluorescence in neat amorphous films are 38% for Spiro-6Φ **22** and as high as $70 \pm 10\%$ for 4-Spiro² **30** [73].

The fluorescence lifetime was determined to be 1124 ps for Spiro-4Φ **21**, 785 ps for Spiro-6Φ **22**, and 831 ps for 4-Spiro² **30** in dichloromethane, whereas in the corresponding amorphous films a nonexponential decay with shorter time constants was observed [82, 83]. These lifetimes are similar to the parent oligophenyls but different from fluorene (10 ns) [84, 85].

The terfluorene unit is also present in the spiro-terfluorenes **41-43**, which show the same long-wavelength absorption band as in 4-Spiro² ($\lambda_{max} = 358 \pm 2$ nm) and similar emission characteristics. Fluorescence quantum yields are reported for doped poly(methylmethacrylate) films (0.01 M) and for neat films. For **41**, the photoluminescence quantum yield is 0.83 in PMMA and 0.28–0.24 in neat film, depending on thermal treatment [86].

For the alkyl-substituted spiro-oligofluorenes **44-48**, solid-state quantum yields have been determined in the range of 0.44 to 0.57 [87]. There is no significant shift in the emission spectrum going from pentafluorene to heptafluorene, indicating an effective conjugation length in the excited state of five fluorene units.

A very important characteristic that holds for all spiro-fluorene compounds discussed in this chapter is that they do not exhibit a long wavelength emission beyond 500 nm, which is usually found in poly(fluorene)s after annealing. This indicates the effective prevention of local aggregates.

When applying oligophenyls as luminescent films, however, it has to be taken into consideration that photooxidation may occur if molecular oxy-

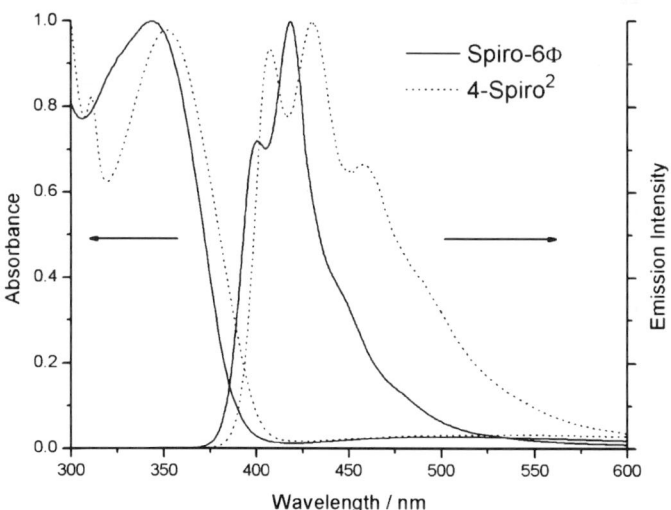

Fig. 20 Absorption and emission spectra of Spiro-6Φ and 4-Spiro² in solid state

gen is present [88, 89]. The proposed pathway for the decomposition is based on the attack of singlet oxygen, which is formed by spin transfer from the excited oligophenyl molecules. Endo- or exoperoxides are formed that may open to carbonyl compounds. This mechanism is supported by the infrared spectroscopic detection of carbonyl groups after irradiation in the presence of oxygen.

5.4
Stimulated Emission

Optical amplification experiments for thin films of spiro-linked oligophenyls are reported in [50, 90, 91]. Spin-coated films were irradiated with a pulsed nitrogen laser at 337 nm. The emitted light that was scattered out at an angle of 45° was detected. At low irradiation intensities the normal fluorescence spectrum appears, whereas at a pump intensity threshold of about $10\,\mu J/cm^2$ in the case of Spiro-6Φ, the spectrum begins to collapse into one narrow line at 419 nm (Fig. 21).

This line corresponds to the 0-1 vibronic transition that is amplified, in contrast to the smaller homolog Spiro-4Φ 21, where the 0-0 transition dominates. The linewidths are as low as 2.9 nm for Spiro-6Φ 22 and 2.2 nm for Spiro-4Φ [90].

The influence of the waveguide structure on the threshold of line narrowing can be deduced from the following observations. First, no gain narrowing

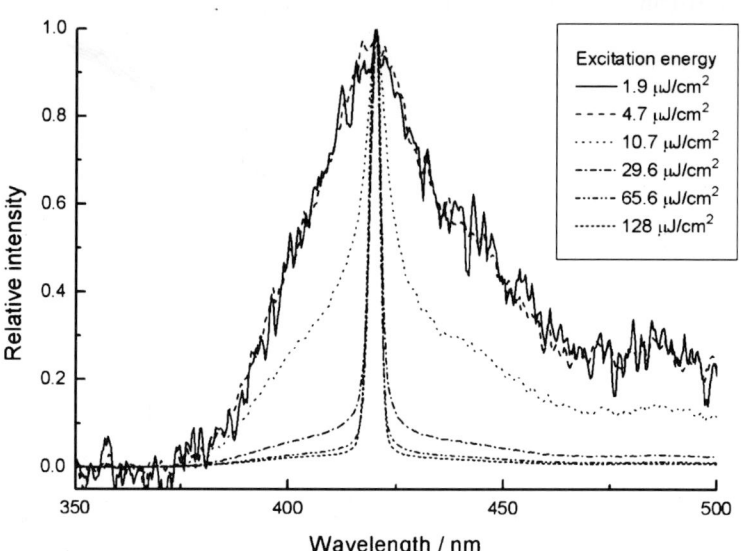

Fig. 21 ASE of a 50-nm film of Spiro-6Φ. At pump intensities around $10\,\mu J/cm^2$, the normal fluorescence spectrum collapses to a *narrow line*

can be observed for film thicknesses of 30 nm, which is below the cutoff thickness for the first mode. Second, if a high-index substrate is used, no narrowing occurs. Third, the lowest threshold is at film thicknesses of around 100 nm [50]. At this thickness, only one mode for each polarization exists, so the optical energy does not have to be divided into different competing modes.

The higher branched derivatives of the spiro compounds, 4-Spiro^2 30 and 4-Spiro^3 32, exhibit also spectral narrowing under pulsed excitation, with emission lines at 428 nm and 443 nm and linewidths of 3.2 nm and 3.9 nm, respectively (Fig. 22) [50]. For the same film thickness, the threshold is lower for 4-Spiro^2 30 than for Spiro-6Φ 22, despite the lower absorption at the pump wavelength, which is attributed to the higher luminescence quantum yield. The colors of the emission lines are very pure and near to the edge of the CIE diagram. The values are $x = 0.12$, $y = 0.16$ for Spiro-6Φ 22, $x = 0.15$, $y = 0.10$ for 4-Spiro^2 30, and $x = 0.16$, $y = 0.03$ for 4-Spiro^3 32.

For Spiro-Octo2 28, the ASE peak maximum was measured at 425.4 nm, and for a tetramethoxy-substituted spiro-sexiphenyl, Spiro-6Φ(MeO)$_4$, it was 427.5 nm. The threshold pump energies were 3 µJ/cm^2 for a 66-nm film of Spiro-Octo2 and 30 µJ/cm^2 for a 50-nm film of Spiro-6Φ(MeO)$_4$ [91].

The first microcavity devices and distributed feedback (DFB) laser with spiro compounds were realized by Benstem et al. [92]. For a microcavity with Spiro-6Φ between a semitransparent silver mirror and a multistack of Ta$_2$O$_5$/SiO$_2$, a spectral halfwidth (FWHM) of 7.2 nm was achieved. The fluorescence lifetime within the microcavity (240 ps) was distinctly lower than without the cavity (345 ps). The emission of the DFB laser (250-nm-

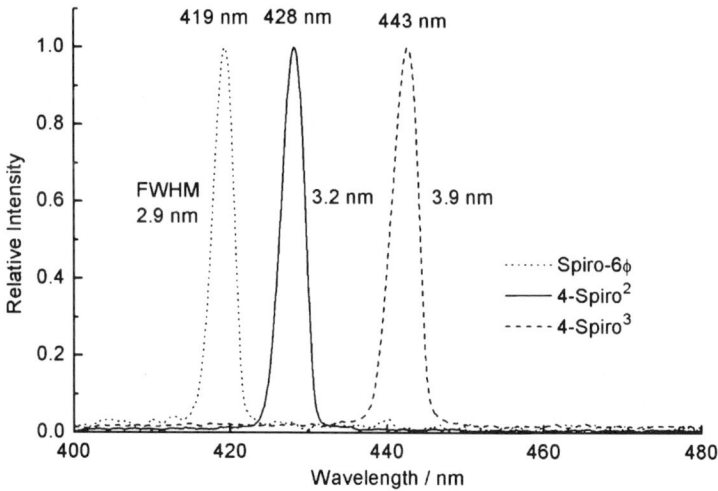

Fig. 22 ASE lines of different blue-emitting spiro compounds. The linewidths are about 3 nm

thick film on top of a 270-nm SiO$_2$ grating with modulation depth 70 nm) was multimodal with a FWHM of 1.85 nm and a threshold of 1.6 mJ/cm^2.

5.5
Electrochemical Properties

Oligophenyls are not as easily oxidized or reduced as special charge transport materials, but reversible oxidation and reduction steps can nevertheless be detected at appropriate conjugation lengths. For the ter(9,9-diarylfluorene) **49** containing three spirobifluorene units in a chain, both oxidation and reduction potentials were measured by Wong et al. [93]. Two oxidation steps were detected at 1.32 V and 1.56°V vs. Ag/AgCl (CH$_2$Cl$_2$, 0.1 M nBu$_4$NPF$_6$) and two reduction steps at -2.01 V and -2.21 V (THF, 0.1 M nBu$_4$NClO$_4$).

An interesting question arises for spiro-oligophenyls with two identical, perpendicular chromophores: whether the double-charged species have their charges on the same branch or on different branches. Crispin et al. [94] performed doping experiments with Li and Na in solid films and concluded that if Spiro-4Φ **21** or Spiro-6Φ **22** is doped with Li, dianions (bipolarons) are formed on the same spiro branch, whereas doping with Na leads to the formation of radical anions (polarons) on each branch. These results are based on photoelectron and optical spectra and indicate that the size and interactions of the dopant play an important role in the formation of the charged species.

5.6
Charge Transport

Despite the unfavorable energy levels for charge injection, charge carrier mobilities can be quite high. Wu et al. measured the hole and electron mobilities of two ter(9,9-diarylfluorenes) **49** and **50** by time-of-flight techniques [76]. Nondispersive bipolar transport was detected. Hole mobilities up to 4×10^{-3} cm^2/Vs at 3.6×10^5 V/cm have been obtained for spiro compound **49**, which is comparable to values for the best arylamine hole-transport materials [76]. The electron mobilities of this compound are one order of magnitude smaller and in the range of 10^{-4}–10^{-3} cm^2/Vs. For **50**, the electron mobility was higher than the hole mobility.

5.7
Electroluminescent Devices

Due to the far-blue emission spectrum of oligophenyls, they can be used as blue electroluminescent emitters, but in order to obtain other colors, dopants are needed.

The hole and electron mobilities provide charge transport even in a single-layer device. An electroluminescent single-layer device with Spiro-6Φ **22** as the active material between ITO and Al : Mg electrodes was described by Salbeck et al. [95]. The electroluminescence spectrum shows a tail on the low energy side compared to the photoluminescence spectrum, resulting in color coordinates of $x = 0.18$ and $y = 0.15$. In a two-layer device made of Spiro-6Φ **22** and Alq$_3$, Spiro-6Φ served as the hole-transport layer, and green emission was obtained from Alq$_3$.

Doping of an emissive oligophenyl with a dye emitting at longer wavelengths also offers the possibility of fine-tuning the dopant concentration in order to obtain mixed colors. Preferred for various display applications is white-light emission. If an emitting host is doped, the color coordinates follow a trajectory in the diagram with increasing dopant concentration. This is displayed in Fig. 23 for a series of electroluminescent devices based on spiro materials as hosts [96].

Using Spiro-4Φ **21** doped with the yellow-emitting rubrene as the emitting layer, the white point is crossed for a dopant concentration of 0.58 wt.-%. The hole-transport layer was Spiro-TAD **56**, and the electron-transport layer was Alq$_3$ in these experiments. Similar results have been obtained by replacing Spiro-4Φ **21** with Spiro-Octo2 **28**.

Electroluminescence devices have also been made with **52** as emitter [36]. For the configuration ITO/TPD/**52**/Alq$_3$/LiF/Al, blue emission was obtained with color coordinates $x = 0.14$ and $y = 0.12$. A maximum luminance of 3125 cd/m^2 at a driving voltage of 12.8 V was measured. A luminous efficiency of 0.9 lm/W and an external quantum efficiency 2.8% were obtained at 100 cd/m^2.

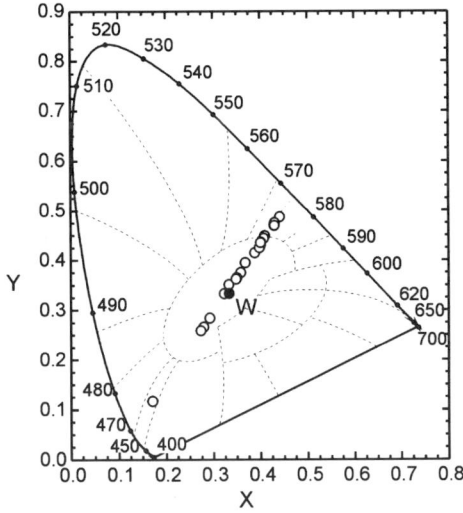

Fig. 23 CIE color diagram for electroluminescent devices, consisting of anode/Spiro-TAD/Spiro-4Φ: Rubrene/Alq$_3$/cathode with different concentrations of rubrene. The spectrum shifts from blue to yellow, crossing the white point (W)

TBSA **38** was used as a blue emitter in the device configuration ITO/CuPc/ a-NPD, TBSA, Alq$_3$/LiF/Al. The color coordinates were $x = 0.14$, $y = 0.08$ with a peak at 442 nm [37].

6
Spiro Compounds with Stilbene and Azobenzene Units

6.1
Synthesis

A spiro compound with stilbene units is Spiro-DPVBi **54**, which was synthesized in 52% yield by Suzuki coupling of 2,2-diphenylvinylboronic acid and 2,2′,7,7′-tetrabromo-9,9′-spirobifluorene (**14**), as shown in Fig. 24 [97].

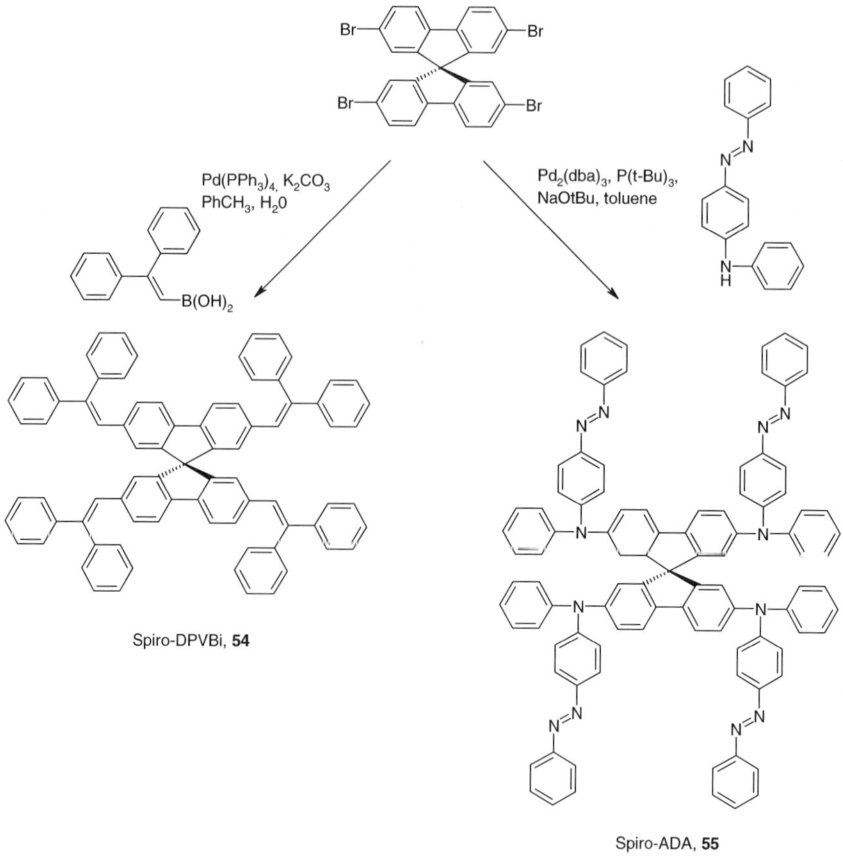

Fig. 24 Synthesis of Spiro-DPVBi **54** and Spiro-ADA **55**

Spiro-ADA **55**, which contains four phenylazodiphenylamine groups, was synthesized by Hartwig–Buchwald coupling of 4-(phenylazo)diphenylamine and 2,2′,7,7′-tetrabromo-9,9′-spirobifluorene (**14**) in 89% yield [98].

6.2
Thermal and Morphological Properties

The glass transition of spiro compounds containing vinylene units is lower than for oligophenyls of a similar size and comparable to spiro-linked arylamines. For instance, the glass transition temperature of Spiro-DPVBi **54** is $T_g = 130\,°C$ [97]. It is substantially increased with respect to the parent compound, 4,4′-bis(2,2′-diphenylethenyl)-biphenyl (DPVBi), which has $T_g = 64\,°C$. The glass transition temperature of Spiro-ADA **55** is $T_g = 147\,°C$ [98], while the parent compound N,N'-bis(4-phenylazo)phenyl)-N,N'-diphenyl-4,4′-diamine (AZOPD) shows a glass transition temperature of $101\,°C$ [99].

6.3
Optical Properties

The parent compound of Spiro-DPVBi **54**, 4,4′-bis(2,2′-diphenylethenyl)-biphenyl (DPVBi), was introduced by Hosokawa et al. as an excellent blue emitter [100]. Spiro-DPVBi absorbs at $\lambda_{max} = 359$ nm and emits at 466 nm in the solid film, exhibiting a large Stokes shift. Amplified spontaneous emission occurs at 475.3 nm, but with high thresholds [91].

Compounds containing azobenzene groups have been studied as photochromic materials. When polarized light at a proper wavelength is irradiated, azobenzene groups show a reversible *trans-cis-trans* isomerization and an orientational distribution perpendicular to the direction of the polarization of the incident laser beam. This means that surface relief gratings (SRG) could be formed through illumination of interference beams on an amorphous film, which was first demonstrated for azo polymers in 1995 [101, 102] and for low-molecular-weight glasses in 1999 [99]. Amorphous films of Spiro-ADA **55** show a very rapid response to linearly polarized light, a high diffraction efficiency of 38.7%, and a large modulation depth of around 320 nm [98].

6.4
Electroluminescent Devices

The blue electroluminescence of Spiro-DPVBi **54** was demonstrated in a device consisting of ITO/CuPc/Spiro-TAD/Spiro-DPVBi/Alq$_3$/Ca [97]. The color coordinates were $x = 0.16$, $y = 0.17$, and slightly more blue was emitted using such a device vs. a corresponding device with the nonspiro analog

DPVBi. Also the efficiencies were better with 4.2 cd/A and 2500 cd/m² at 10 V. But most important, the stability of the device was demonstrated for temperatures up to 120 °C, whereas diffusion and intermixing were obtained at 90 °C for DPVBi devices.

7
Spiro Compounds Based on Arylamines

7.1
Synthesis

Due to their ease and reversibility of oxidization, aromatic amines are widely studied and used as hole-transporting compounds in organic electronics. Electronic characteristics like redox potentials and charge carrier mobilities can be adjusted by the substitution pattern of the aryl substituents. Spiro-linked arylamines can be prepared in high yields either by Hartwig–Buchwald coupling [103] of secondary amines and halogenated spiro precursors or by the copper-promoted Ullmann coupling reaction [104]. The second pathway is especially useful for the synthesis of arylamines with different substituents on the two phenyl rings, since it allows a two-step protocol by the use of an acyl-protected amine in the first step. The synthesis of spiro-linked tetraamines is outlined in Fig. 25. Figures 26 and 27 give an overview of the different compounds published so far [32].

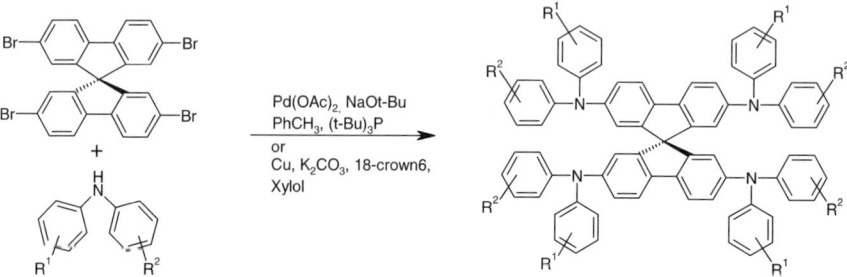

Fig. 25 Synthesis of spiro-linked arylamines

7.2
Thermal and Morphological Properties

For the arylamines, the glass transition temperatures are lower than for spiro-linked oligophenyls. The T_g of Spiro-TAD **56** is 133 °C, whereas the flexible methoxy groups in Spiro-MeO-TAD **62** decrease T_g further to 121 °C.

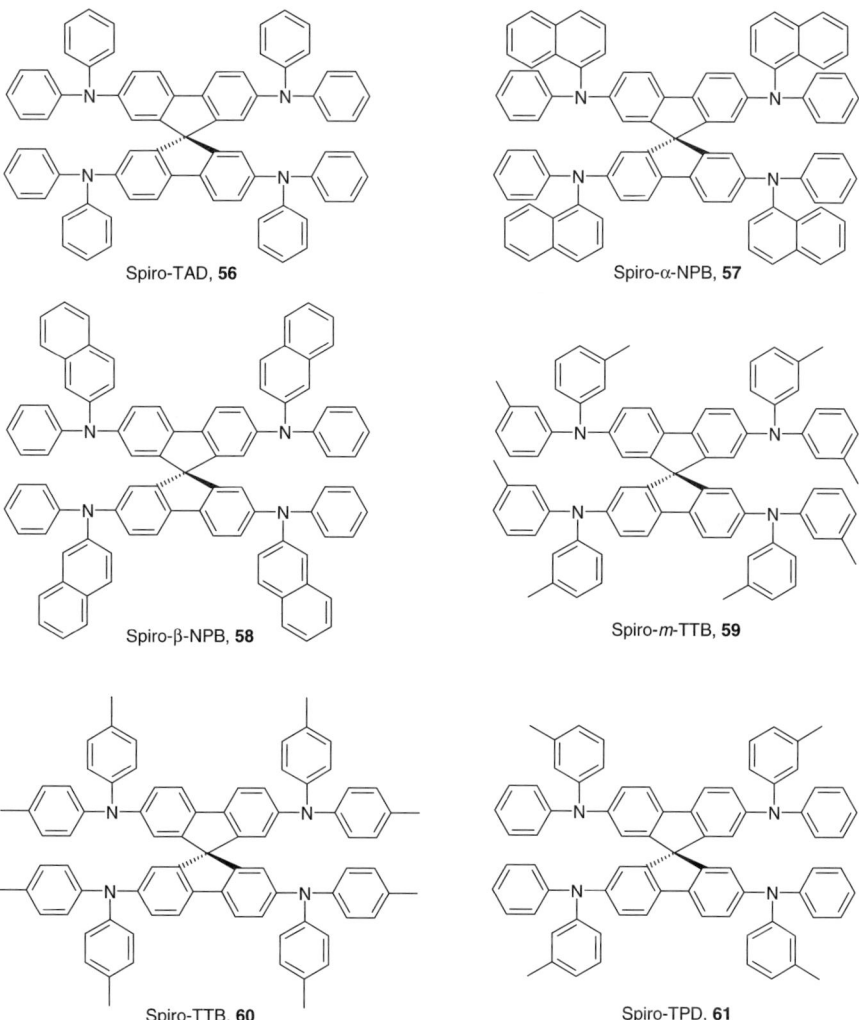

Fig. 26 Substituted spiro-linked arylamines

7.3
Optical Properties

Arylamines are not frequently used as emitters; the quantum yields are low, and the materials are subjected to photodegradation in air. For Spiro-TAD **56**, we measured the solid-state absorption band at 382 nm and the main emission at 405 nm. The carbazole Spiro-Carb **65** absorbs at $\lambda_{max} = 346$ nm and emits in the deep violet range with a peak at 382 nm. All these values are given for neat amorphous films.

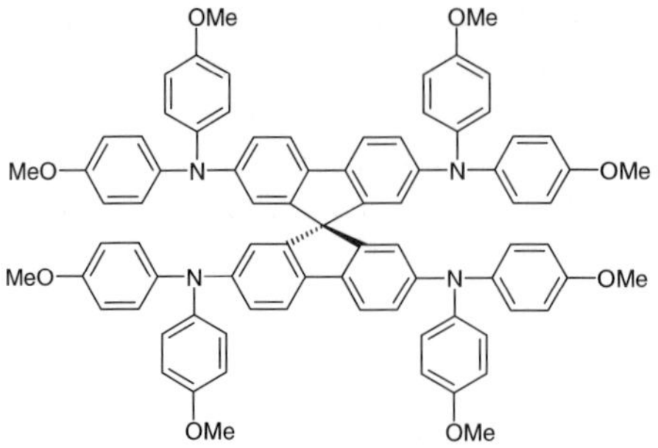

Spiro-MeO-TAD, **62**

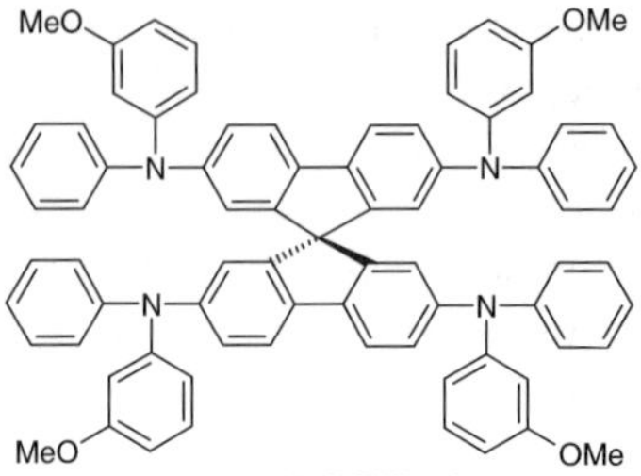

Spiro-MeO-TPD, **63**

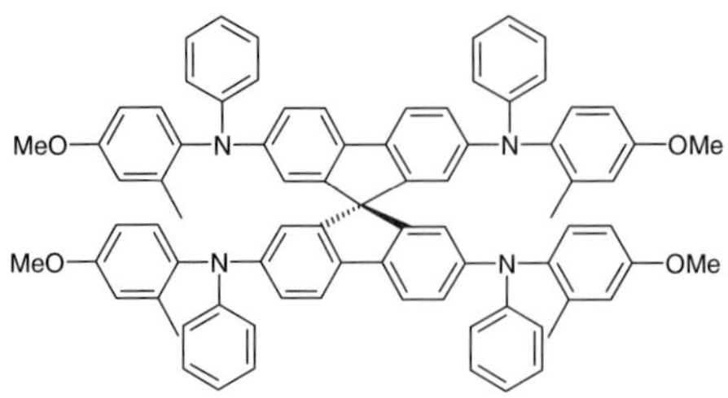

Spiro-MeMeO-TAD, **64**

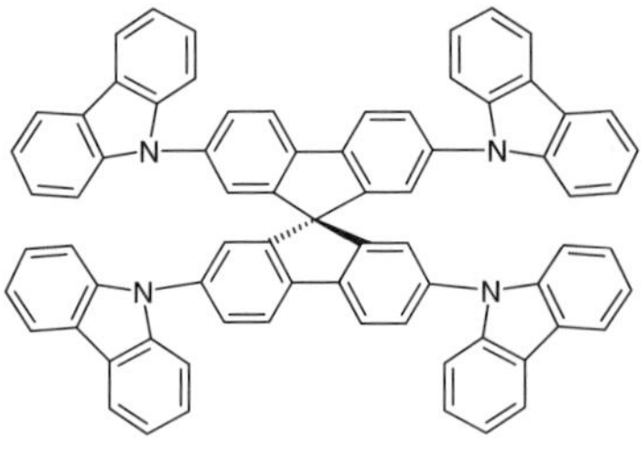

Spiro-Carb, **65**

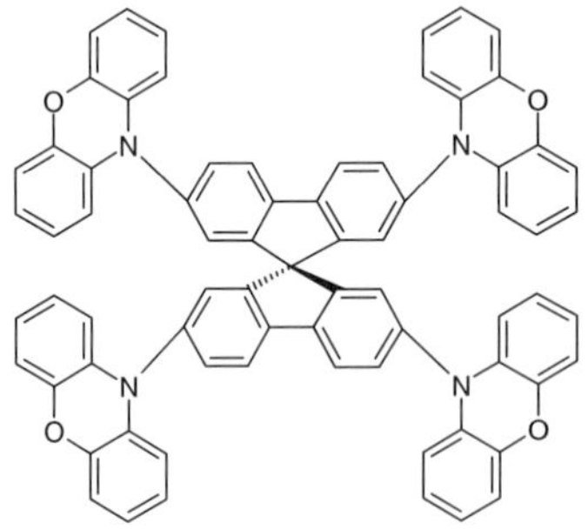

Spiro-PhOx-TAD, **66**

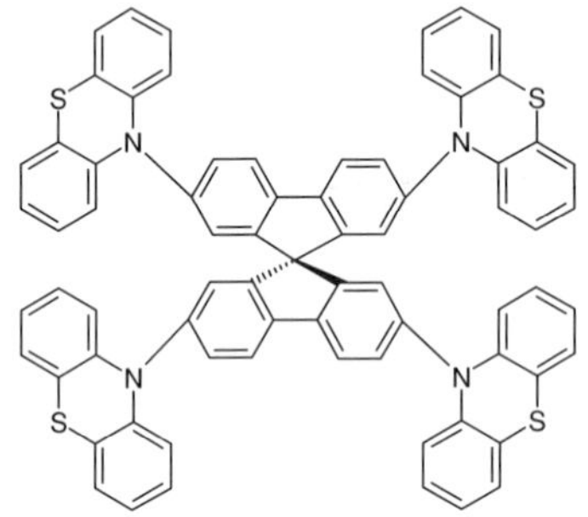

Spiro-PhTh-TAD, **67**

Fig. 27 Substituted spiro-linked arylamines (cont.)

Amplified spontaneous emission (ASE) was measured for both Spiro-TAD **56** and Spiro-Carb **65** [91]. Whereas the threshold of Spiro-TAD **56** is quite high, the ASE peak being at 404.7 nm, Spiro-Carb **65** exhibits reasonable good ASE at 400.3 nm with a threshold of 50 $\mu J/cm^2$ for a 117-nm-thick film.

7.4 Electrochemical Properties

In order to give a comparison of the energy levels, the values determined by electrochemistry should be seen in relation to the standard hole conductor TPD. For TPD, measurements are available from different groups that allow a direct comparison of different experimental setups. The ionization potential, which corresponds to the HOMO level under the assumptions mentioned above, was measured with photoelectron spectroscopy to be 5.34 eV [105]. Anderson et al. identified the onset of the photoelectron spectrum with the ionization potential and the first peak with the HOMO energy, and reported separate values of 5.38 eV and 5.73 eV, respectively [106]. The cyclic voltammetry data reveal a first oxidation wave at 0.34 V vs. Fc/Fc$^+$ in acetonitrile (J. Uebe and J. Salbeck, unpublished results), and 0.48 V vs. Ag/0.01 Ag$^+$ in dichloromethane [107]. The oxidation proceeds by two successive one-electron oxidations, the second one being located at 0.47 V vs. Fc/Fc$^+$.

Relative to TPD, the spiro derivative Spiro-TAD **56** has a lower first oxidation potential, which can be explained by the better resonance stabilization of the radical cation [71]. The material exhibits two successive one-electron oxidations (0.23 V and 0.38 V vs. Fc/Fc$^+$) and one subsequent formal two-electron oxidation (0.58 V) to the tetracation. The same behavior can be found in the case of Spiro-α-NPB **57**. The cyclovoltammogram, shown in Fig. 28, exhibits two successive one-electron oxidations at 0.20 V and 0.35 V and one subsequent formal two-electron oxidation at 0.60 V vs. Fc/Fc$^+$.

Phenothiazine **67** (0.27 V vs. Fc/Fc$^+$) and phenoxazine **66** (0.29 V vs. Fc/Fc$^+$) have higher oxidation potentials than Spiro-TAD. The carbazole Spiro-Carb **65** has an even higher oxidation potential, but in this case the oxidation is not reversible [32]. There are some hints, however, of bipolar behavior: the parent compound 4,4'-bis(N-carbazolyl)-biphenyl was applied as electron-transport layer in electroluminescent devices [108].

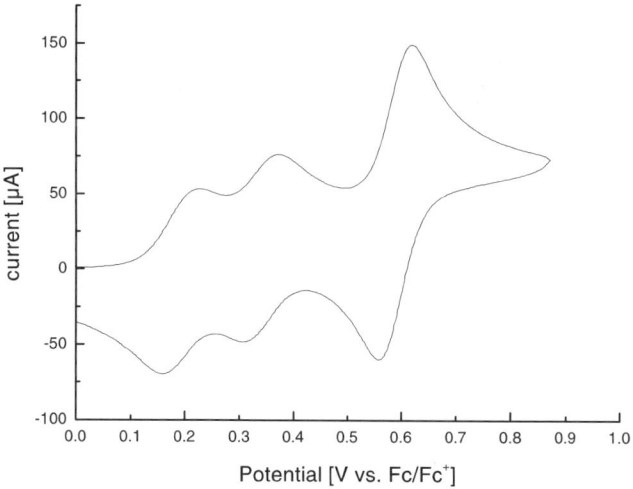

Fig. 28 Cyclovoltammogram for Spiro-α-NPB **57**. The oxidation proceeds in two one-electron waves and one formal two-electron wave. Solvent: Dichloromethane/TBAHFP 0.1 M, scan rate 100 mV/s

7.5
Charge Transport

Again, the values for spiro compounds will be compared with mobility data for TPD. For TPD, the hole mobility was measured by Heun and Borsenberger at fields between 40 and 400 kV/cm and in a wide temperature range from 213 K up to 345 K [109]. For high field and ambient temperature, the mobility is in the range of 10^{-3} cm^2/Vs with $\mu_0 = 3.0 \times 10^{-2}$ cm^2/Vs, $\sigma = 0.077$ eV, and $\Sigma = 1.6$.

Time-of-flight data for Spiro-TAD **56** and Spiro-mTTB **59** were reported by Bach et al. [110]. Spiro-TAD exhibits a hole mobility of 3×10^{-4} cm^2/Vs at 200 kV/cm with the model parameters $\mu_0 = 1.6 \times 10^{-2}$ cm^2/Vs, $\sigma = 0.08$ eV, and $\Sigma = 2.3$. The values for m-TTB are in the same order of magnitude with $\mu_0 = 1.0 \times 10^{-2}$ cm^2/Vs, $\sigma = 0.08$ eV, and $\Sigma = 1.2$. Within the description by the disorder model, the spiro linkage reduces the prefactor mobility μ_0 by a factor of approximately 2 and decreases the positional disorder Σ. The energetic disorder parameter σ is not influenced. The mobility values are lower than for TPD, but these differences should not be overestimated since the values fall into the same order of magnitude and there might also be some differences due to the sample preparation procedures.

For Spiro-MeO-TAD **62**, the mobility was measured by Poplavskyy and Nelson with three different techniques: time-of-flight (TOF), dark-injection

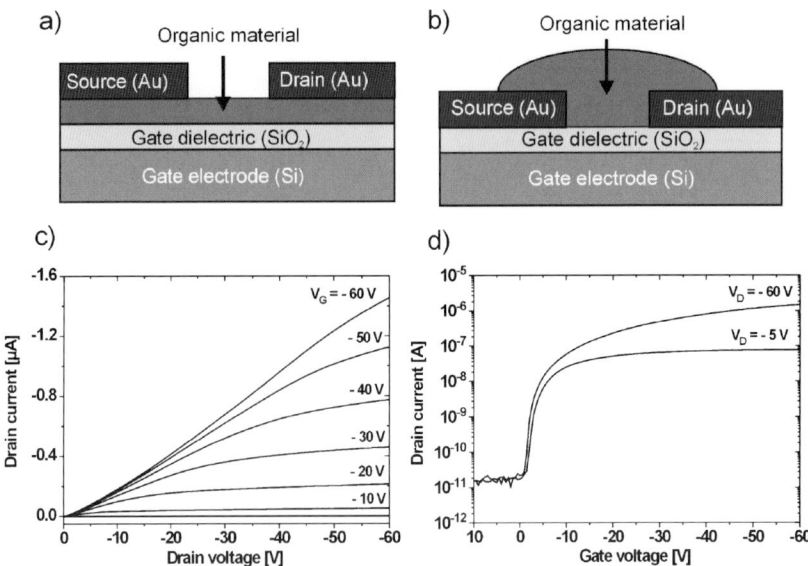

Fig. 29 a Top contact and **b** bottom contact field-effect transistor structures for determination of charge-transport properties of organic semiconductors. **c** Typical output characteristics and **d** transfer characteristics of an organic field-effect transistor with Spiro-TAD **61** as active material (measurements by T.P.I. Saragi)

space-charge-limited current, and steady-state current-voltage characteristics [111]. The values of the different methods agree well and give mobilities of 2×10^{-4} cm^2/Vs at room temperature and high field (3.6×10^5 V/cm). An analysis of the TOF data by the disorder model gives $\mu_0 = 4.7 \times 10^{-2}$ cm^2/Vs, $\sigma = 0.10$ eV, and $\Sigma = 2.3$.

The field effect transistor method has also been applied for the determination of hole mobilities. For Spiro-TAD **56** in a top-contact structure field effect transistor (Fig. 29a), we measured an apparent mobility of 8×10^{-4} cm^2/Vs by evaluating the transfer characteristic in the saturation regime [112].

For bottom-contact structures (Fig. 29b), a mobility of 7×10^{-5} cm^2/Vs was obtained [113]. The field-effect mobility of Spiro-TPD **61** does not differ much from the values for Spiro-TAD. All field-effect transistors are characterized by a high on/off switching ratio [113].

7.6
Electroluminescent Devices

Spiro-TAD **56** has been used as hole-transport material in combination with Alq$_3$ as emitting layer, replacing TPD or α-NPB, which have been used in the original KODAK structures, but have substantially lower T_gs than Spiro-TAD. Consequently, the devices are more stable against thermal treatment (Fig. 30). Whereas the performance of the TPD device breaks down at 70 °C, which cor-

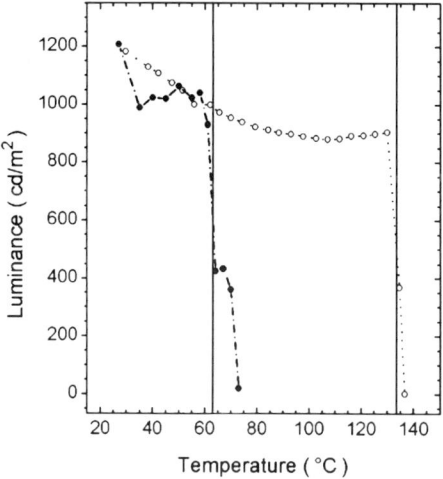

Fig. 30 Breakdown of device performance at T_g: the luminance for a constant current density of 25 mA/cm^2 is plotted for two light-emitting diodes made with hole-transport materials of different glass transition temperatures. *Filled circles*: 75 nm TPD/65 nm Alq$_3$; *open circles*: 75 nm Spiro-TAD **56**/65 nm Alq$_3$. The *vertical lines* mark the respective T_gs: TPD 63 °C, Spiro-TAD 133 °C. At these temperatures the materials become soft, which results in a steep decrease in efficiency

responds to the glass transition temperature of TPD, the Spiro-TAD device is stable up to 130 °C [96].

For a device consisting of anode/CuPc/Spiro-TAD **56**/Alq$_3$/Ca, an onset voltage of 2.2 V and a luminance of 5530 cd/m^2 at 10 V, detected through the semitransparent metal cathode, were obtained [97].

Spiro-TAD was also used in sophisticated OLEDs combining the concepts of (a) doped hole-transport layers in order to increase the conductivity and (b) quantum well structures for better recombination properties due to charge-carrier confinement [114]. The devices consist of ITO/TDATA: F$_4$-TCNQ (82 nm)/Spiro-TAD (8 nm)/[Alq$_3$ (3 nm)/Spiro-TAD (3 nm)]$_n$/ Alq$_3$ (54–42 nm)/LiF (1 nm)/Al, n being 0-3 double layers. The turn-on voltages were ca. 2.5 V for obtaining a luminance of 1 cd/m^2. Except for the triple quantum well structure, the operating voltages were less than 4 V for obtaining a brightness of 100 cd/m^2. For the double quantum well structure, a best luminous efficiency of 5 cd/A at 72 mA/cm^2 was obtained.

7.7
Solar Cells

The conversion of light to electric current in photovoltaic devices represents the direct inversion of the electroluminescent process in OLEDs; thus it is not surprising that spiro compounds have also been exploited for the realization of solar cells, namely, in the dye-sensitized solar cell.

The dye-sensitized solar cell, commonly referred to as the Grätzel cell, is based on the photoinduced electron transfer from a dye to mesoporous TiO$_2$ [115]. The network of connected TiO$_2$ nanoparticles forms a large surface area that is active for light absorption and electron transfer. As active dye, ruthenium dyes such as Ru(II)L$_2$(SCN)$_2$, where L is 4,4'-dicarboxy-2,2'-bipyridyl ligands, are very suitable because of the good adhesion of the carboxy groups to the TiO$_2$ surface. If the dye is excited, electrons are injected into the conducting band of TiO$_2$, leaving a ruthenium (III) species.

$$[\text{Ru(II)L}_2(\text{SCN})_2]^* \rightarrow [\text{Ru(III)L}_2(\text{SCN})_2]^+ + e^-(\text{TiO}_2) \tag{13}$$

The electrons are collected by an F-doped SnO$_2$ electrode, and the electrical circuit is closed via a gold electrode by a mediator that transports the charges needed for the back reduction of the dye (electrons from the gold electrode to the LUMO of the dye, or, in an equivalent picture, holes from the dye to the gold electrode). The original cell suffers from the use of a liquid electrolyte with the redox system I_2/I_3^- as mediator, which causes some problems due to leakage and solvent evaporation if not sealed properly.

Thus all-solid-state variations of this cell have been developed by replacing the liquid electrolyte by hole-transporting molecular glasses [116]. The tetramethoxy derivative Spiro-MeO-TAD **62** was used, with Ru(II)L$_2$(SCN)$_2$

as sensitizing dye. The structure and energy level diagram of this cell is displayed in Fig. 31.

The internal photon-to-electron-conversion efficiency (IPCE) was determined to be 5%, which is still lower than that for comparable liquid cells. However, by doping the hole-transport material by $N(PhBr)_3SbCl_6$ and $Li[(CF_3SO_2)_2N]$, which increases the charge-carrier concentration by partial oxidation, an IPCE of 33% was measured. The overall efficiency of the cell was 0.74% at white-light irradiation with 9.4 mW/cm^2. Under full sunlight (air mass 1.5, 100 mW/cm^2), short-circuit photocurrents of 3.18 mA/cm^2 have been achieved. In a later report [117], the performance was improved by blending Spiro-MeOTAD **62** with a combination of 4-*tert*-butylpyridine and $Li[(CF_3SO_2)_2N]$. Open-circuit voltages of 910 mV and short-circuit current densities of 4.8 mA/cm^2 were obtained, yielding a power efficiency of 2.56%.

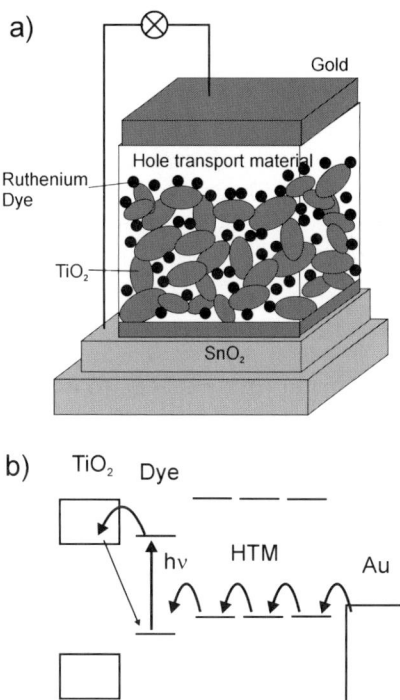

Fig. 31 **a** Organic solar cell with molecular glass Spiro-MeO-TAD **62** as solid-state hole conductor. The photosensitive ruthenium dye is attached as a monolayer to TiO$_2$ nanoparticles, thus forming a large active area for photoinduced electron transfer. **b** Energy-level scheme. Photoinduced electron transfer takes place from photoexcited ruthenium dye into TiO$_2$ conduction band. The recombination directly back to dye must be suppressed. Instead, current is directed through circuit to counter electrode and hole conductor that brings electrons back via hopping transport. HTM: hole-transport material

8
Thiophene Compounds

8.1
Synthesis

Spiro-linked compounds containing heterocyclic units have been prepared for many applications. In general, there are two synthetic pathways to build up heterocyclic spiro compounds. On the one hand, cross-coupling reactions like the Negishi, Kharash, Stille, Suzuki, or Sonogashira coupling reaction can be utilized to connect the heterocyclic subunit with the central spiro core [118]. On the other hand, the heterocycle can be built up from spiro precursors containing heteroatoms.

The first spiro compounds containing oligothienylene or oligothienyleneethinylene subunits as branches connected to a center core of spirobifluorene were synthesized by Tour and coworkers [19, 31, 119], who attempted to realize a concept from Aviram [120], dealing with the theoretical background

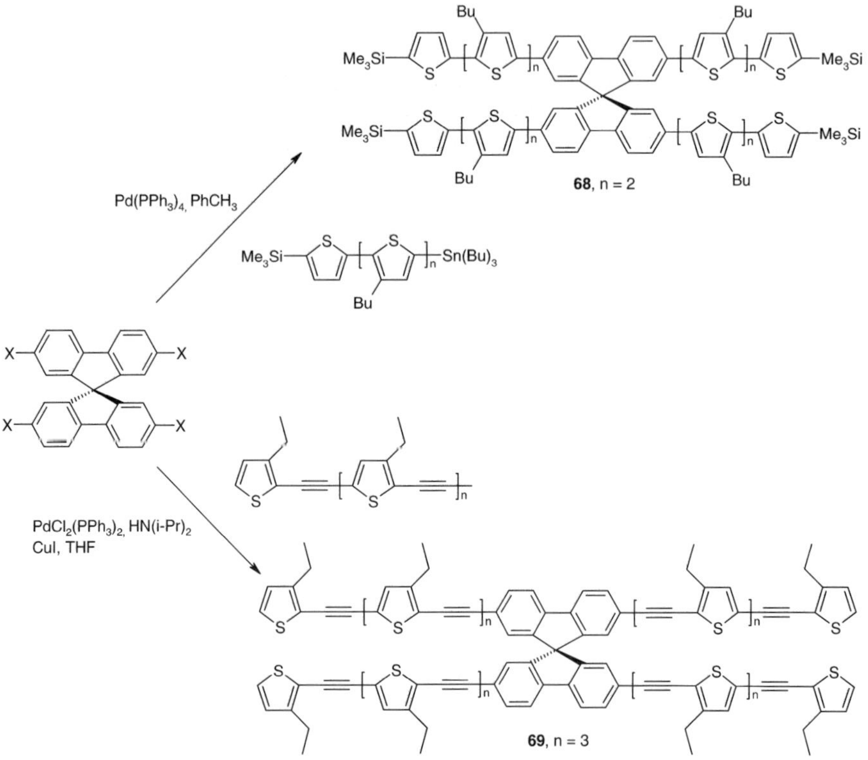

Fig. 32 Synthesis of oligothiophenes with central core of spirobifluorene

for the design of molecular devices for memory, logic, and amplification purposes. These compounds were synthesized by Stille and Sonogashira coupling reactions from thienylene precursors and a tetrahalogenated spiro core, as shown in Fig. 32.

The spiro-oligothiophenes shown in Fig. 33 are similar to **68**. These compounds were reported by Pei and coworkers, who utilized the Suzuki cross-coupling reaction to connect the precursor thiophene chains and the halogenated spiro cores [28, 121].

Unsymmetric spirobifluorenyl bridged oligothiophenes **98** and **99** with a mixed spiro core were reported by Mitschke and Bäuerle [122]. In these compounds not only the branches but also one half of the central core consists of a bithiophene unit as shown in Fig. 34. The central building block for the synthesis of the spirobifluorenyl-bridged oligothiophenes **98** and **99** is the spiro compound **95**, available starting from 4H-cyclopenta[2,1-b:3,4-b']dithiophen-4-one and 2-lithiobiphenyl, which was prepared by lithiation of 2-bromobiphenyl. The spiro core was formed by an intramolecular condensation reaction induced by either concentrated hydrochloric acid in acetic acid or boron trifluoride-diethyl ether in dichloromethane. In contrast to the synthesis of spirobifluorene (**1**), the α-thienyl position is more reactive, so that also "dimer" **96** can be identified as the main byproduct. After bromination at the α-thienyl positions with two equivalents of NBS, the synthesis of target compounds **98** and **99** was achieved by transition metal cross-coupling reactions and the precursors. **98** and **99** show solubilities in dichloromethane of 12 g/l and 3.8 g/l, while their non-spirobifluorenyl-bridged counterparts

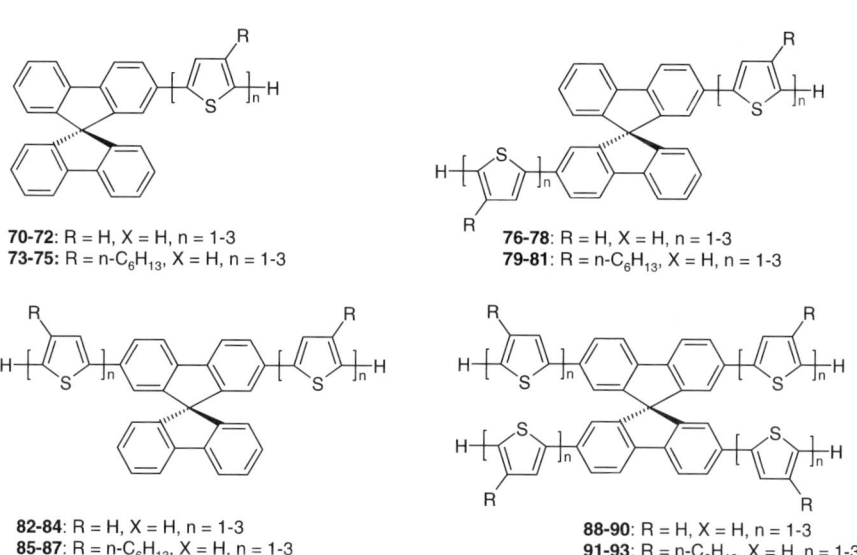

70-72: R = H, X = H, n = 1-3
73-75: R = n-C$_6$H$_{13}$, X = H, n = 1-3

76-78: R = H, X = H, n = 1-3
79-81: R = n-C$_6$H$_{13}$, X = H, n = 1-3

82-84: R = H, X = H, n = 1-3
85-87: R = n-C$_6$H$_{13}$, X = H, n = 1-3

88-90: R = H, X = H, n = 1-3
91-93: R = n-C$_6$H$_{13}$, X = H, n = 1-3

Fig. 33 Other spirobifluorene-bridged oligothiophenes

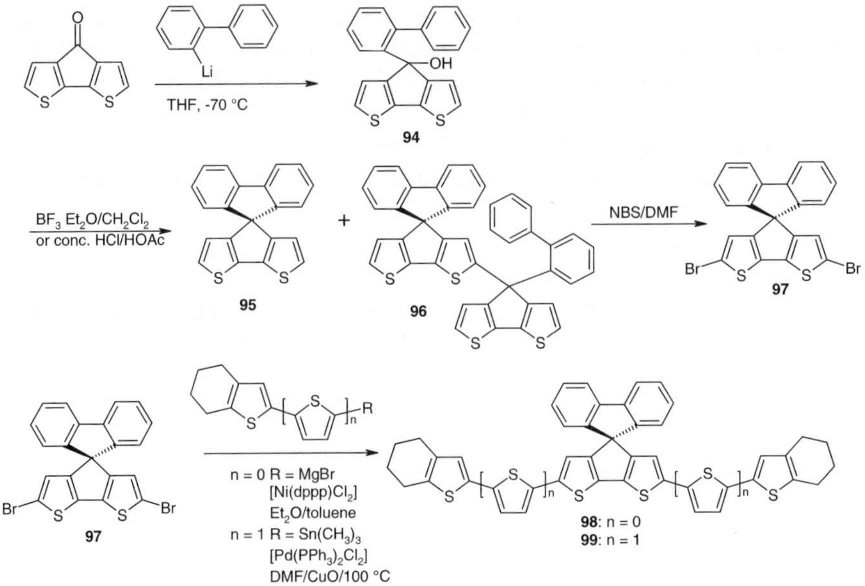

Fig. 34 Synthesis of unsymmetric spirobifluorenyl-bridged oligothiophenes

only have values of 4.6 g/l and 1.1×10^{-3} g/l. A different route to the precursor was recently reported by Ong, utilizing 3-bromo-2,2′-bithiophene and 9-fluorenone [123].

8.2
Optical Properties

Incorporating an increasing number of thiophene rings into molecular glasses allows a wide tunability of the optical properties. For the unsubstituted α-polythiophenes, denoted usually as $H-T_n-H$, the absorption wavelengths reach from 302 nm ($n = 2$) to 432 nm ($n = 6$) (in CHCl$_3$) [124]. The corresponding emission wavelengths vary from 362 nm to 510 nm (in dioxane/acetonitrile) [125].

For spirofluorenyl-bridged oligothiophenes, absorption and emission are bathochromically shifted with respect to the unbridged oligothiophenes: **98** absorbs at 431 nm and emits at 484 nm/511 nm, **99** absorbs at 472 nm and emits at 536 nm with a shoulder at 570 nm [126].

8.3
Electrochemical Properties

The oligothiophene-functionalized 9,9′-spirobifluorenes **70–93** exhibit reversible oxidation waves but show poor reversibility in reduction [28, 121].

For spirobifluorene functionalized with four unsubstituted mono-, bi-, and terthiophene units, **88–90**, onset oxidation potentials of 1.20 V, 0.86 V, and 0.65 V vs. SCE were reported, respectively. For the compounds **91–93**, in which the thiophenes are substituted with alkyl chains in the 3 position, the onset potentials shift to 0.99 V, 0.88 V, and 0.85 V, respectively.

The oxidation potentials of spirofluorenyl-bridged oligothiophene **98** are 0.18 V vs. Fc/Fc$^+$ for the first oxidation and 0.57 V for the second oxidation. For the higher homolog **99**, the values are 0.16 V and 0.37 V, respectively. A third, quasireversible, oxidation step was detected at 1.09 V. Reduction takes place at −2.58 V and −2.78 V (irreversible) for **98**, and at −2.32 V and −2.43 V for **99** [122]. In comparison with unbridged oligothiophenes, these materials have a higher effective conjugation length, which can be seen in smaller energy gaps and lower oxidation potentials.

8.4
Electroluminescent Devices

To date, there have been no data on electroluminescent thin film devices with spirothiophenes available, but Bäuerle et al. examined the applicability of these compounds by electrogenerated chemiluminescence experiments. They found for the quaterthiophene **98** an intense but unstable electrochemiluminescence because of the instability of the radical anion. On the other hand, the sexithiophene **99** showed a more intense and more stable electrochemiluminescence. The emission spectra center around 506 nm and 560 nm for **98** and **99**, respectively.

9
Oxadiazole Compounds

Another class of compounds for use in organic electroluminescence are substances containing oxadiazole heterocyles. These systems are well known for their good electron-transporting capabilities but often suffer from low glass transition temperatures and low morphologic stability. Oxadiazoles can be prepared in high yields by the reaction of carboxyl chlorides with tetrazoles [127–129]. This reaction was utilized for the preparation of a spirolinked version **102** of the electron transporting compound PBD [71, 95].

9.1
Thermal and Morphological Properties

For Spiro-PBD **102**, T_g is at 163 °C and the melting point of crystals at 331 °C. The amorphous state shows no recrystallization upon heating at 10 K/min but decomposition at temperatures between 300 and 350 °C (Fig. 2b).

9.2
Optical Properties

The emission region of oxadiazoles are at the violet end of the visible spectrum. Typical emission maxima are located around 370 nm, with absorption bands around 300–330 nm. For the spiro-oxadiazole, Spiro-PBD **102**, a detailed study of the electronic structure and optical properties was published [81]. The vibronic structure of the lowest energy absorption band is well resolved, in solution as well as in amorphous film, with a 0-0 transition at 351 nm (3.53 eV), and as strongest absorption peaks the 0-1 and 0-2 phonon bands at 336 nm (3.69 eV) and 318 nm (3.90 eV). The fluorescence spectrum of this compound is symmetrical to the absorption spectrum with a Stokes shift of 43 nm in solution. In neat films, the second vibronic emission band is more pronounced (λ_{Abs} = 334 nm, λ_{Em} = 406 nm). Amplified spontaneous emission (ASE) was measured with a peak emission at 387 nm, but with high threshold [91].

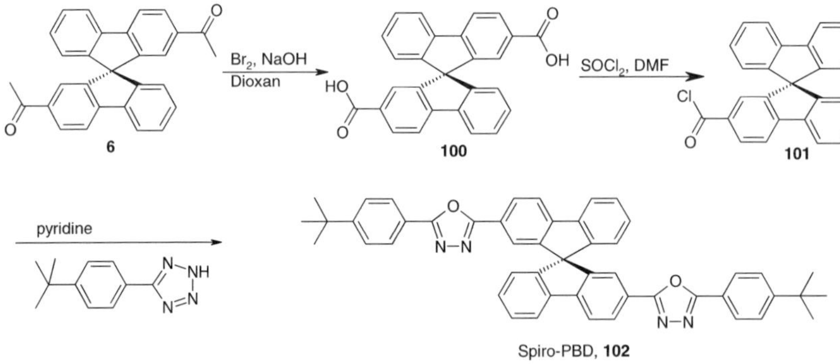

Fig. 35 Synthesis of Spiro-PBD **102**

The applications of oxadiazoles in devices are dominated by the electron-transporting and hole-blocking properties. In multilayer devices comprising oxadiazoles, usually other layers with fluorophores emitting at longer wavelengths act as emission layers. In these devices, oxadiazoles are used effectively as exciton barrier. The excited states of chromophores can be interpreted as Frenkel excitons in terms of semiconductor physics, and exciton diffusion corresponds to resonant energy transfer. This means that as exciton barrier, a layer with higher photoexcitation energy like the oxadiazoles can be used since exciton transfer does not proceed upwards in energy.

9.3
Electrochemical Properties

The oxadiazoles shown have a lower tendency toward reduction than the standard electron-transport material Alq_3 and thus a higher barrier for elec-

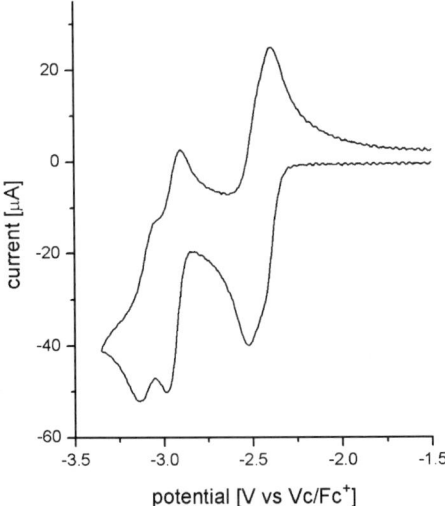

Fig. 36 Cyclovoltammogram for the electron-transporting material Spiro-PBD. The first reduction is a merged wave with an overall transfer of two electrons. Solvent: THF/TBAHFP 0.1 M, scan rate 100 mV/s

tron injection. Spiro-PBD **102** can accept four electrons, the first electron transfer (merged wave for two electrons) taking place at -2.46 eV vs. Fc/Fc$^+$ (Fig. 36) [71]. Since the HOMO-LUMO gap is more than 1 eV larger than for Alq$_3$, the hole-blocking properties are better for the oxadiazoles.

10
Other *N*-Containing Heterocycles

The blue-emitting pyrimidine-containing compound TBPSF (**103**) was synthesized by Wu and coworkers by Suzuki coupling of 2,7-spirobifluorene boronic diacid and 5-bromo-2-(4-tert-butylphenyl)-pyrimidine in the presence of Pd(PPh$_3$)$_4$ and t-Bu$_3$P [130].

The spiro-bridged bis(phenanthroline) ligand **104** for the synthesis of transition-metal complexes were developed by Juris and Ziessel [131, 132]. The synthetic pathway involves the Friedländer condensation of 2,2'-diacetylspirobifluorene (**6**) and 8-amino-7-quinolinecarbaldehyde outlined in Fig. 38.

Di- and tetra-9,9'-spirobifluorene porphyrins **105-107**, shown in Fig. 39, were synthesized by Poriel et al. [22]. These compounds show a hindered rotation about the porphyrin-spiro bond, which leads to the occurrence of atropisomers. Manganese and iron complexes of these spiro-porphyrins have been studied with respect to their properties as oxidation catalysts in heterogeneous catalysis [133, 134].

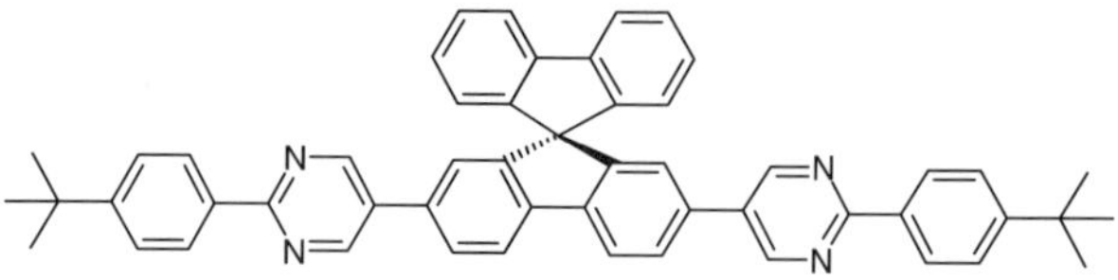

TBPSF, **103**

Fig. 37 Chemical structure of TBPSF **103**

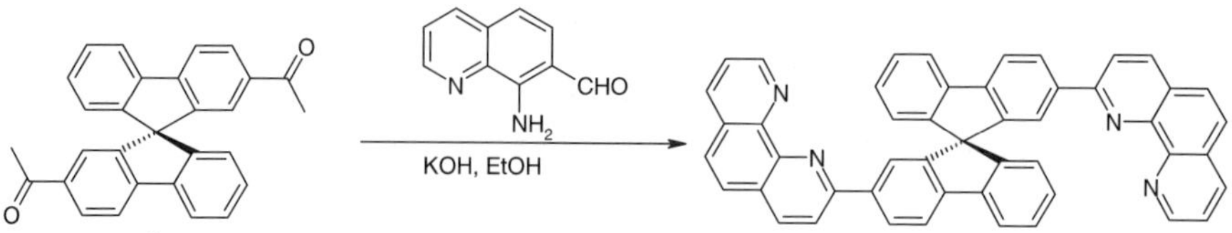

6 **104**

Fig. 38 Synthesis of spiro-bridged bis(phenanthroline) ligands

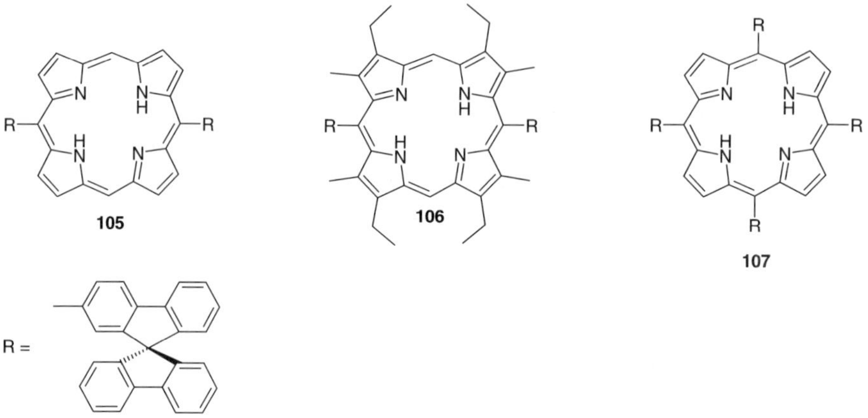

105 **106** **107**

Fig. 39 Di- and tetra-9,9′-spirobifluorene porphyrins

11
Spiro Compounds with Mixed Chromophores

As already mentioned, most emitting spiro-linked compounds are systems whose emission color covers the blue part of the electromagnetic spectrum. The electronic structure of most fluorescence dyes emitting in the green or red part of the electromagnetic spectrum consists of a conjugated system with donor and acceptor groups attached [135]. A combination between the spiro concept and this design strategy leads to a new general structure for vertical, or "left–right", unsymmetric spiro compounds with a 2,2′-A-7,7′-B substitution pattern (Fig. 40, left).

Fig. 40 General structures of "left–right" and "top–down" unsymmetric spiro compounds

The other possible class of mixed spiro compounds with a horizontal, or "top–down", unsymmetric, 2,7-A-2′,7′-B substitution pattern is interesting due to the possibility of combining different functions (emitting properties, charge-transporting properties) in one molecule, which should lead to a reduced number of necessary layers in an OLED device.

11.1
Synthesis

11.1.1
"Left–Right" Unsymmetric Spiro Compounds

Based on the concept of "left–right" unsymmetric substitution, a series of spiro-bridged emitters equipped with electron-donating arylamine groups and electron-accepting oxadiazole moieties have been reported by Pudzich and Salbeck [136]. The synthesis of these compounds is sketched in Fig. 41, starting with 2,2′ diacetyl-7,7′-dibromo-9,9′-spirobifluorene (**108**), which was first prepared by Diederich et al. [137]. The oxidation of the acetyl groups led to the corresponding carboxylic acid **109**, which is converted to the oxadiazole heterocyclus in the intermediate compounds **110a** and **110b** [128, 129, 138]. The target compounds Spiro-AMO-tBu **111** and Spiro-AMO-CN **112** were synthesized by the palladium-catalyzed Hartwig–Buchwald reaction [139]; for the synthesis of Spiro-AMPO-tBu **113** and Spiro-AMPO-CN **114** the Suzuki coupling reaction was used [72].

11.1.2
"Top–Down" Unsymmetric Spiro Compounds

Spiro compounds based on a "top–down" unsymmetric substitution pattern were initially synthesized to combine independent emitting and/or charge-transporting moieties in one molecule. A series of compounds containing combinations of hole-transporting arylamine with emitting oligophenyl (Spiro-DPSP **115**), electron-transporting oxadiazole with emitting oligophenyl (Spiro-SPO **116**), and hole-transporting arylamine with electron-transporting oxadiazole moieties (Spiro-DPO **117** and Spiro-MeO-DPO **118**) were synthesized by Pudzich [140]. Recently, Chien et al. reported

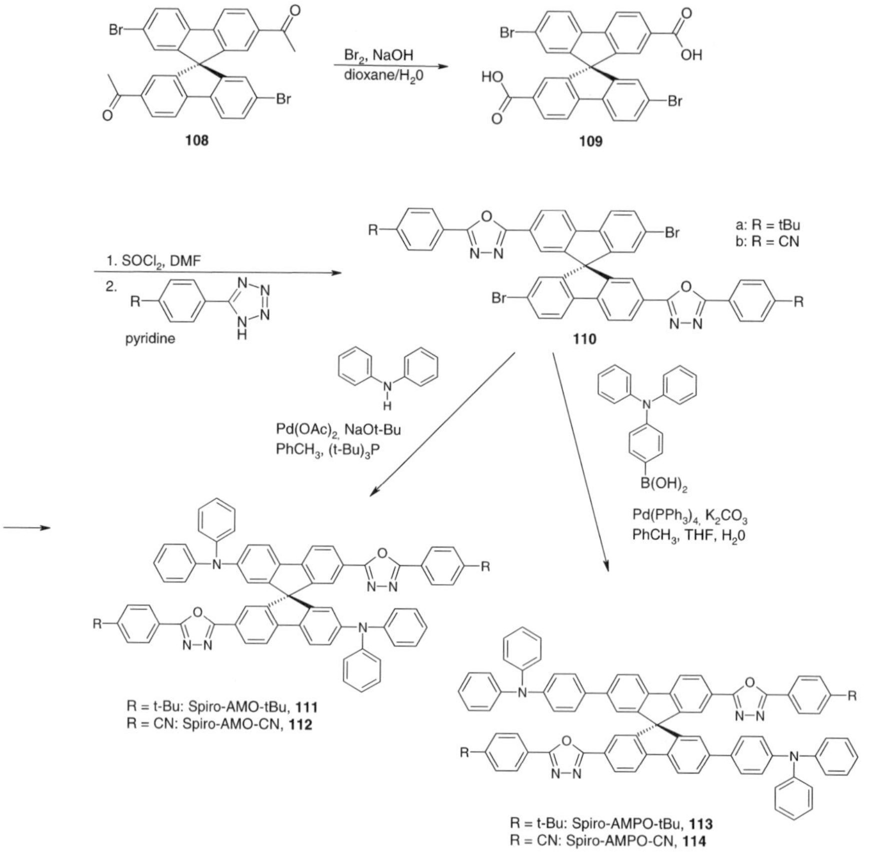

Fig. 41 Synthesis of spiro-linked, left–right unsymmetric oxadiazoleamines

on the synthesis of compound **119**, which is also a combination of an electron-transporting with a hole-transporting half [141].

The synthesis of Spiro-DPSP **115** starts with 2,7-dibromospirobifluorene (**9**), which is iodinated in the 2′ and 7′ position by the procedure of Merkushev [35]. After Suzuki coupling [72] of **120** with biphenylboronic acid, the target compound is formed by Hartwig–Buchwald coupling [103] of **121** with diphenylamine (Fig. 43).

As in the case of Spiro-DPSP **115**, the synthesis of the oxadiazole-containing compounds Spiro-SPO **116**, Spiro-DPO **117**, and Spiro-MeO-DPO **118** starts from 2,7-dibromospirobifluorene (**9**), which is first converted into the 2-acetyl-2′,7′-dibromospirobifluorene (**122**). After oxidation of the acetyl groups, the resulting carboxylic acid **123** is used to form the oxadiazol heterocycle by reaction with *t*-butylphenyltetrazole. Again, the target compounds are built by use of transition-metal-catalyzed cross-coupling reactions as shown in Fig. 44. Chien et al. used a different route to build the "dou-

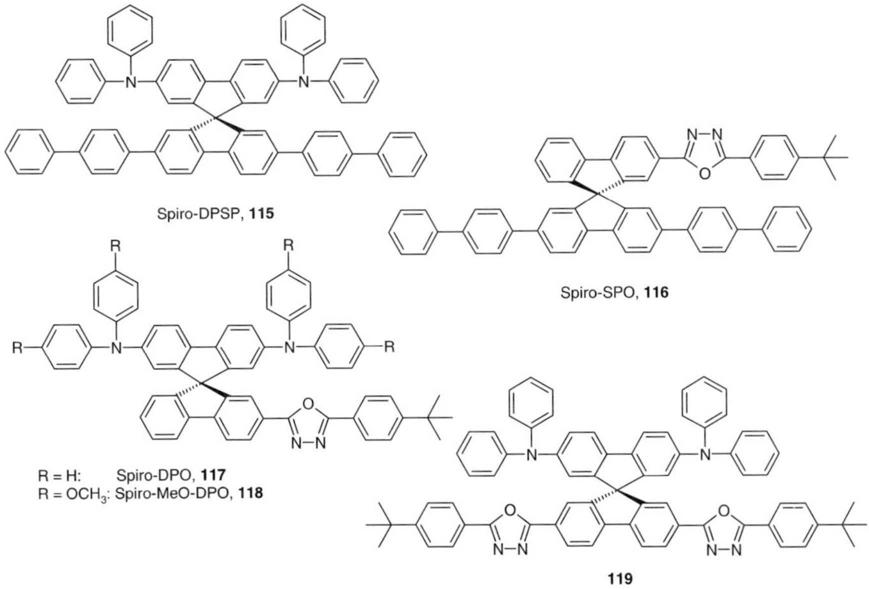

Fig. 42 Molecular structures of "top–down" unsymmetric spiro compounds

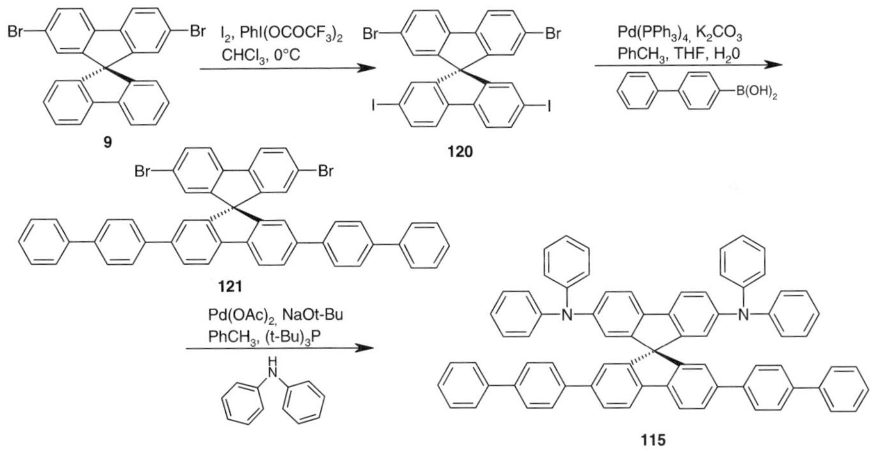

Fig. 43 Synthesis of Spiro-DPSP 115

ble" oxadiazole compound **119**. Instead of introducing two carboxylic acid moieties into the one half of the spiro core, they formed the bistetrazole **128** from the corresponding dicyano precursor **127**. This route is outlined in Fig. 45.

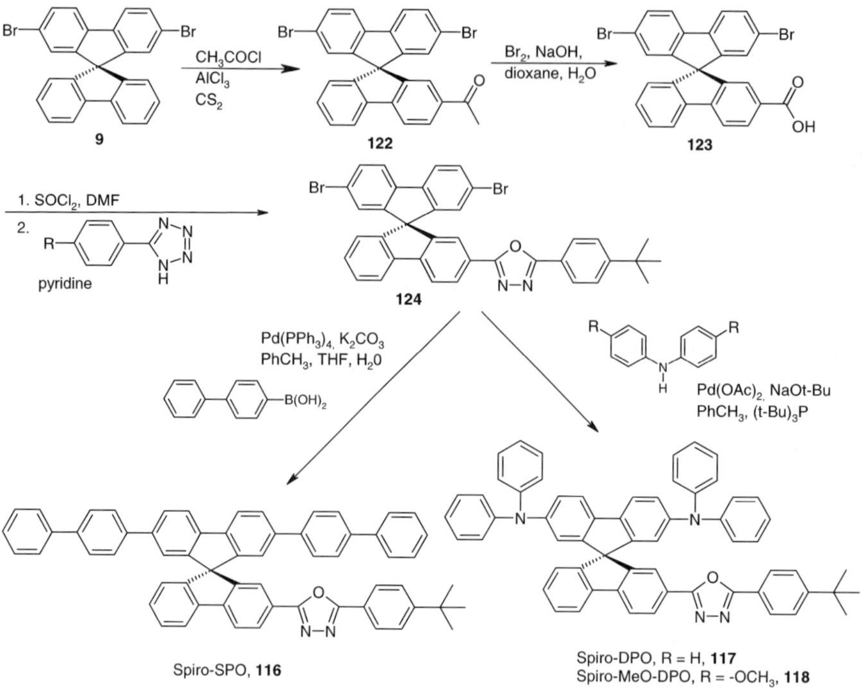

Fig. 44 Synthesis of Spiro-SPO **116**, Spiro-DPO **117**, and Spiro-MeO-DPO **118**

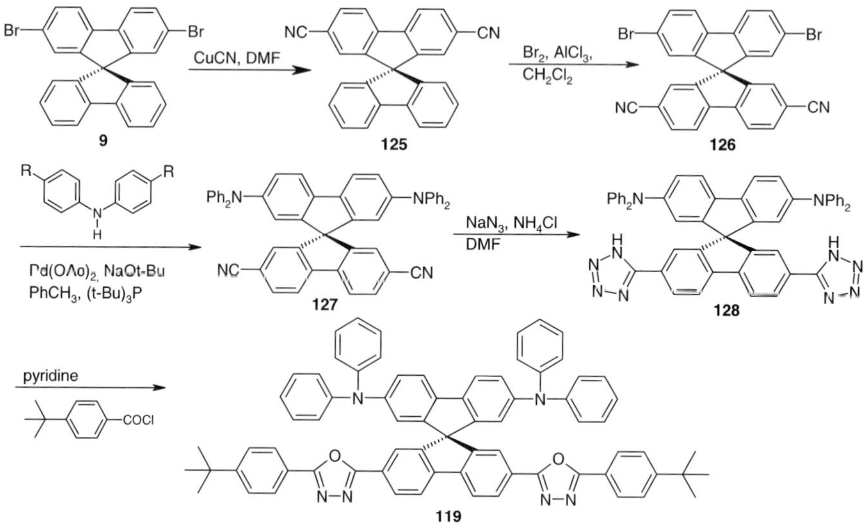

Fig. 45 Synthesis of **119**

11.2
Thermal and Morphological Properties

The glass transition temperatures of the "left–right" unsymmetric substituted spiro-oxadiazoleamines Spiro-AMO-tBu **111**, Spiro-AMPO-tBu **113**, Spiro-AMO-CN **112**, and Spiro-AMPO-CN **114** show high values with $T_g = 165\,°C$, $177\,°C$, $168\,°C$, and $212\,°C$, respectively. Note that a recrystallization was not observed in the case of the *tert*-butyl-substituted compounds, while the cyano-substituted compounds showed both recrystallization and melting signals in the time scale of the DSC experiment. Obviously, the bulky *tert*-butyl groups reduce the kinetics of crystallization compared to the rigid cyano-group [136].

The "top–down" unsymmetric substituted spiro compounds Spiro-DPSP **115**, Spiro-SPO **116**, Spiro-DPO **117**, and Spiro-MeO-DPO **118** show glass transition temperatures with values between the "symmetric" substituted "parent" spiro compounds. Spiro-DPSP **115**, for example, has $T_g = 152\,°C$, while the "symmetric" substituted Spiro-6Φ **22** has $T_g = 212\,°C$, and Spiro-TAD **56** has $T_g = 133\,°C$ (R. Pudzich and J. Salbeck, unpublished results).

11.3
Optical Properties

The absorption spectra of the "left–right" unsymmetrically substituted spiro-oxadiazoleamines are all characterized by two peaks. The most intense absorption maximum in dichloromethane solution was determined to be 389 nm for Spiro-AMO-tBu **111**, 374 nm for Spiro-AMPO-tBu **113**, 400 nm for Spiro-AMO-CN **112**, and 383 nm for Spiro-AMPO-CN **114**. All compounds show intense fluorescence in solution as well as in the solid state (amorphous film spin-coated on glass substrate) and exhibit large Stokes shifts. The values for the emission maximum in solution are 477 nm (Spiro-AMO-tBu), 497 nm (Spiro-AMPO-tBu), 541 nm (Spiro-AMO-CN), and 540 nm (Spiro-AMPO-CN).

The "top–down" unsymmetric substituted spiro compounds behave differently with regard to their fluorescence properties. In the case of Spiro-SPO **116**, the absorption and fluorescence characteristics are mainly determined by the sexiphenyl chain, with an absorption maximum at 331 nm and an intense fluorescence peak with a maximum at 380 nm in dichloromethane. For this compound, the threshold for ASE in neat films is quite low [91]. For a 122-nm-thick film, a threshold of $1\,\mu J/cm^2$ was measured, with a peak wavelength of 419.0 nm. We attribute the low threshold to an efficient absorption by the oxadiazole moieties that have a higher extinction coefficient than the sexiphenyl moieties at the pump wavelength (337 nm) and a subsequent energy transfer to the radiative sexiphenyl chains.

The stronger electronically unsymmetric spiro compounds Spiro-DPSP **115**, Spiro-DPO **117**, Spiro-MeO-DPO **118** and **119** show a distinct dependence of their optical properties on the polarity of the environment. Spiro-DPSP, for example, shows an absorption maximum at 342 nm in dichloromethane, which is only slightly shifted in hexane ($\lambda_{abs} = 338$ nm), or acetonitrile ($\lambda_{abs} = 343$ nm). The fluorescence properties, however, undergo a dramatic shift if the polarity of the solvent is changed, as shown in Fig. 46. In hexane, Spiro-DPSP **115** shows fluorescence with two maxima at 394 nm and 419 nm and a quantum yield of 7%. By addition of dichloromethane, the fluorescence intensity decreases, and the maximum is shifted to longer wavelengths. In pure dichloromethane, only a broad and weak fluorescence maximum at 521 nm can be detected.

The similarity of the absorption spectra and the strong solvent dependence of the emission spectra show that the two halves of the molecule demonstrate negligible interaction in the ground state but strong interaction in the vibrationally relaxed exited state. These effects can be understood in terms of a photoinduced intramolecular electron transfer reaction. A possible explanation can be given by the effect of "spiro conjugation" (see next section) [142–145]. For compound **119**, Chien et al. determined the free energy of electron transfer in different solvents to a value of $\Delta G \geq -0.63$ eV [141].

The effect of photoinduced electron transfer can be applied in organic phototransistors, as we demonstrated recently [146]. Spiro-DPSP **115** is a hole-transport material with a low charge carrier concentration in the dark. Under illumination, electron–hole pairs are generated that increase the number of positive charges in the film, contributing to the conductivity. As a result, the transfer characteristic curve shifts toward lower switch-on voltages.

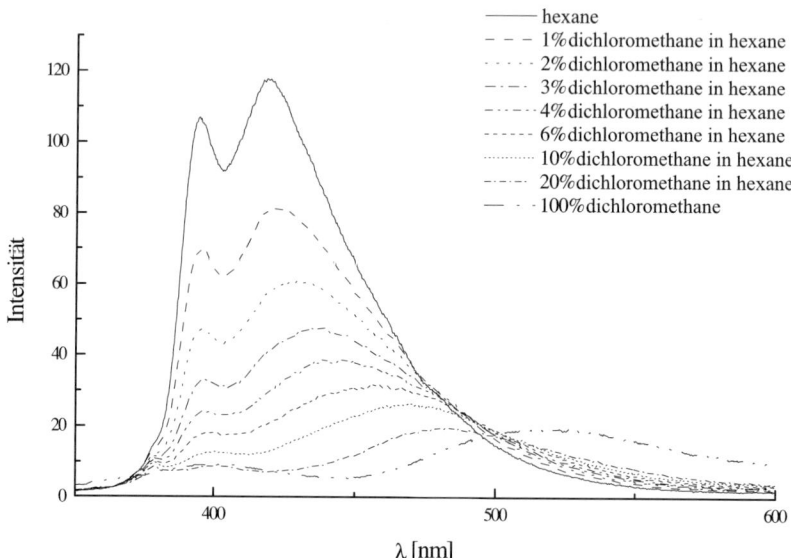

Fig. 46 Fluorescence of Spiro-DPSP with increasing solvent polarity

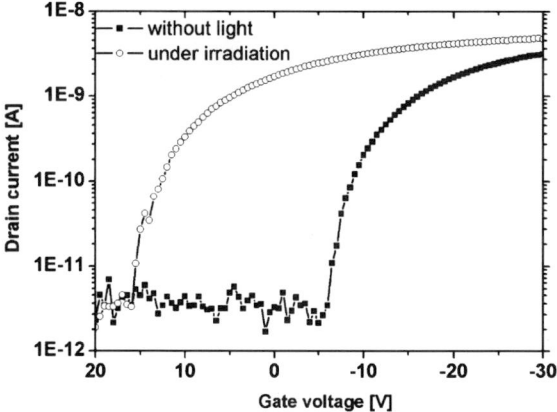

Fig. 47 Transfer characteristics of organic phototransistor with Spiro-DPSP as active material. Irradiation at 375 nm shifts switch-on voltage to positive values (after [146])

12
Chirality in Spiro Compounds

If the spirobifluorene core in low-molecular-weight compounds is totally symmetrically substituted, the point group of the molecules is D_{2d} comprising an S_4 improper rotation axis, and chirality cannot be observed. The symmetry can be broken by bridging the 2 and 2' positions, and the 7 and 7' positions, respectively. Resulting chiral molecules with the point group D_2, the vespirenes (Fig. 48), have been synthesized and characterized by Haas and Prelog [25].

If spirofluorene is equally substituted only in the 2 and 2' positions, the symmetry is C_2, and the compounds are also chiral. This is the case for a series of materials discussed in previous sections, but usually racemic mixtures are obtained and characterized. In a few cases, however, the enantiomers were separated, and the chiroptical properties of the individual enantiomers

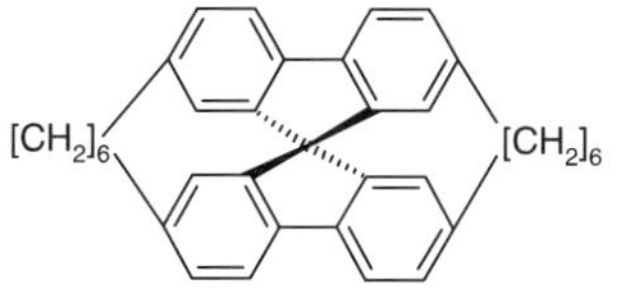

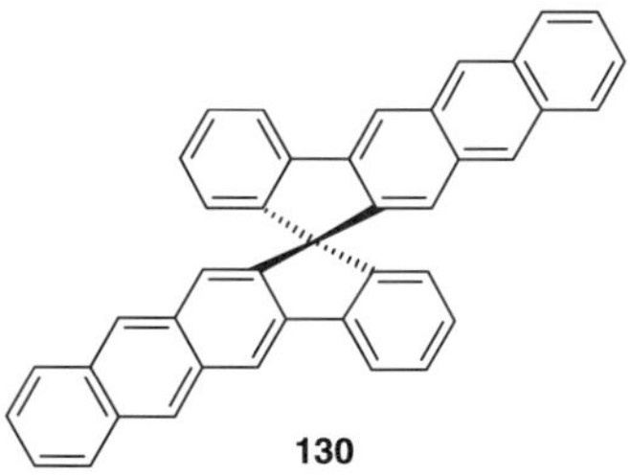

Fig. 48 Chiral spiro compounds with assigned stereochemistry. *Left*: (R)-(−)-[6,6]-vespiren (**129**), *right*: (R)-(+)-spiro-bis(anthracene) (**130**)

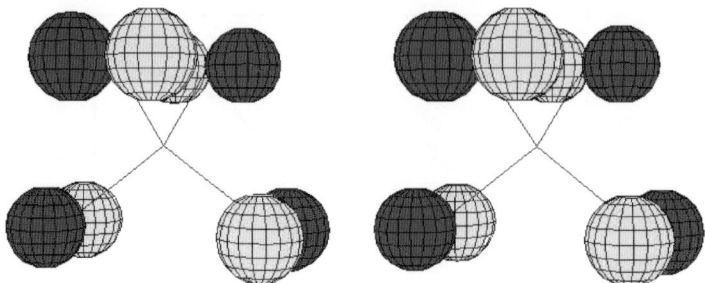

Fig. 49 Orbital symmetry at central spiro-carbon atom in case of spiro-binding interaction (stereo picture)

were characterized [24, 147–149]. Separation of the enantiomers was done from a carboxylic acid precursor by condensation with chiral dehydroabietyl amine and separation of the diastereomeric amides. In the circular dichroism (CD) spectra, the main UV absorption bands are split into a couplet with positive and negative signs, respectively [24, 147]. These results show a substantial coupling between the two chromophores. For the explanation of coupling between two spiro-bridged chromophores, different models can be applied: deviation of the chromophore orientation from orthogonality [24], vibronic coupling [150, 151], or spiro conjugation [152]. Spiro conjugation means the allowed coupling of higher orbitals with two nodal planes in each chromophore (Fig. 49). From the sign of the CD bands of the coupled chromophores the absolute stereochemistry of the spiro center can be determined [24, 147].

13
Spiro Polymers

Even though the spiro concept has been used mainly to improve the morphological stability of low-molecular-weight materials for optoelectronic applications, it has also proven its benefits in the development of polymeric materials.

13.1
Synthesis

Spiro-functionalized polyfluorenes based on a 2,2′-substitution pattern as well as on a 2,7-substitution pattern at the central core of spirobifluorene have been reported [27, 29, 153]. These materials were synthesized by Suzuki coupling of the halogenated spiro core and bis-2,7-(9,9′-octylfluorene) boronic acid, as shown in Fig. 50. Other transition-metal coupling reactions like the

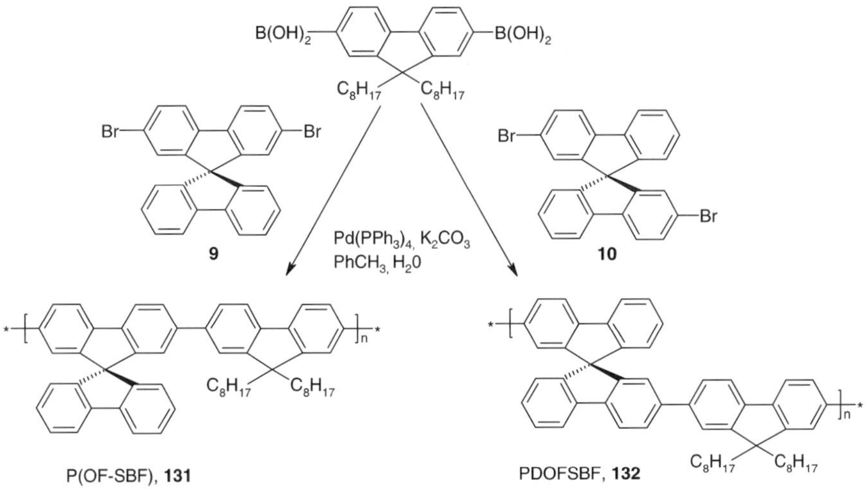

Fig. 50 Synthesis of spiro-conjugated polyfluorenes

Ni(0)-promoted Yamamoto coupling have also been applied for the synthesis of spiro-functionalized polyfluorenes [153].

Spiro-conjugated polyamides [154], polyimides [21], and polyquinolines [155] have also been reported in the literature.

13.2
Electroluminescent Devices

Spiro-functionalized polyfluorenes have been used for the fabrication of blue polymer LEDs [27, 153]. A device was reported using ITO/PDOFSBF 132/Ca that exhibited a maximum external quantum efficiency of 0.12%. Turn-on voltage and efficiency were improved by inserting copper phthalocyanine (CuPc) between anode and polymer (external quantum efficiency 0.54%).

Müller et al. reported three-color OLEDs based on the deposition of spiro polymers with cross-linkable side chains [156]. The devices consist of ITO/PEDOT (20 nm)/spiro polymer (80 nm)/Ca. The polymers are based on a poly(phenyl) backbone that is responsible for the color in the blue device ($x = 0.16$, $y = 0.19$). For the green-emitting polymer, an oligo(phenylene)vinylene moiety was incorporated ($x = 0.31$, $y = 0.58$). For the red one, isobenzothiadiazoles were used ($x = 0.67$, $y = 0.33$). For non-cross-linked devices, efficiencies of 2.9 cd/A have been measured for the blue, 7 cd/A for the green, and 1 cd/A for the red, respectively. Upon cross-linking, the turn-on voltages were reduced slightly, and the efficiencies remained the same, with the exception of the green device in which the efficiency was reduced by 8%.

14
Conclusion

In this article we have presented an overview of the spiro concept as a way to enhance the morphological properties of low-molecular-weight compounds that are of interest for optoelectronic applications due to their electrical and optical properties. Synthetic pathways to precursors as well as target compounds and their important characteristics were discussed. Today, the range of applications for thin-film devices based on these materials spans the whole field of interaction between light and electricity, beginning from the emission of light in light-emitting diodes and lasers up to the response to light in photovoltaic and photochromic systems. However, this does not limit the range of applications. In particular, the application of electrical properties (e.g., for organic transistors) offers a huge potential for future devices.

References

1. Destriau G (1936) J Chem Phys 33:587
2. Holonyak N, Bevacqua SF (1962) Appl Phys Lett 1:82
3. Pope M, Kallmann HP, Magnante P (1963) J Chem Phys 38:2042
4. Helferich W, Schneider WG (1965) Phys Rev Lett 14:229
5. Tang CW, Van Slyke SA (1987) Appl Phys Lett 51:913
6. Adachi C, Tsutsui T, Tokito S, Saito S (1988) Jpn J Appl Phys 28:L269
7. Adachi C, Tsutsui T, Saito S (1990) Appl Phys Lett 56:799
8. Burroughes JH, Bradley DDC, Brown AR, Marks RN, Mackay K, Friend RH, Burns PL, Holmes AB (1990) Nature 347:539
9. Gustavson G, Cao Y, Treacy GM, Klavetter F, Colaneri N, Heeger AJ (1992) Nature 357:477
10. Kolosov D, English DS, Bulovic V, Barbara PF, Forrest SR, Thompson ME (2001) J Appl Phys 90:3242
11. Ke L, Chen P, Chua SJ (2002) Appl Phys Lett 80:697
12. McElvain J, Antoniadis H, Hueschen MR, Miller JN, Roitman DM, Scheats JR, Moon RL (1996) J Appl Phys 80:6002
13. Naito K, Miura A (1993) J Phys Chem 97:6240
14. Naito K (1994) Chem Mater 6:2343
15. Shirota Y, Kuwabara Y, Okuda D, Okuda R, Ogawa H, Inada H, Wakimoto T, Nakada H, Yonemoto Y, Kawami S, Imai K (1997) J Lumin 72–74:985
16. Wang S, Oldham WJ, Hudack RA, Bazan GC (2000) J Am Chem Soc 122:5695
17. Salbeck J (1996) In: Mauch RH, Gumlich H-E (eds) Inorganic and organic electroluminescence (EL96). Wissenschaft & Technik, Berlin, p 243
18. Clarkson RG, Gomberg M (1930) J Am Chem Soc 52:2881
19. Tour JM, Wu R, Schumm JS, Pearson DL (1996) J Org Chem 61:6906
20. Sutcliffe FK, Shahidi HM, Patterson D (1978) J Soc Dyers Colour 94:306
21. Chou C-H, Reddy DS, Shu C-F (2002) J Polym Sci Part A Polym Chem 40:3615
22. Poriel C, Ferrand Y, Juillard S, Le Maux P, Simonneaux G (2004) Tetrahedron 60:145
23. Weisburger JH, Weisburger EK, Ray FE (1950) J Am Chem Soc 72:4253
24. Prelog V, Bedekovic D (1979) Helv Chim Acta 62:2285

25. Prelog V, Haas G (1969) Helv Chim Acta 52:1202
26. Mattiello L, Fioravanti G (2001) Synth Commun 31:2645
27. Yu W-L, Pei J, Huang W, Heeger AJ (2000) Adv Mater 12:828
28. Pei J, Ni J, Zhou XH, Cao XY, Lai YH (2002) J Org Chem 67:4924
29. Wu F-I, Rajasekhar D, Reddy DS, Shu C-F (2002) J Mater Chem 12:2893
30. Moore JS, Weinstein EJ, Wu Z (1991) Tetrahedron Lett 32:2465
31. Tour JM, Wu R, Schumm JS (1990) J Am Chem Soc 112:5662
32. Weissörtel F (1999) Synthese und Charakterisierung spiroverknüpfter niedermolekularer Gläser für optoelektronische Anwendungen. Universität Regensburg
33. Lützen A, Thiemann F, Meyer S (2002) Synthesis 2771
34. Wu X-m, Chen X-c, Cao X-p, Pan X-f (2001) Ganguang Kexue Yu Guang Huaxue 19:161
35. Merkushev EB (1988) Synthesis 923
36. Lee H, Oh J, Chu HY, Lee J-I, Kim SH, Yang YS, Kim GH, Do L-M, Zyung T, Lee J, Park Y (2003) Tetrahedron 59:2773
37. Kim Y-H, Shin D-C, Kim S-H, Ko C-H, Yu H-S, Chae Y-S, Kwon S-K (2001) Adv Mater 13:1690
38. Treacher K, Becker H, Stoessel P, Spreitzer H, Falcou A, Parham A, Buesing A (2002) PCT Int Appl (Covion Organic Semiconductors GmbH, Germany), WO 02077060
39. Stoessel P, Spreitzer H, Becker H, Drott J (2002) PCT Int Appl (Covion Organic Semiconductors GmbH, Germany), WO 02051850
40. Hung LS, Tang CW, Mason MG (1997) Appl Phys Lett 71:1762
41. Shaheen SE, Jabbour GE, Morrell MM, Kawabe Y, Kippelen B, Peyghambarian N, Nabor MF, Schlaf R, Mash EA, Armstrong NR (1998) J Appl Phys 84:2324
42. Brown TM, Friend RH, Millard IS, Lacey DJ, Burroughes JH, Cacialli F (2000) Appl Phys Lett 77:3096
43. Furukawa K, Terasaka Y, Ueda H, Matsumura M (1997) Synth Met 91:99
44. Nuesch F, Forsythe EW, Lee QT, Gao Y, Rothberg LJ (2000) J Appl Phys 87:7973
45. Van Slyke SA, Chen CH, Tang CW (1996) Appl Phys Lett 69:2160
46. Carter SA, Angelopoulos M (1997) Appl Phys Lett 70:2067
47. Aminaka E, Tsutsui T, Saito S (1996) J Appl Phys 79:8808
48. Yeh P (1988) Optical waves in layered media. Wiley, New York
49. McGehee MD, Gupta R, Veenstra S, Miller EK, Diaz-Garcia MA, Heeger AJ (1998) Phys Rev B 58:7035
50. Salbeck J, Schörner M, Fuhrmann T (2002) Thin Solid Films 417:20
51. Heinze J (1984) Angew Chem 96:823
52. Ishii H, Sugiyama K, Ito E, Seki K (1999) Adv Mater 11:605
53. Bard AJ, Faulkner LR (2001) Electrochemical methods, fundamentals and applications. Wiley, New York
54. Pommerehne J, Westweber H, Guss W, Mahrt RF, Bässler H, Porsch M, Daub J (1995) Adv Mater 7:551
55. Horowitz G (2000) In: Hadziioannou G, van Hutten PF (eds) Semiconducting polymers. Chemistry, Physics and Engineering. Wiley, Weinheim, p 463
56. Nelson SF, Lin Y-Y, Gundlach J, Jackson TN (1998) Appl Phys Lett 72:1854
57. Katz HE, Lovinger AJ, Johnson J, Kloc C, Siegrist T, Li W, Lin Y-Y, Dodabalapur A (2000) Nature 404:478
58. Adam D, Schumacher P, Simmerer J, Häußling L, Siemensmeyer K, Etzbach K-H, Ringsdorf H, Haarer D (1994) Nature 371:141
59. Bässler H (1993) Phys Stat Sol (b) 175:15
60. Facci JS, Stolka M (1986) Philos Mag B 54:1

61. Stephan J, Schrader S, Brehmer L (2000) Synth Met 111:353
62. Brown AR, Jarrett CP, de Leeuw DM, Matters M (1997) Synth Met 88:37
63. Borsenberger PM, Weiss DS (1993) Organic photoreceptors for imaging systems. Marcel Dekker, New York
64. Forrest SR, Bradley DDC, Thompson ME (2003) Adv Mater 15:1043
65. American Society for Testing and Materials (1998) G159–198
66. Brabec CJ, Sariciftci NS (2000) In: Hadzioannou G, van Hutten PF (eds) Semiconducting polymers. Wiley, Weinheim, p 528
67. Wirth HO (1961) In: Proceedings of the International Conference on Luminescence of organic and inorganic materials. New York, p 226
68. Wirth HO, Herrmann FU, Herrmann G, Kern W (1968) Mol Cryst 4:321
69. Ried W, Freitag D (1968) Angew Chem 80:932
70. Salbeck J (1996) Ber Bunsen-Ges Phys Chem Chem Phys 100:1667
71. Salbeck J, Weissortel F, Bauer J (1997) Macromol Symp 125:121
72. Miyaura N, Suzuki A (1995) Chem Rev 95:2457
73. Salbeck J, Weinfurthner K-H, Weissörtel F, Harmgarth G (1998) Proc SPIE 3476:40
74. Minato A, Tamao K, Hayashi T, Suzuki K, Kumada M (1980) Tetrahedron Lett 21:845
75. Shen W-J, Dodda R, Wu C-C, Wu F-I, Liu T-H, Chen H-H, Chen CH, Shu C-F (2004) Chem Mater 16:930
76. Wu C-C, Liu T-L, Hung W-Y, Lin Y-T, Wong K-T, Chen R-T, Chen Y-M, Chien Y-Y (2003) J Am Chem Soc 125:3710
77. Wong K-T, Chien Y-Y, Chen R-T, Wang C-F, Lin Y-T, Chiang H-H, Hsieh P-Y, Wu C-C, Chou HC, Yuhlong OS, Lee G-H, Peng S-M (2002) J Am Chem Soc 124:11576
78. Davydov AS (1948) 18:515
79. Berlman IB (1968) Mol Cryst 4:157
80. Brackmann U (2000) Lambdachrome Laser Dyes. Lambda Physik, Goettingen
81. Johansson N, dos Santos DA, Guo S, Cornil J, Fahlman M, Salbeck J, Schenk H, Arwin H, Bredas JL, Salanek WR (1997) J Chem Phys 107:2542
82. Milota F (1999) Master Thesis, University of Vienna
83. Milota F, Warmuth C, Tortschanoff A, Sperling J, Fuhrmann T, Salbeck J, Kauffmann HF (2001) Synth Met 121:1497
84. Birks JB (1970) Photophysics of aromatic molecules. Wiley, London
85. Berlman IA (1971) Handbook of fluorescence spectra of aromatic molecules. Academic, New York
86. Geng Y, Katsis D, Culligan SW, Ou JJ, Chen SH, Rothberg LJ (2002) Chem Mater 14:463
87. Katsis D, Geng YH, Ou JJ, Culligan SW, Trajkovska A, Chen SH, Rothberg LJ (2002) Chem Mater 14:463
88. Schartel B, Damerau T, Hennecke M (2000) Phys Chem Chem Phys 2:4690
89. Schörner M (2000) Master's thesis, Kassel University, Germany
90. Johansson N, Salbeck J, Bauer J, Weissortel F, Broms P, Andersson A, Salaneck WR (1998) Adv Mater 10:1136
91. Spehr T, Pudzich R, Fuhrmann T, Salbeck J (2003) Org Electron 4:61
92. Benstem T (2002) Lumineszenz-Dynamik und stimulierte Emission von organischen Dünnschichten. Cuvillier, Göttingen
93. Wong K-T, Chien Y-Y, Chen R-T, Wang C-F, Lin Y-T, Chiang H-H, Hsieh P-Y, Wu C-C, Chou HC, Yuhlong OS, Lee G-H, Peng S-M (2002) J Am Chem Soc 124:11576
94. Crispin A, Crispin X, Fahlman M, dos Santos DA, Cornil J, Johansson N, Bauer J, Weissortel F, Salbeck J, Bredas JL, Salaneck WR (2002) J Chem Phys 116:8159
95. Salbeck J, Yu N, Bauer J, Weissortel F, Bestgen H (1997) Synth Met 91:209

96. Steuber F, Staudigel J, Stossel M, Simmerer J, Winnacker A, Spreitzer H, Weissortel F, Salbeck J (2000) Adv Mater 12:130
97. Salbeck J, Spreitzer H, Schenk H, Weissortel F, Riel H, Riess W (1999) Proc SPIE 3797:316
98. Chun C, Kim M-J, Vak D, Kim DY (2003) J Mater Chem 13:2904
99. Fuhrmann T, Tsutsui T (1999) Chem Mater 11:2226
100. Hosokawa C, Higashi H, Nakamura H, Kusumoto T (1995) Appl Phys Lett 67:3853
101. Rochon P, Batalla E, Natansohn A (1995) Appl Phys Lett 66:136
102. Kim DY, Li L, Kumar J, Tripathy SK (1995) Appl Phys Lett 66:1166
103. Hartwig JF (1998) Angew Chem 110:2154
104. Gauthier S, Fréchet JM (1987) Synthesis 383
105. Fujikawa H, Tokito S, Taga Y (1997) Synth Met 91:161
106. Anderson JD, McDonald EM, Lee PA, Anderson ML, Ritchie EL, Hall HK, Hopkins T, Mash EA, Wang J, Padias A, Tayumanavan S, Barlow S, Marder SR, Jabbour GE, Shaheen S, Kippelen B, Peyghambarian N, Whightman RM, Armstrong NR (1998) J Am Chem Soc 120:9646
107. Shirota Y (2000) J Mater Chem 10:1
108. Kanai H, Ichinosawa S, Sato Y (1997) Synth Met 91:195
109. Heun S, Borsenberger PM (1995) Chem Phys 200:245
110. Bach U, De Cloedt K, Spreitzer H, Grätzel M (2000) Adv Mater 12:1060
111. Poplavskyy D, Nelson J (2003) J Appl Phys 93:341
112. Saragi TPI, Pudzich R, Fuhrmann T, Salbeck J (2002) MRS Proc 725:85
113. Saragi TPI, Fuhrmann-Lieker T, Salbeck J (2005) Synth Met 148:267
114. Huang J, Pfeiffer M, Blochwitz J, Werner A, Salbeck J, Liu S, Leo K (2001) Jpn J Appl Phys 40:6630
115. O'Regan B, Grätzel M (1991) Nature 353:737
116. Bach U, Lupo D, Comte P, Moser JE, Weissortel F, Salbeck J, Spreitzer H, Gratzel M (1998) Nature 395:583
117. Krüger J, Plass R, Cevey L, Piccirelli M, Grätzel M (2001) Appl Phys Lett 79:2085
118. Stanforth SP (1998) Tetrahedron 54:263
119. Tour JM (1994) Adv Mater 6:190
120. Aviram A (1988) J Am Chem Soc 110:5687
121. Pei J, Ni J, Zhou X-H, Cao X-Y, Lai Y-H (2002) J Org Chem 67:8104
122. Mitschke U, Bäuerle P (2001) J Chem Soc Perkin Trans 1:740
123. Ong T-T, Ng S-C, Chan HSO, Vardhanan RV, Kumura K, Mazaki Y, Kobayashi K (2003) J Mater Chem 13:2185
124. van Pham C, Burckhardt A, Shabana R et al. (1989) Phosphor Sulfur Silicon 46:153
125. Becker RS, Melo JSD, Macanita AL, Elisei F (1995) Pure Appl Chem 67:9
126. Bäuerle P, Mitschke U, Mena-Osteritz E, Sokolowski M, Müller D, Groß M, Meerholz K (1998) Proc SPIE 3476:32
127. Huisgen R (1963) Angew Chem 75:604
128. Huisgen R, Seidl H (1965) Chem Ber 98:2966
129. Brown HC, Kassal RJ (1967) J Org Chem 32:1871
130. Wu CC, Lin YT, Chiang HH, Cho TY, Chen CW, Wong KT, Liao YL, Lee GH, Peng SM (2002) Appl Phys Lett 81:577
131. Wu F, Riesgo EC, Thummel RP, Juris A, Hissler M, El-ghayoury A, Ziessel R (1999) Tetrahedron Lett 40:7311
132. Juris A, Prodi L, Harriman A, Ziessel R, Hissler M, El-ghayoury A, Wu F, Riesgo EC, Thummel RP (2000) Inorg Chem 39:3590

133. Poriel C, Ferrand Y, Le Maux P, Raul-Berthelot J, Simonneaux G (2003) Chem Commun 1104
134. Poriel C, Ferrand Y, Le Maux P, Rault-Berthelot J, Simonneaux G (2003) Tetrahedron Lett 44:1759
135. Maeda M (1984) Laser Dyes. Academic, Orlando, FL
136. Pudzich R, Salbeck J (2003) Synth Met 138:21
137. Diederich F, Alcazar V, Moran JR (1992) Isr J Chem 32:69
138. Hetzheim A (1994) Houben-Weyl, Methoden der organischen Chemie (Houben-Weyl, Methoden der organischen Chemie), vol E8c. Georg Thieme, Stuttgart, p 525
139. Hartwig JF (1998) Angew Chem-Int Edit Engl 37:2046
140. Pudzich R (2002) Synthese und Charakterisierung spiroverknüpfter Emitter- und Ladungstransportmaterialien mit kombinierten Funktionalitäten. PhD thesis, Universität Kassel, Germany
141. Chien Y-Y, Wong K-T, Chou P-T, Cheng Y-M (2002) Chem Commun 2874
142. Maslak P, Chopra A, Moylan CR, Wortmann R, Lebus S, Rheingold AL, Yap GPA (1996) J Am Chem Soc 118:1471
143. Maslak P (1994) Adv Mater 6:405
144. Maslak P, Chopra A (1993) J Am Chem Soc 115:9331
145. Maslak P, Augustine MP, Burkey JD (1990) J Am Chem Soc 112:5359
146. Saragi TPI, Pudzich R, Fuhrmann T, Salbeck J (2004) Appl Phys Lett 84:2334
147. Harada N, Ono H, Nishiwaki T, Uda H (1991) J Chem Soc Chem Commun 1753
148. Alcazar V, Diederich F (1992) Angew Chem 104:1503
149. Cuntze J, Diederich F (1997) Helv Chim Acta 80:897
150. Dantzig NA, Levy DH, Vigo C, Piotrowiak P (1995) J Chem Phys 103:4894
151. Shain AL, Ackerman JP, Teague MW (1969) Chem Phys Lett 3:550
152. Schweig A, Weidner U, Hellwinkel D, Krapp W (1973) Angew Chem 85:360
153. Kreuder W, Lupo D, Salbeck J, Schenk H, Stehlin T (1997) US Patent, US5621131
154. Wu S-C, Shu C-F (2003) J Polym Sci Part A Polym Chem 41:1160
155. Chiang C-L, Shu C-F (2002) Chem Mater 14:682
156. Müller CD, Falcou A, Reckefuss N, Rojahn M, Wiederhirn V, Rudati P, Frohne H, Nuyken O, Becker H, Meerholz K (2003) Nature 421:829

Editor: Ullrich Scherf

Charge Transport and Catalysis by Molecules Confined in Polymeric Materials and Application to Future Nanodevices for Energy Conversion

Masayuki Yagi[1] · Masao Kaneko[2] (✉)

[1] Faculty of Education and Human Sciences, and Center for Transdisciplinary Research, Niigata University, 8050 Ikarashi-2, 950-2181 Niigata, Japan
yagi@ed.niigata-u.ac.jp

[2] Faculty of Science, Ibaraki University, 2-1-1 Bunkyo, 310-8512 Mito, Japan
mkaneko@mx.ibaraki.ac.jp

1	Introduction	145
2	Charge Transport by Redox Molecules in Polymeric Solid Materials	146
2.1	In a Nafion Film	146
2.1.1	Measurements and Analyses of Charge Transport Processes	147
2.1.2	Mechanism of Charge Transport and Its Influencing Factors in a Nafion Film	153
2.2	In Polysaccharide Solids Containing Excess Liquid	159
2.2.1	Polysaccharide Solids Containing Excess Liquid and Their Application	159
2.2.2	Characteristics of Polysaccharide Solids Containing Excess Water	160
2.2.3	Transport of Charges and Molecules in Polysaccharide Solids Containing Excess Water	160
2.2.4	Transport of Ions in Polysaccharide Solids (Ionic Conductivity)	164
3	Molecular Catalysts for Multielectron Reactions	165
3.1	Water Oxidation Catalysts	165
3.1.1	Manganese Complexes	166
3.1.2	Ruthenium Complexes	169
3.2	Proton Reduction Catalysts	173
3.3	CO_2 Reduction Catalysts	176
4	Application to Future Energy Conversion Devices	178
4.1	Dye-sensitized Solar Cell	178
4.2	Artificial Photosynthesis	183
5	Conclusion and Future Directions	185
	References	185

Abstract Polymeric materials confining functional molecules are one of the most promising materials for designing nanodevices for energy conversion, e.g., solar cells, fuel cells, and artificial photosynthetic devices that are expected to provide a renewable energy resource. Charge transport (CT) and catalysis by redox molecules in polymeric solid materials are reviewed with a focus on a polyanion film, typically Nafion and other polymeric materials, containing excess water. CT in a polyanion film is evaluated based on the physical displacement (physical diffusion) and charge hopping mechanisms between redox molecules. The mechanism of CT is exhibited to depend on the structure and redox reaction of the center molecules, and the influencing factor on CT is discussed. For the polymeric solid reactor containing excess water, the physical data of CT and molecular transport in the bulk matrix are summarized to demonstrate that the electrochemical reaction in the solid reactor occurs similarly as in an aqueous solution. Recent progress in molecular catalysis for multielectron redox reactions with a focus on water oxidation, reduction of proton, and carbon dioxide is introduced, and the catalytic activity and mechanism in solution and polymeric matrixes are reviewed. A dye-sensitized solar cell was fabricated using polymeric solid materials containing excess organic solution as an electrolyte layer, and its performance similar to a liquid-type solar cell is discussed based on the physicochemical data in the polymeric solid materials. Recent approaches toward construction of an artificial photosynthetic system are reviewed, and, finally, concluding remarks and directions for future research are given.

Keywords Charge transport · Catalysis · Polymeric solid materials · Artificial photosynthesis · Dye-sensitized solar cell

Abbreviations

A	electrode area
A_0	absorbance at time zero
A_t	absorbance at time t
bpp$^-$	3,5-bis(2-pyridyl)pyrazolate
bpy	2,2'-bipyridine
bpz	2,2'-bipyrazine
btpyan	1,8-bis(2, 2' : 6', 2''-terpyridyl)anthracene
t-Bu$_2$qui	3,6-di-*tert*-butyl-1,2-benzoquinone
t-Bu$_2$sq	3,6-di-*tert*-butyl-1,2-semiquinone
c_0	initial concentration of a redox molecule in a film or solution (mol dm^{-3} or mol cm^{-3})
C_{dl}	capacitance in a double layer
CT	charge transport
CV	cyclic voltammogram
D_{app}	apparent diffusion coefficient (cm^2 s^{-1})
Ea	activation energy
EIS	Electrochemical impedance spectra
ff	fill factor
F	Faraday constant
Fc	ferrocene
i_p	peak current
$i(t)$	current density at time t
ITO	indium tin oxide
J_{sc}	short circuit photocurrent
k_{app}	apparent charge transport rate constant (s^{-1})
k_c	second-order rate constant (M^{-1} s^{-1}) for charge transport by charge hopping
k_p	first-order rate constant (s^{-1}) for charge transport by physical displacement
L	film thickness

M	mol dm^{-3}
MV^{2+}	methylviologen
n	number of electrons involved in reaction
N3 dye	*cis*-bis(thiocyanato)bis(4,4′-dicarboxyl-2,2′-bipyridine)ruthenium(II)
NPV	normal pulse voltammetry
OEC	oxygen evolving center
Pc	phthalocyanine
PSCA	potential-step chronoamperometry
PSCC	potential-step chronocoulometry
PSCCS	potential-step chronocoulospectrometry
PSCAbs	potential-step chronoabsorptometry
pz	pyrazine
R_{ct}	charge transfer resistance at the electrode/solution interface
R_s	solution resistance
Ru-red	$[(NH_3)_5Ru^{III}(\mu-O)Ru^{IV}(NH_3)_4Ru^{III}(\mu-O)(NH_3)_5]^{6+}$
Ru-brown	$[(NH_3)_5Ru^{IV}(\mu-O)Ru^{III}(NH_3)_4Ru^{IV}(\mu-O)(NH_3)_5]^{7+}$
$[Ru^{II}-Ru^{II}]^{4+}$	$[(NH_3)_5Ru^{II}(\mu-pz)Ru^{II}(NH_3)_5]^{4+}$
Ru^{III}ORu^{III}	$[(bpy)_2(H_2O)Ru^{III}(\mu-O)Ru^{III}(H_2O)(bpy)_2]^{4+}$
SCE	saturated calomel reference electrode
S$_2$R	azadithiolate
t	reaction time (s)
$t_{1/2}$	half-life period
terpy	2,2′:6′,2″-terpyridine
TPP	tetraphenylporphyrin
v	scan rate for potential sweep
v_{CT}	initial charge transport rate (M s^{-1})
V_{oc}	open circuit photovoltage
ε	molar absorption coefficient (M^{-1} cm^{-1})
Φ_c	fraction of the contribution of charge hopping to v_{CT} ($\Phi_c = k_c c_0/(k_p c_0 + k_c c_0^2)$)
η	light-to-electricity conversion efficiency
ω	resistance for charge transport

1
Introduction

Nanodevices are attracting a great deal of attention; they are investigated, e.g., for energy conversion, solar cells, fuel cells, and artificial photosynthetic devices that are expected to provide a renewable energy resource [1–4]. Polymeric solid materials are promising as matrixes for functional molecules and as electrolytes in these nanodevices. In both cases a rapid charge and molecular transport in the polymeric solid materials is necessary for the devices to work efficiently. The development of polymeric solid materials to establish rapid charge and molecular transport is an urgent subject for scientists in related research fields. Chemical energy conversion devices such as fuel cells and artificial photosynthetic devices would require in the future molecular catalysts for multielectron redox reactions such as water oxidation, dioxygen reduction, proton reduction, dihydrogen oxidation, carbon dioxide reduction, etc. [5, 6]. The development of molecular catalysts that show high activity and stability is essential for achieving breakthroughs and for creating nanodevices in this field.

The present article reviews recent progress in CT in solid and quasisolid polymeric materials as well as the characteristics of molecule-based catalysts for multielectron reactions in polymeric materials. This mainly covers work published in the last 5 years from 2000 to 2004, as well as significant work in the last 10 years. Following the introduction (Sect. 1), mechanistic features of CT in polymer films and related polymeric materials confining functional redox molecules are discussed in Sect. 2 with a focus on recent progress in the present authors' groups. In Sect. 3, our attention is particularly focused on molecular catalysts for water oxidation, proton reduction, and CO_2 reduction that are important for designing an artificial photosynthetic system. Various molecular catalysts are introduced, and their molecular functions in polymeric matrixes are described. Finally, applications of the polymeric materials to a dye-sensitized solar cell and an artificial photosynthetic system are discussed to suggest directions for future research.

2
Charge Transport by Redox Molecules in Polymeric Solid Materials

2.1
In a Nafion Film

A wide range of successful applications of polymer-modified electrodes [7, 8] to electroanalysis [8], electrocatalysis [5, 9–11], photoelectrochemistry [12], and solar energy conversion [13, 14] have given impetus to develop these kinds of modified electrodes. A polymer film confining functional molecules is one of the most attractive new nanodevice design materials. The under-

$$\left[\left(CF_2-CF \right) - \left(CF_2-CF_2 \right)_m \atop \left[\begin{array}{c} O \\ | \\ CF_2 \\ | \\ CFCF_3 \\ | \\ O \\ | \\ CF_2CF_2SO_3H \end{array} \right]_k \right]_n$$

Nafion

standing of CT in such a polymer film is an important fundamental topic for the development of new nanodevices because CT is a key process in the efficient functioning of such devices [7, 8, 15]. Pioneering work on the nature of CT by redox molecules taking place in polymeric films has been reported [16–26]. However, its mechanism remains poorly understood.

The Nafion (perfluorosulfonated ionomer) films have prompted the development of many chemical industries commercializing electrocatalytic technologies and fuel cells. Nafion has been studied mainly for its structure and properties as well as its application in heat-resistant cation exchange membranes. The chemical structure of Nafion consists of a poly(tetrafluoroethylene) backbone with perfluorinated pendant chains terminated by sulfonate groups with either acidic ($-SO_3H$) or anionic form ($-SO_3^-$). A Nafion film is known to involve two fundamentally distinctive structural regions: a hydrophobic region formed by the perfluorinated polymer backbone and a hydrophilic ionic cluster region comprising sulfonate groups, counter cations, and water molecules. The neighboring ionic clusters are interconnected through channels that enable transport of ions and water molecules. A Nafion film can easily incorporate cationic functional molecules by cation exchange from its solution, which allows one to develop molecular electronic and photonic devices using a Nafion film, leading to a variety of applications. In order to design Nafion film-based devices it is of particular importance to understand the CT kinetics and mechanism in Nafion films that incorporate functional molecules.

In this section, recent progress in studies on the CT mechanism in a Nafion film incorporating functional redox molecules is summarized. This highlights the characteristics of the CT in the Nafion film depending on the structure and redox properties of redox center molecules and also helps us to understand the general principle of CT in polymer films. Measurements and analyses of the CT process for the purpose of elucidating the mechanism will be described first, followed by a discussion of CT features including its influencing factors.

2.1.1
Measurements and Analyses of Charge Transport Processes

In most of the traditional studies on CT in polymer films incorporating functional redox molecules, conventional electrochemical techniques such as cyclic voltammogram (CV), normal pulse voltammetry (NPV), potential-step chronoamperometry (PSCA), and potential-step chronocoulometry (PSCC) have been used. However, in some cases electrochemical information such as amperometry and coulometry does not quantitatively correspond to real changes of redox center molecules due to unfavorable side reactions [27, 28], showing that electrochemical techniques combined with absorption spectroscopic techniques should be adopted in order to clarify CT by redox reactions of the center molecules in a film. Several electrospectroscopic techniques using UV-visible absorption, emission, and EPR spectroscopies have been reported [27–32].

The potential-step chronoabsorptometry (PSCAbs) with UV-visible absorption spectroscopy is applicable to various electrochemical reactions. An example of a PSCAbs measurement is shown in Fig. 1, in which the absorbance by $[(NH_3)_5Ru^{II}(\mu-pz)Ru^{II}(NH_3)_5]^{4+}$ (abbreviated to $[Ru^{II}-Ru^{II}]^{4+}$ and pz = pyrazine) incorporated in an electrode coated Nafion film decreased at $\lambda_{max} = 537$ nm due to $[Ru^{II}-Ru^{II}]^{4+}$ by oxidation to $[Ru^{II}-Ru^{III}]^{5+}$ in a potential step from – 0.2 V to 0.3 V vs. SCE.

The apparent diffusion coefficient, D_{app} (cm^2 s^{-1}), is generally used to compare the CT rate for various systems. D_{app} in the film can be obtained from the absorbance decrease using the modified Cottrell Eq. 1:

$$A_0 - A_t = 2 c_0(\varepsilon_{II,II} - \varepsilon_{II,III})(D_{app} t/\pi)^{1/2}, \qquad (1)$$

where A_0 and A_t are the absorbances at 537 nm of the film at time zero and time t, respectively, c_0 (M) is the initial concentration of $[Ru^{II}-Ru^{II}]^{4+}$ in the film, $\varepsilon_{II,II}$ and $\varepsilon_{II,III}$ (M^{-1} cm^{-1}) are the molar absorption coefficients of $[Ru^{II}-Ru^{II}]^{4+}$ and $[Ru^{II}-Ru^{III}]^{5+}$, respectively, and t (s) is the reaction time. D_{app} can be obtained from the slope of the plots of $A_0 - A_t$ vs. $t^{1/2}$ according to Eq. 1, as shown in the inset of Fig. 1.

We proposed the analysis of CT with a kinetic parameter of the initial CT rate (v_{CT}/M s^{-1}) [27, 28, 33–36]. v_{CT} can conveniently be used for comparison of CT including in cases that cannot be analyzed by a simple diffusion process. v_{CT} is obtained from the UV-Visible absorption spectral data according

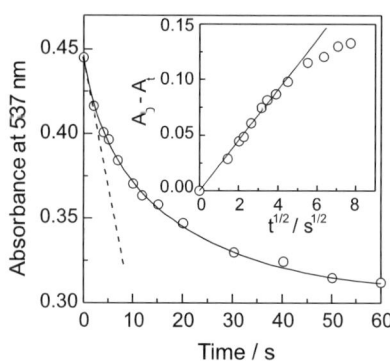

Fig. 1 Absorbance change at 537 nm with time in potential-step chronoabsorptometry from – 0.2 to 0.3 V vs. SCE. Inset shows plots of $A_0 - A_t$ vs. $t^{1/2}$ according to Eq. 1. The complex concentration in the film is 0.26 M (reprinted with permission from American Chemical Society [39])

to Eq. 2:

$$v_{CT} = \frac{\{d(A_0 - A_t)/dt\}_{initial}}{(\varepsilon_{II,II} - \varepsilon_{II,III})L},\qquad(2)$$

where L is the film thickness (usually 1.0 μm). The v_{CT} value is estimated from the slope of the initial absorbance change (537 nm) at time zero as shown by the dotted line in Fig. 1.

There are two kinds of CT mechanisms in a nonconductive polymer film incorporating redox molecules: (1) physical displacement (physical diffusion) of the redox molecules and (2) charge hopping between the redox molecules, as shown in Fig. 2. CT by physical displacement of the redox molecules in a film is a kind of molecular diffusion process and regarded as a unimolecular process that involves diffusions of both the oxidized and reduced species of the redox couple. On the other hand, CT by charge hopping takes place by the self-exchange of charges between the redox couple, which is a bimolecular process.

These two mechanisms can therefore be discriminated from each other by measuring the concentration dependence of v_{CT}.

When CT takes place by both physical displacement and charge hopping, v_{CT} can be represented by Eq. 3, which is a combination of the first-order

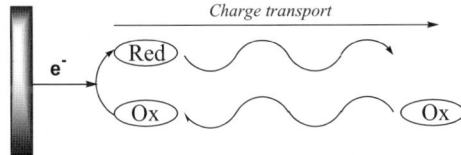

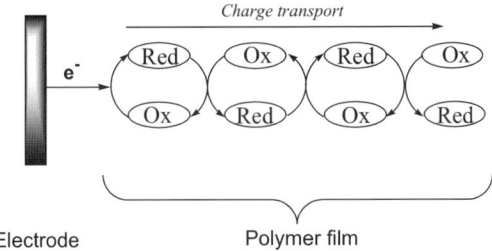

Fig. 2 Illustration for charge transport (CT) in a polymer film by **a** physical displacement and **b** charge hopping mechanisms (Reprinted with permission from American Chemical Society [33])

physical displacement and the second-order charge hopping [27, 28, 33–36]:

$$v_{CT} = k_p c_0 + k_c c_0^2, \quad (3)$$

where k_p (s^{-1}) and k_c (M^{-1} s^{-1}) are the first-order rate constant for CT by physical displacement and the second-order rate constant by charge hopping, respectively. To make the contribution of each mechanism to v_{CT} clearer, the apparent CT rate constant, k_{app} (s^{-1}), is defined by Eq. 4:

$$k_{app} = v_{CT}/c_0 = k_p + k_c c_0. \quad (4)$$

The plots of k_{app} vs. c_0 show several types depending on the CT mechanism, as shown in Fig. 3.

When the plots exhibit a straight line with an intercept with a slope of 0, CT takes place by physical displacement, and the k_p value is obtained from the intercept. When the plots show a straight line passing through the origin, a charge hopping mechanism is activated, and the k_c value is obtained from the slope. If CT occurs by a combined mechanism of physical displacement and charge hopping, the plots provide a straight line with intercept and slope demonstrating the contribution of both k_p (from intercept) and k_c (from slope). In a few cases the k_{app} value by physical displacement may depend on c_0 (e.g., the diffusion is suppressed by high concentration), or the k_{app} value by charge hopping is not of the first order with respect to c_0 (for instance by some interaction between the redox center molecule and the polymer matrix); in these cases the plots may exhibit some curvature, not a straight line. The details of such CT processes are described in the literature [15]. The mechanisms of CT in various redox molecules/Nafion film system have been investigated by analysis of the k_{app} vs. c_0 plots and summarized in Table 1, classified according to mechanism.

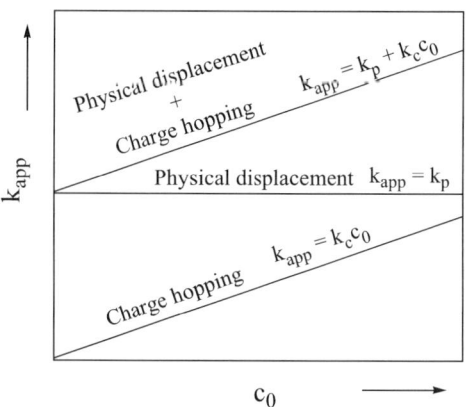

Fig. 3 Typical concentration dependence of apparent rate constant (k_{app}/s^{-1}) for CT

Table 1 Summary of CT mechanism of redox molecule/Nafion film system

Mechanism	v_{CT}, k_{app} vs. c_0	Redox molecule	Redox reaction	Kinetic parameters (25 °C)	Ref.
1) Physical displacement	$v_{CT} = k_p c_0$ $k_{app} = k_p$	MV^{2+}	$MV^{2+} \rightarrow MV^{\cdot+}$	$k_p = 1.4 \times 10^{-1}$ s^{-1}	[37]
		$[Co^{III}TPP]^+$	$Co^{III} \rightarrow Co^{II}$	$k_p = 6.2 \times 10^{-2}$ s^{-1}	[38]
		$Zn^{II}Pc$	$Pc \rightarrow Pc^{\cdot+}$	$k_p = 3.3 \times 10^{-3}$ s^{-1} (for monomer) $k_p = \sim 0$ (for dimer)	[34]
		$[(NH_3)_5Ru^{II}(\mu-pz)_5Ru^{II}(NH_3)_5]^{4+}$	$Ru^{II} - Ru^{II} - Ru^{II} \rightarrow Ru^{II} - Ru^{II} - Ru^{III}$	$D_{app} = 1.3 \times 10^{-10}$ cm^2 s^{-1} $k_p = 1.1 \times 10^{-1}$ s^{-1}	[39]
		$[(NH_3)_5Ru^{II}(\mu-pz)_5Ru^{III}(NH_3)_5]^{5+}$	$Ru^{II} - Ru^{II} - Ru^{III} \rightarrow Ru^{II} - Ru^{III} - Ru^{III}$	$D_{app} = 1.0 \times 10^{-11}$ cm^2 s^{-1} $k_p = 2.1 \times 10^{-2}$ s^{-1}	[39]
		$[Ru^{III}(bpy)_3]^{3+}$	$Ru^{III} \rightarrow Ru^{II}$	$D_{app} = 1 \times 10^{-10}$ cm^2 s^{-1} $k_p = 7.5 \times 10^{-2}$ s^{-1}	[41]
		$[(bpy)_2(H_2O)Ru^{III}(\mu-O)-Ru^{III}(H_2O)(bpy)_2]^{4+}$	$Ru^{III} - Ru^{III} \rightarrow Ru^{II}$ monomer	$k_p = 1.1 \times 10^{-2}$ s^{-1}	[36]
2) Charge hopping	$v_{CT} = k_c c_0^2$ $k_{app} = k_c c_0$	$[Ru^{II}(bpy)_3]^{2+}$	$Ru^{II} \rightarrow Ru^{III}$	$k_c = 4.4 \times 10^{-2}$ M^{-1} s^{-1} (mixture casting method) $k_c = 2.5 \times 10^{-1}$ M^{-1} s^{-1} (adsorption method)	[28]
		$[(bpy)_2(H_2O)Ru^{III}(\mu-O)-Ru^{III}(H_2O)(bpy)_2]^{4+}$	$Ru^{III} - Ru^{III} \rightarrow Ru^{III} - Ru^{IV}$	$k_c = \sim 0.2$ M^{-1} s^{-1} (depending on c_0)	[44]

Table 1 (continued)

Mechanism	v_{CT}, k_{app} vs. c_0	Redox molecule	Redox reaction	Kinetic parameters (25 °C)	Ref.
3) Physical displacemet + Charge hopping	$v_{CT} = k_p c_0 + k_c c_0^2$ $k_{app} = k_p + k_c c_0$	$[(NH_3)_5Ru^{III}Ru^{III}(\mu-O)Ru^{IV}\text{-}(NH_3)_4Ru^{III}(\mu-O)(NH_3)_5]^{6+}$	$Ru^{III} - Ru^{IV} - Ru^{III} \rightarrow Ru^{IV} - Ru^{III} - Ru^{IV}$	$D_{app} = 0.68 \sim 8.3 \times 10^{-11}$ cm^2 s^{-1} $k_p = 5.7 \times 10^{-3}$ s^{-1} $k_c = 1.2 \times 10^{-1}$ M^{-1} s^{-1}	[35]
		$[Ru^{II}(bpz)_3]^{2+}$	$bpz \rightarrow bpz^{·-}$	$D_{app} = 7.6 \times 10^{-10}$ cm^2 s^{-1} (at 0.2 M) $k_p = 9.0 \times 10^{-2}$ s^{-1} $k_c = 1.5$ M^{-1} s^{-1}	[33]

2.1.2
Mechanism of Charge Transport and Its Influencing Factors in a Nafion Film

The mechanism of CT in a Nafion film strongly depends on the kind of redox molecule, as shown in Table 1. What is controlling the CT mechanism? To better understand the controlling factors, we studied the activation energy and other activation parameters for each CT mechanism. In this section, the features and influencing factors for each mechanism will be discussed.

(A) Physical Displacement Mechanism

CT in methylviologen (MV^{2+})/Nafion was studied using a potential-step chronoamperospectrometry (PSCAS) technique, and the results were compared with those in a poly(styrene)-pendant MV^{2+} system. The CT by reduction of MV^{2+} to $MV^{\bullet+}$ was found to occur by physical displacement of the center molecule with $k_p = 1.4 \times 10^{-1}$ s^{-1} [37]. By contrast, CT took place by a charge hopping mechanism in the poly(styrene)-pendant MV^{2+} system. The redox couple of MV^{2+} and $MV^{\bullet+}$ is considered to diffuse in a hydrophilic ion channel of Nafion, but in a poly(styrene)-pendant MV^{2+} system, the diffusion should be difficult in the film by the covalent attachment of the MV^{2+} center to the polymer matrix, which would result in a charge hopping mechanism [37].

CT by reduction of a macrocyclic complex, tetraphenylporphyrin cobalt(III) ([$Co^{III}TPP$]$^+$), in a Nafion film was investigated using a PSCAS [38]. The CT mechanism was found to be a physical displacement of [$Co^{III}TPP$]$^+$ with $k_p = 6.2 \times 10^{-2}$ s^{-1}. It was surprising to see CT by such a large and hydrophobic molecule taking place by physical diffusion in Nafion. The reduced species of hydrophobic [$Co^{II}TPP$] had been considered to be unfavorable for diffusion in the hydrophilic ion channel of Nafion. The physical displacement mechanism would suggest that the complexes ([$Co^{III}TPP$]$^+$ and [$Co^{II}TPP$]) diffuse in the interfacial region between hydrophobic polyfluorocarbon clusters and the hydrophilic ion channel.

The CT by oxidation of phthalocyanine zinc(II) ($Zn^{II}Pc$) in a Nafion film was also studied using a potential-step chronocoulospectrometry (PSCCS) technique [34]. Absorption spectra of $Zn^{II}Pc$/Nafion indicated the formation of the $Zn^{II}Pc$ dimer, and the equilibrium constant between its monomer and dimer in the Nafion film was 75 M^{-1}. The plots of bulk CT rate vs. total $Zn^{II}Pc$ concentration gave a downwardly deviating curve, and the process was analyzed considering the equilibrium between the monomer and the dimer. The analysis result showed that CT takes place by physical displacement of the $Zn^{II}Pc$ monomer with $k_p = 3.3 \times 10^{-3}$ s^{-1} and not by charge hopping, and that the contribution of the dimer to the CT is negligible. The macrocyclic complexes such as [$Co^{III}TPP$]$^+$ and $Zn^{II}Pc$ are known to be an active electrocatalyst. The obtained CTs by [$Co^{III}TPP$]$^+$ and $Zn^{II}Pc$ should be taken into account in the electrocatalytic activity of these macrocyclic complexes for proton reduction (Sect. 2.2).

$[(NH_3)_5Ru^{II}(\mu-pz)Ru^{II}(NH_3)_5]^{4+}$ ($[Ru^{II}-Ru^{II}]^{4+}$, pz = pyrazine) underwent 1-electron oxidation to $[Ru^{II}-Ru^{III}]^{5+}$ in a Nafion film at 0.09 V vs. SCE, and the formed $[Ru^{II}-Ru^{III}]^{5+}$ underwent further 1-electron oxidation to $[Ru^{III}-Ru^{III}]^{6+}$ at 0.46 V. A double potential-step chronoabsorptometry (DPSCAbs) technique was used to investigate the CT by the two steps of reversible oxidation, $[Ru^{II}-Ru^{II}]^{4+}/[Ru^{II}-Ru^{III}]^{5+}$ and $[Ru^{II}-Ru^{III}]^{5+}/[Ru^{III}-Ru^{III}]^{6+}$, exhibiting the influence of charge of the complexes on the CT in the film [39]. For both steps charge was transported by a physical displacement mechanism. However, the CT rate constant for the first step was 5.2 times higher than that for the second step at 25 °C. The activation energy (Ea) and other activation parameters at 25 °C are summarized in Table 2.

The CTs in both steps are entropy controlled in the activation at room temperature ($-T\Delta S^{\neq} > \Delta H^{\neq}$). The $\Delta H^{\neq}$ (12 kJ mol^{-1}) for the physical displacement in the second step was lower than that (24 kJ mol^{-1}) for the first step, showing that the physical displacement for the second step is enthalpically favorable compared with that for the first step. The corresponding $\Delta S^{\neq}$ (-234 J K^{-1} mol^{-1}) for the second step is lower than that ($\Delta S^{\neq} = -184$ J K^{-1} mol^{-1}) for the first step, making the physical displacement for the second step more entropically difficult than that for the first step. Although the detachment of the cationic complexes from the Nafion sulfonate anionic groups should contribute positively to $\Delta S^{\neq}$, the reorganization of the solvent, as determined in Fig. 4, would cause totally the negative $\Delta S^{\neq}$.

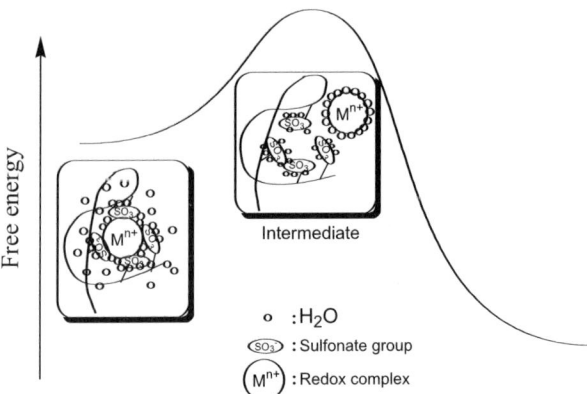

Fig. 4 Imaged illustration for dissociation of a positively charged complex from the sulfonate groups of Nafion in a physical displacement mechanism. The disorder and randomness of solvents decrease in dissociation (reprinted with permission from American Chemical Society [39])

Table 2 Summary of activation energy (E_a) and activation parameters at $T = 25\,°C$

Complex	Redox reaction	Mechanism	Ea kJ mol⁻¹	$\Delta G^{\neq a}$ kJ mol⁻¹	$\Delta H^{\neq b}$ kJ mol⁻¹	$\Delta S^{\neq c}$ J K⁻¹ mol⁻¹	$-T\Delta S^{\neq}$ kJ mol⁻¹	Refs.
[(NH₃)₅Ru^II(μ−pz)- Ru^II(NH₃)₅]⁴⁺	Ru^II−R^II/Ru^II−R^III	Physical displacement	26	79	24	−184	55	[39]
	Ru^II−R^III/Ru^III−R^III	Physical displacement	14	83	12	−234	71	[39]
[Ru^II(bpz)₃]²⁺	bpz/bpz·⁻	Physical displacement	78	79	75	−13	3.8	[33]
		Charge hopping	40	72	37	−116	35	
[Ru^II(bpy)₃]²⁺	Ru^II/Ru^III	Physical displacement	−	−	−	−	−	[28]
		Charge hopping	55	76	53	−79	24	
[(NH₃)₅Ru^III(μ−O)Ru^IV- (NH₃)₄Ru^III(μ−O)(NH₃)₅]⁶⁺	Ru^III Ru^IV Ru^III / Ru^IV Ru^III Ru^IV	Physical displacement	13	86	11	−252	75	[35]
		Charge hopping	55	78	53	−86	26	

ᵃ $\Delta G^{\neq}$ was calculated by $k = kT/h\,exp(-\Delta G^{\neq}/RT)$, where k, h, and R are rate constant (k_p or k_c), Boltzmann constant, Planck constant and gas constant, respectively. ᵇ $\Delta H^{\neq}$ was calculated by $\Delta H^{\neq} = E_a - RT$ (RT is zero-point energy.). ᶜ $\Delta S^{\neq}$ was calculated based on $\Delta G^{\neq} = \Delta H^{\neq} - T\Delta S^{\neq}$

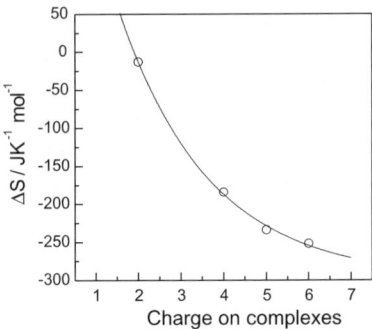

Fig. 5 Plots of $\Delta S^{\neq}$ vs. positive charge on the complex. The plots include the data of [Ru(bpz)$_3$]$^{2+}$ [33] [Ru^{II} – Ru^{II}]$^{4+}$ [39] and [Ru^{II} – Ru^{III}]$^{5+}$ [39] and [(NH$_3$)$_5$ Ru(μ – O)Ru(NH$_3$)$_4$ (μ – O)Ru(NH$_3$)$_5$]$^{6+}$ [35] for 2+, 4+, 5+, and 6+ charges (reprinted with permission from American Chemical Society [39])

The lower $\Delta S^{\neq}$ (– 234 J K^{-1} mol^{-1}) for the second step than that (– 184 J K^{-1} mol^{-1}) for the first step could be explained by the higher degree of solvation required in the detachment of the [Ru^{II} – Ru^{III}]$^{5+}$/[Ru^{III} – Ru^{III}]$^{6+}$ redox pair from the Nafion sulfonate groups than that for the [Ru^{II} – Ru^{II}]$^{4+}$/[Ru^{II} – Ru^{III}]$^{5+}$ pair due to the highly positive charges. Figure 5 shows the plots of the $\Delta S^{\neq}$ for physical displacement vs. the positive charges of the complexes. The decrease of the $\Delta S^{\neq}$ suggests that the highly positively charged complex is entropically unfavorable for physical displacement, presumably due to a higher degree of solvation of the complexes in their detachment from the sulfonate groups.

(B) Charge Hopping Mechanism

A [Ru(bpy)$_3$]$^{2+}$ (bpy = 2, 2$'$-bipyridine)/Nafion film can be prepared by two methods: (1) the complex is adsorbed from its aqueous solution into a Nafion film precoated on an electrode (adsorption method) or (2) an alcoholic mixture solution of Nafion and the complex is cast and air-dried on an electrode (mixture casting method). CT by oxidation of Ru^{II} to Ru^{III} in the film prepared by a different method was compared [28]. CT was found to take place by a charge hopping mechanism for the films prepared by both the adsorption and mixture casting methods. However, k_c (2.5 × 10^{-1} M^{-1} s^{-1}) for the former was 5.7 times higher than that (4.4 × 10^{-2} M^{-1} s^{-1}) for the latter [28]. The higher k_c value is attributable to the higher $\Delta S^{\neq}$ of CT by charge hopping, which was explained by the higher local complex concentration in the film prepared by the adsorption method since the complex is incorporated only in the hydrophilic region of the Nafion. The degree of localization of the complex was estimated to be 5.1 by the comparison of the electrochemical reactivity in both films [40]. CT by reverse reduction of Ru^{III} to Ru^{II} in the film prepared by a mixture casting method was investigated; charge was transferred by a physical displacement mechanism with $k_p = 7.5 \times 10^{-2}$ s^{-1}, in contrast to CT by the oxidation of Ru^{II} to Ru^{III}, showing that the CT mechanism depends on the direction of the redox reaction (oxidation or reduction), even though the redox couple involved in CT is the same [41].

For charge hopping, the distance is also an important parameter in addition to k_c. A charge hopping distance between the [Ru(bpy)$_3$]$^{2+}$ complexes in the Nafion film was found to be 1.3 ~ 1.6 nm including the distance of the bounded motion[1] of the complex that is dependent on the time scale of the reaction [42, 43]. Here the bounded motion is a kind of local oscillation of the redox molecules in the incorporated matrix.

[(bpy)$_2$(H$_2$O)Ru^{III}(μ – O)Ru^{III}(H$_2$O)(bpy)$_2$]$^{4+}$(Ru^{III}ORu^{III}) is an attractive molecule since it is an active water oxidation catalyst applicable to an electrocatalysis system (Sect. 3.1.2). CT by oxidation of Ru^{III}ORu^{III} to Ru^{III}ORu^{IV} occurred by a charge hopping mechanism with k_c depending on the complex concentration [44], i.e., the plots of k_{app} vs. c_0 gave an upwardly deviating curve. The k_c depending on the complex concentration was explained by a restricted short bounded motion of the complex. In this case, the k_c was either diffusion controlled (or bounded motion controlled) or activation controlled, which was different from the k_c for the [Ru(bpy)$_3$]$^{2+}$/Nafion film that was solely activation controlled. Also, CT by reduction of Ru^{III}ORu^{III} to Ru^{II} monomeric species was reported to occur by a physical displacement mechanism [36], demonstrating again that the CT mechanism depends on the direction (oxidation or reduction) of the redox reaction.

(C) Parallel Mechanism of Physical Displacement and Charge Hopping

[(NH$_3$)$_5$Ru^{III}(μ – O)Ru^{IV}(NH$_3$)$_4$Ru^{III}(μ – O)(NH$_3$)$_5$]$^{6+}$ (Ru-red) is an active electrocatalyst for water oxidation [10], and the mechanism of CT by oxidation of Ru-red to Ru-brown (Ru^{IV} – Ru^{III} – Ru^{IV}) in a Nafion film was examined using a PSCAS technique [35, 45]. The charge was found to be transported by a parallel mechanism of physical displacement and charge hopping. The k_c increased by one order of magnitude when the temperature increased from 5 to 35 °C, with k_p increasing by only 1.6 times with the same temperature change [35]. Activation parameters showed that the physical displacement mechanism is entropy controlled ($-T\Delta S^{\neq} = 75$ kJ mol^{-1} > $\Delta H^{\neq} = 11$ kJ mol^{-1}), but the charge hopping mechanism is enthalpy controlled ($\Delta H^{\neq} = 53$ kJ mol^{-1} > $-T\Delta S^{\neq} = 26$ kJ mol^{-1}) at 25 °C (Table 2). The

---

[1] When a redox molecule is strongly attached on polymer matrices, it can not diffuse freely in the matrixes. However, it can move slightly to aid charge hopping reaction between the molecules. This movement of the molecule is defined as "bounded motion."

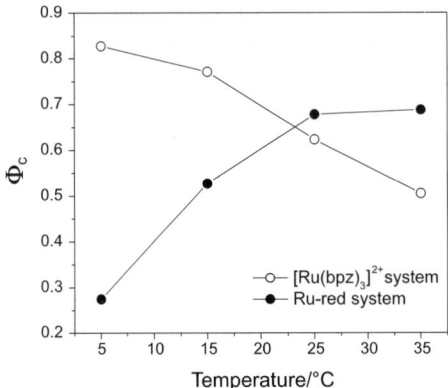

Fig. 6 Plots of Φ_c vs. temperature for Nafion/[Ru(bpz)$_3$]$^{2+}$ (○) and Nafion/Ru-red systems (●). The complex concentration is 0.1 M (reprinted with permission from American Chemical Society [33])

entropy-controlled physical displacement could be contributed by the high charges (6+) on Ru-red, as described in the physical displacement section (A). The activation parameters for the charge hopping mechanism reflect the characteristics of the self-exchange electron transfer reaction between a redox pair involving the redox equilibrium, counter ion movement, and reorganization of solvent in the film.

The parallel mechanism of physical displacement and charge hopping was also observed in CT by reduction (bpz → bpz$^{•-}$) of [Ru(bpz)$_3$]$^{2+}$ (bpz = 2,2′-bipyrazine) in a Nafion film [33]. However, k_p increased by 25 times when the temperature increased from 5 to 35 °C, with k_c increasing by 5 times with the same temperature change. The physical displacement mechanism is enthalpy controlled ($\Delta H^{\neq} = 75$ kJ mol^{-1} > $-T\Delta S^{\neq} = 3.8$ kJ mol^{-1}) at 25 °C, in contrast to the entropy-controlled physical displacement for the Ru-red/Nafion system (Table 2). The enthalpy-controlled physical displacement could suggest a stronger attachment of [Ru(bpz)$_3$]$^{2+}$ to the Nafion film than that of Ru-red by hydrophobic interaction of bpz ligands with the Nafion fluorocarbons as well as the electrostatic interaction. A striking difference in temperature dependence of the fraction, $\Phi_c = k_c c_0^2/(k_p c_0 + k_c c_0^2)$, of the contribution of charge hopping to v_{CT} was shown between the [Ru(bpz)$_3$]$^{2+}$/Nafion and Ru-red/Nafion systems. The Φ_c decreased with the increase in temperature for the Nafion/[Ru(bpz)$_3$]$^{2+}$ system but increased with the temperature increase for the Nafion/Ru-red system. This is illustrated as temperature dependences of Φ_c (at 0.1 M complex concentration) in Fig. 6.

2.2
In Polysaccharide Solids Containing Excess Liquid

2.2.1
Polysaccharide Solids Containing Excess Liquid and Their Application

In polymeric materials CT is usually slow in comparison with that in a solution as mentioned in the previous section, which is the drawback in using polymeric materials for practical use. If CT can take place in a solid phase in the same way as in a liquid phase, such solid materials can be applied to various purposes including practical uses such as sensors and other devices.

It is well known that hydrophilic polymers form a hydrogel that contains excess water [46]. However, the diffusion of molecules and ions in these gels has scarcely been studied due to the lack of suitable methodology. We have found that a tight and elastic polysaccharide solid containing excess water can be used as a solid medium for electrochemical measurements in the same way as liquid water, and that diffusion of molecules and ions takes place in this solid in the same way as in a liquid [47–51]. This allows the solid to be used not only as a medium for electrochemistry, but also as a solid reactor for various chemical reactions. It is of further interest that, in this solid bulk, natural convection does not exist [51], meaning that molecular diffusion can be discriminated from bulk natural convection that should always exist on the earth due to gravity.

Recently a dye-sensitized solar cell has been attracting a great deal of attention for converting solar energy into electricity [52]. Since this cell uses a redox electrolyte solution, it is important to solidify the liquid in order to stabilize the cell, but the task is not easy to achieve. To overcome this problem, solidification of the organic redox electrolyte solution by molten salts and gelator [53] or by polymer film [54, 55] has been achieved. We have successfully used the polysaccharide solid to solidify the electrolyte solution [56, 57], as later described in Sect. 4.1.

In the present Sect. 2.2 the fundamental properties of polysaccharide solids containing excess water are at first explained (Sect. 2.2.2). In Sect. 2.2.3 the characteristics of the polysaccharide solids as media for electrochemistry will

Agarose

κ-Carrageenan

then be described in detail. A novel property of the polysaccharide solid will be shown wherein bulk natural convection does not exist (Sect. 2.2.4).

2.2.2
Characteristics of Polysaccharide Solids Containing Excess Water

The typical polysaccharides used are agarose and κ-carrageenan. It is well known that polysaccharides form a tight and elastic solid containing excess water [46, 58–60]. On cooling its hot aqueous solution, the polysaccharide chains form a helical structure which then aggregate to double helix. This double helical structure further aggregates to a bundled structure. The double and/or bundled helical structures act as a bridging point for a three-dimensional (3D) network that can contain excess water within the network [46, 58–60]. Since κ-carrageenan involves sulfonate anionic groups, the helical structures are bundled by the presence of cations that induce aggregation of helixes by electrostatic interaction. Very tight polysaccharide solids containing excess water could be obtained by applying a microwave very carefully when preparing its aqueous solution [47].

The hardness of a 2 wt % agarose solid was one third of a conventional rubber eraser and that of a 3 wt % κ-carrageenan containing 0.1 M KCl (M = mol dm^{-3}) half of a rubber eraser. It is interesting that the solid surface is superhydrophilic: the contact angle with water is almost zero degrees. The water inside the solid evaporates much like liquid water, and after standing under ambient conditions the solid loses all the water to become a very hard dry solid.

2.2.3
Transport of Charges and Molecules in Polysaccharide Solids Containing Excess Water

The electrochemical behavior of redox molecules in polymer films and gels has been investigated [4, 7, 18, 23, 61–66], but such behavior has usually been studied by using a modified electrode coated with a polymer film or gel in the presence of an outer electrolyte solution. In a few examples, entirely solid-state voltammetry was also achieved, but by using a microelectrode array [65, 66] composed of working, counter, and reference electrodes because of the slow ionic or molecular diffusion in the solid matrices. The apparent diffusion coefficient (D_{app}) of a redox substrate in the films or solids coated on an electrode was very small [4, 7, 18, 23, 62–66], usually of the order of less than 10^{-7} cm^2 s^{-1}. Another example of solid state votammetry is a report on the electrochemistry of Prussian blue in silica sol-gel electrolytes [67], but only Pt gauze working and counter electrodes for a ~ 1 mm-thick silica solid were used. Moreover, it is well known that solid electrolytes have been used on various sensors, electrochromic devices, etc. [68, 69]. However, in spite of these activities in solid electrolytes, there has been almost no work, to our knowledge, using a solid medium for conventional electrochemical measurement with an ordinary three-electrode system. There has been only one report [70] that used agarose gel (1 wt %) for an electrochemical measurement; in this report only a gold wire working electrode was used, and a much lower diffusion coefficient of ferricyanide in the agarose gel was obtained compared to that in an aqueous solution. If electrochemical measurements using ordinary electrodes could be carried out in a solid medium in the same way as in a pure liquid, a new kind of electrochemistry and electrochemical measurements could be initiated.

In the present section, tight and elastic polysaccharide solids were used as solid electrolyte media for very conventional three-electrode electrochemical measurements. CT by molecules and its diffusion coefficient in the polysaccharide solids will be presented.

The CV of $K_4[Ru(CN)_6]$ in a 2 wt % agarose solid, in a 2 wt % κ-carrageenan solid, and in an aqueous solution containing 0.1 M KNO_3 are shown in Fig. 7 [49].

The CVs in the agarose and κ-carrageenan solids show very similar features as in liquid water including the redox potential and peak currents, but with a slightly larger peak separation for the solid systems. These polymeric materials were so tight and stable that CV could be measured without any outer cell or vessel.

The D_{app} of redox molecules was estimated either from the peak current of the CV using the Randles–Sevcik equation (Eq. 5) [59] or with a potential-

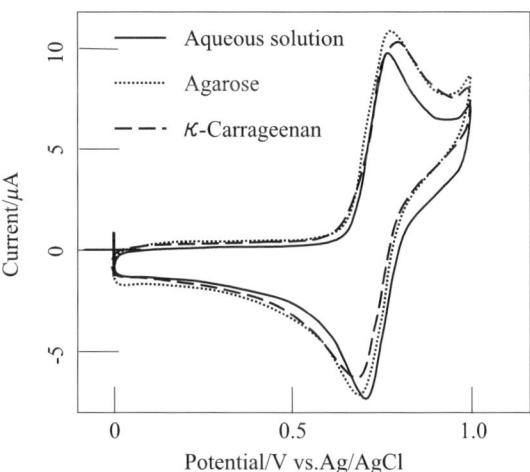

Fig. 7 CV of $K_4[Ru(CN)_6]$ (0.1 mM) in the solids of agarose (2 wt %), κ-carrageenan (2 wt %), and aqueous solution containing 0.1 M KNO_3. Working electrode is ITO, and counter electrode Pt. scan rate, 20 mV s^{-1} (reprinted with permission from Elsevier [49])

step method by using Cottrell's equation (Eq. 6) [59]:

$$i_p = (2.69 \times 10^5) A\, n^{3/2} D_{app}^{1/2} c_0 v^{1/2}, \tag{5}$$

where i_p is the peak current, A the electrode area, n the number of electrons for the reaction, c_0 the initial concentration (mol cm^{-3}) of redox compound, and v the scan rate;

$$i(t) = n F c_0 D_{app}^{1/2} (\pi t)^{-1/2}, \tag{6}$$

where $i(t)$ is current density at time t and F is Faraday constant.

As shown in Fig. 8, the D_{app} values in the solids are very similar to those in liquid water, meaning that molecular diffusion of the complex takes place in the solids, much like in an aqueous solution.

Electrochemical impedance spectra (EIS) of a – 5 mM [Fe(CN)$_6$]$^{3-}$ were measured in 2 wt % agarose and κ-carrageenan solids and in an aqueous solution containing 0.5 M KCl at the rest potentials [48]. All the spectra showed linear-relation characteristics for a diffusion-controlled process in the bulk phase at low frequencies (< 30 Hz). However, at high frequencies semicircles were clearly seen in the κ-carrageenan solid and in an aqueous solution, while the impedance spectrum in the agarose solid was slightly different, showing a larger semicircle. These semicircles at high frequencies are determined both by the capacitance (C_{dl}) in the double layer and charge transfer resistance (R_{ct}) at the electrode/solution interface. The estimated equivalent circuit for the impedance spectra was analyzed based on the Randles circuit, and the solution resistance (R_s) was determined from the equivalent circuit, and then D_{app} was calculated. The concentration dependence of D_{app} of [Fe(CN)$_6$]$^{3-}$ determined by EIS showed a trend similar to that of the above K$_4$[Ru(CN)$_6$]. On the other hand, the D_{app} of [Ru(bpy)$_3$]$^{2+}$ in the 2 wt %

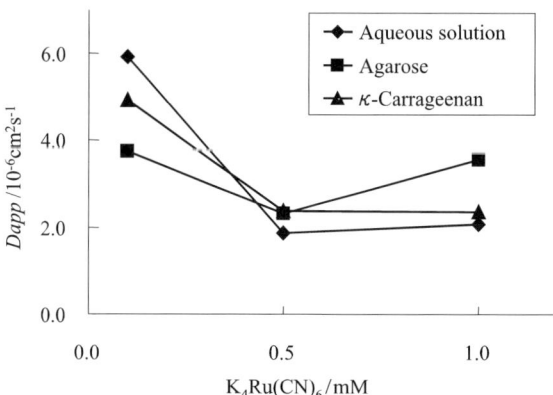

Fig. 8 Concentration dependence of D_{app} for K$_4$[Ru(CN)$_6$] in the solids of agarose (2 wt %), κ-carrageenan (2 wt %), and aqueous solution containing 0.1 M KNO$_3$ (reprinted with permission from Elsevier [49])

κ-carrageenan solid (4.5×10^{-6} cm^2 s^{-1}) was about 70% of the D_{app} in the 2 wt % agarose solid (6.7×10^{-6} cm^2 s^{-1}) [49]. This result indicates that the diffusion of [Ru(bpy)$_3$]$^{2+}$ in the κ-carrageenan solid is slightly suppressed by the anionic groups of the κ-carrageenan.

The contact angle of water on the present solid surface was almost zero, showing that the surface is superhydrophilic. The concentration dependence of the charge transfer resistance (R_{ct}) at the electrode/solid interface for the [Fe(CN)$_6$]$^{3-}$ obtained by EIS is shown in Fig. 9a [48].

It is remarkable that the R_{ct} in the 2 wt % κ-carrageenan solid was almost the same as that in the aqueous solution, although the R_{ct} in the 2 wt % agarose solid was much larger than that in the aqueous solution. Despite the high hydrophilicity of the solid surface, there could be some contact problems between the electrode and the agarose solid. As for the κ-carrageenan solid, the sulfonate anionic groups would improve the contact between the electrode and the solid.

The dependence of the double-layer capacitance (C_{dl}) on the [Fe(CN)$_6$]$^{3-}$ concentration is shown in Fig. 9b. In an aqueous solution, C_{dl} tends to increase with the [Fe(CN)$_6$]$^{3-}$ concentration, but for the solid system C_{dl} values are only weakly dependent on the [Fe(CN)$_6$]$^{3-}$ concentration. At 5 mM [Fe(CN)$_6$]$^{3-}$ concentration, C_{dl} values of the solids are similar to those in an aqueous solution.

The charge transfer resistance (R_{ct}) vs. polysaccharide concentration for the 10 mM [Fe(CN)$_6$]$^{3-}$ in the agarose and κ-carrageenan solids was investigated [49]. The R_{ct} in the κ-carrageenan solid is of the same order of magnitude as in an aqueous solution and even lower in the 4 wt % κ-carrageenan. A high concentration of the $-SO_3^-$ groups would facilitate a charge transfer between the electrode and the redox compound. It is surprising that the

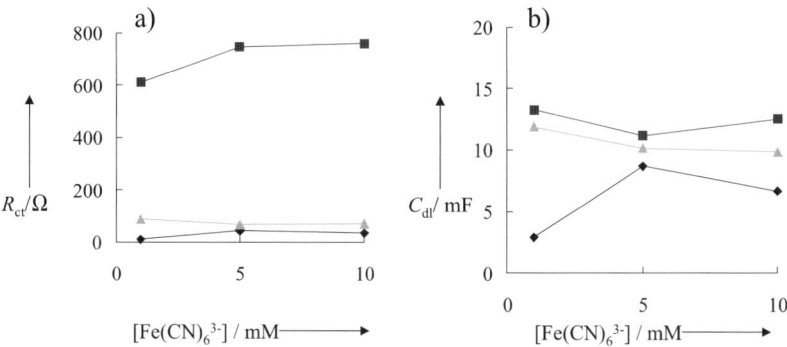

Fig. 9 Concentration dependence of **a** the charge transfer resistance (R_{ct}) and **b** the double layer capacitance (C_{dl}) of [Fe(CN)$_6$]$^{3-}$ in 2 wt % agarose (*squares*), 2 wt % κ-carrageenan (*triangles*) and aqueous solution (*diamonds*) containing 0.5 M KCl at the rest potential (reprinted with permission from Elsevier [48])

R_{ct} decreased with increasing polysaccharide concentration in the solids. One reason might be a technical problem; since a higher polysaccharide concentration causes a higher viscosity of the polysaccharide solution when it is warm, the insertion of the electrodes into this solution could give a better and more stable contact between the electrodes and the medium. It was inferred by other experiments that the increase of the polysaccharide concentration induces the growth of the chain aggregation (bundle formation) rather than increasing the crosslinked structures that can be a resistance to molecular/ionic diffusion.

2.2.4
Transport of Ions in Polysaccharide Solids (Ionic Conductivity)

Electrochemical processes in solid electrolyte can be used for various devices such as sensors, energy conversion apparatus, etc. Many groups, including ours, have reported CT by redox molecules in polymer thin films modified on an electrode [4, 7, 18, 23, 61–67]. However, in redox polymer solid phases, CT is usually very slow, showing that other heterogeneous media should be developed where CT takes place in the same way as in a solution. It was shown in Sect. 2.2.3 that molecular diffusion takes place in tight and elastic polysaccharide solids containing excess water, much like in liquid water. This type of solid can offer not only a new solid medium for electrochemical measurement but also an excellent medium for electrochemical sensors and devices. For these purposes it is important to investigate the fundamental ionic conductivity of such a solid as a medium.

An ionically conductive polymer is applied to the lithium battery, the fuel cell, etc. These polymer electrolytes have a fairly high ionic conductivity. However, general electrochemical measurement could not be performed in such polymer electrolytes. Electrochemical measurement in such polymer electrolytes has been possible only by special electrode systems described earlier. The reason is that the diffusion of ions or redox molecules and the rate of electron transfer are slow in these polymer electrolytes.

In the current section, it is shown that polysaccharide solids containing excess water can be investigated by a conventional three-electrode system because of the high ionic conductivity of the solids.

Impedance spectra of agarose and κ-carrageenan solids containing an excess 0.1 M KNO_3 aqueous solution and of a 0.1 M KNO_3 aqueous solution were measured. The conductivity of polysaccharide solids containing various electrolyte solutions obtained from the impedance spectra are summarized in Table 3 [50]. The conductivity in the 2 wt% agarose solids and 2 wt% κ-carrageenan solids is almost the same as that in an aqueous solution, showing that the transport of ions in the solid takes place in the same way as in liquid water.

Table 3 Conductivity of 2 wt % agarose and 2 wt % κ-carrageenan solids containing 0.1 M KCl, 0.1 M KNO$_3$, 0.1 M KI, 0.1 M CsCl, or 0.1 M CaCl$_2$ (reprinted with permission from Elsevier [50])

	Conductivity/mS cm^{-1}				
	0.1 M KCL	0.1 M KNO$_3$	0.1 M KI	0.1 M CsCl	0.1 M CaCl$_2$
Aq. solution	12.90	12.00	13.05	13.12	20.50
2 wt % agarose	11.35	15.20	13.13	13.27	20.46
2 wt % κ-carrageenan	13.16	13.22	13.17	15.08	21.08

Double-layer capacitance values of the solids were similar to those in an aqueous solution for every electrolyte. The double-layer capacitance values estimated from impedance spectra were almost independent of the polysaccharide concentration. These results clearly showed that the present solids could be used as an ionic conductive solid as a medium for electrochemistry in the same manner as an aqueous solution.

3
Molecular Catalysts for Multielectron Reactions

3.1
Water Oxidation Catalysts

Many scientists have been trying to activate water for both O_2 and H_2 production, which is indispensable for the construction of an artificial photosynthetic device to develop a renewable energy resource. A development of a molecular catalyst for water oxidation to evolve O_2 (Eq. 7) is a key process to achieve breakthrough for water activation:

$$2 H_2O \xrightarrow{\text{water oxidation catalyst}} O_2 + 4 H^+ + 4 e^-. \qquad (7)$$

The detection and analysis of O_2 evolved are necessary to assure the ability of the molecular catalyst for water oxidation, for which the activity (turnover rate) and stability (maximum turnover number) of the catalyst should be estimated based on the amount of produced O_2. At the same time, attention must be paid to the source of the O atom for O_2 evolution in a system including other compounds involving O atoms except water. In this section the activity and stability of the catalysts in a solution or in a polymer film, as well as the mechanism for O_2 formation including identification of the O atom source, will be reviewed.

3.1.1
Manganese Complexes

Mn complexes have been studied in relation to the structure and function of a photosynthetic oxygen evolving center (OEC), whose structure was recently proposed by X-ray diffraction analysis [71, 72]. [(terpy)(H$_2$O)Mn(μ – O)$_2$Mn(terpy)(H$_2$O)]$^{3+}$ (terpy = 2,2′:6′,2″-terpyridine) (**1**) was synthesized and structurally characterized by Limburg et al. [73] and Collomb et al. [74]. Subsequently, Limburg et al. reported O$_2$ evolution by the reaction between **1** and sodium hypochlorite (NaClO) or potassium peroxymonosulfate (KHSO$_5$), both of which are highly active oxygen atom transfer agents [73, 75]. **1** was decomposed to MnO$_4^-$ during the catalytic cycle, leading in the end of the cycle after about 6 h to the maximum turnover of the catalyst of 4 using NaClO. The kinetic analysis of O$_2$ evolution suggested that the catalysis follows Michaelis–Menten kinetics with first-order dependence with respect to **1**. Limburg et al. demonstrated by an ^{18}O isotope-labeling experiment that the O atom comes from water and not from the oxidant. However, the isotope exchange between ClO$^-$ and H$_2$(^{18}O) was confirmed to be fast ($t_{1/2} \ll 1$ h) in solution by Raman spectroscopic measurements. Although HSO$_5^-$ is slow to exchange O atoms with water ($t_{1/2} \gg 1$ h) in solution, an isotope exchange is missed for HSO$_5^-$ coordinated on a hypothesized intermediate of di-μ-oxo Mn$_2$ dimer. Their ^{18}O isotope-labeling experiment is consequently incomplete, so that it could not be excluded that the O atom comes from the oxidant (ClO$^-$ or HSO$_5^-$). They proposed the catalytic cycle for O$_2$ evolution in Fig. 10, which involves a hypothesized intermediate of di-

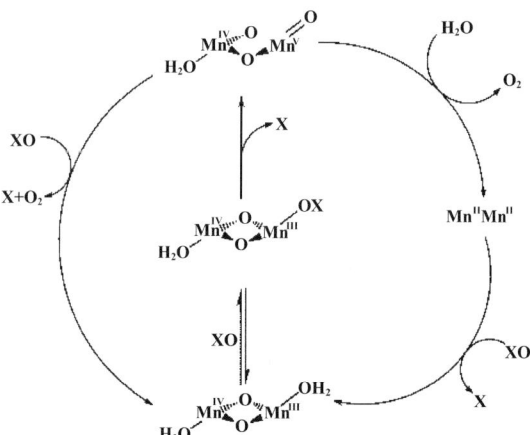

Fig. 10 Proposed mechanism for the reaction between **1** and oxygen atom transfer reagents, XO (reprinted with permission from American Chemical Society [75])

μ-oxo $Mn^{IV} - Mn^{V}$ dimer with terminal manganyl (Mn = O) that reacts with water to produce O_2 and di-μ-oxo-$Mn^{II} - Mn^{III}$ dimer.

However, the mechanism of O_2 formation in this system is completely unclear. It could involve the reaction of terminal Mn = O or oxo-bridge with OH^-, or even disproportionation of $2 ClO^- \rightarrow O_2 + 2 Cl^-$ or $SO_5^{2-} \rightarrow 1/2 O_2 + SO_4^{2-}$ could be involved. The former disproportionation is known to be catalyzed by Mn(II) and other Lewis acids [76], the latter being fast in basic solution. To exclude this possibility, O_2 evolution experiments should be conducted using oxidizing agents that do not contain any oxygen atoms.

Recently, reaction of 1 with a Ce(IV) oxidant was demonstrated in an aqueous solution. 1 was found to decompose to MnO_4^- without O_2 evolution in an aqueous solution (Fig. 11a) [77], which contrasts markedly with the O_2 evolution using 1 and either NaClO or $KHSO_5$ as an oxidant [73, 75]. A detailed reexamination of the proposed O_2 evolution mechanism should be required in the system using NaClO or $KHSO_5$. Nevertheless, the reaction of 1 with a Ce(IV) oxidant catalytically produced O_2 from water when 1 is adsorbed on clay compounds (Fig. 11c) [77]. The schematic illustration of the reaction of 1 with a Ce(IV) oxidant in solution and on clay compounds is shown in Fig. 12. The ^{18}O-labeling experiments showed that the oxygen atoms in O_2 originate exclusively from water. The kinetic analysis of O_2 evolution suggests that catalysis requires cooperation of two equivalents of 1 adsorbed on clay compounds. 1 was also found to produce catalytically O_2 from water in

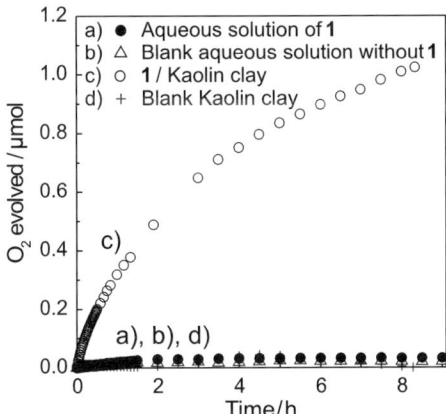

Fig. 11 Time courses of the amount of O_2 evolved in reaction of 1 and a 50 mM Ce^{IV} oxidant. **a** aqueous solution of 1 (0.84 µmol; 0.42 mM), **b** aqueous solution without 1 for a blank experiment, **c** aqueous suspension of Kaolin clay (75 mg) adsorbing 1 (0.72 µmol), **d** aqueous suspension of Kaolin clay (75 mg) without 1 for a control experiment; liquid volume, 2.0 ml; pH = 1.0 (reprinted with permission from American Chemical Society [77])

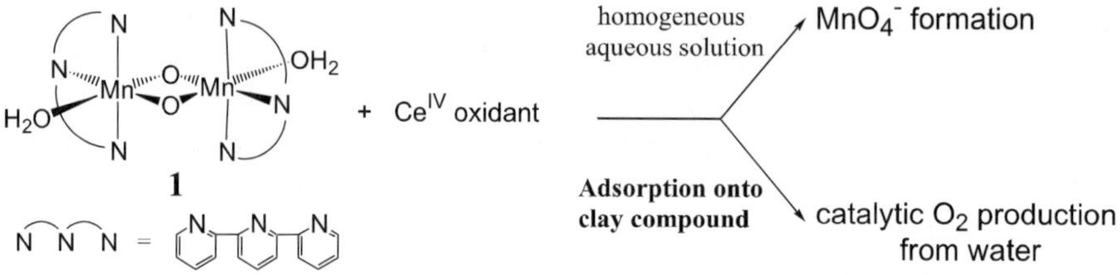

Fig. 12 Schematic illustration of reaction of **1** and a Ce(IV) oxidant in solution and on clay compounds (reprinted with permission from American Chemical Society [77])

a Nafion film using a Ce(IV) oxidant, ensuring the catalytic activity of **1** induced by adsorption onto solid matrixes [78]. Moreover, electrocatalysis was demonstrated using a Nafion-coated electrode incorporating **1**. The CV of the **1**/Nafion film-coated electrode dipped in a 0.1 M KNO$_3$ aqueous solution exhibited the catalytically anodic current above 1.05 V vs. SCE accompanying O$_2$ evolution. Such a catalytic current was not observed in an aqueous solution (1 mM) of **1**.

Naruta et al. reported that O$_2$ is evolved by electrochemical oxidation of Mn tetraarylporphyrin dimers linked by a 1,2-phenylene bridge at 1.2 vs. Ag/Ag$^+$ in a CH$_3$CN solution with a 5% water content, although it auto-oxidizes rapidly and its turnover number is low [79]. Recently the OH – Mn^V = O (– OH$^-$ and = O are axial ligands for a Mn tetraarylporphyrin

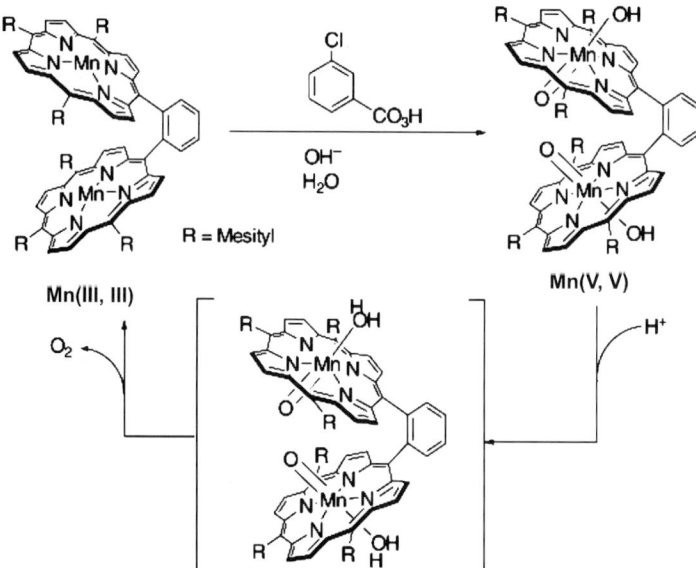

Fig. 13 Structure of Mn$_2$ porphyrin dimer and a reaction pathway for O$_2$ formation (reprinted with permission from WILEY-VCH [80])

unit) was characterized as an intermediate for the O_2 evolution. The addition of CF_3SO_3H to a CH_2Cl_2/CH_3CN solution of $OH-Mn^V=O$ species with a 10% water content led to O_2 evolution (but noncatalytic) with the formation of a $Mn^{III,III}$ dimer [80]. The authors hypothesized that protonation of $OH-Mn^V=O$ by CF_3SO_3H leads to $OH_2-Mn^V=O$ that would be attacked either by water or by the other $Mn^V=O$ unit in the dimer to yield the $O-O$ bond formation (Fig. 13).

3.1.2
Ruthenium Complexes

Ru ammine complexes were reported to be active catalysts for water oxidation in either homogeneous or heterogeneous systems [10, 11, 81–89]. Catalysis by $[(NH_3)_5Ru^{III}(\mu-O)Ru^{IV}(NH_3)_4(\mu-O)Ru^{III}(NH_3)_5]^{6+}$ (Ru-red) has been well studied [10, 11, 81, 90]. The O_2 evolution rate in reaction of Ru-red with a large excess of Ce(IV) is first-order kinetics with respect to Ru-red, suggesting a unimolecular mechanism of O_2 evolution by Ru-red. Its maximum rate (5.1×10^{-2} s^{-1}) was obtained in its homogeneous solution when the concentration was low (~ 0.1 mM). In high concentrations (~ 1 mM) the O_2 evolution rate decreased due to bimolecular decomposition of Ru-red with N_2 evolution derived from its ammine ligands. However, the decomposition can be remarkably suppressed by incorporation of Ru-red into a polymer film while keeping its intrinsic catalytic activity (4.5×10^{-2} s^{-1}) [81]. The turnover rate of Ru-red in a Nafion film decreased with its concentration by the enhanced bimolecular decomposition. The turnover rate was analyzed based on its molecular distribution in a Nafion film to suggest the importance of isolating the catalyst molecules with each other for optimizing the catalytic activity in the film [81]. Electrocatalysis for water oxidation was reported using a Ru-red/Nafion-film-coated electrode [10]. It was suggested that the turnover rate was contributed by CT via hopping between the catalysts and by bimolecular decomposition. The analysis of the turnover rate based on its molecular distribution revealed that a difficult problem arises as to the intermolecular distance between catalysts. (A shorter intermolecular distance is required for the efficient CT and a longer intermolecular distance for keeping from the bimolecular decomposition.) For a detailed discussion of the catalysis by Ru ammine complexes, the reader is referred to [5].

$[(bpy)_2(H_2O)Ru^{III}(\mu-O)Ru^{III}(H_2O)(bpy)_2]^{4+}$ (**2**) [91, 92] and its derivatives [93, 94] are known to work as water oxidation catalysts, and there is a considerable body of literature on the redox and structural chemistry of the complex for understanding the mechanism of water oxidation [5, 95, 96]. Meyer's group reported the mechanism of water oxidation by **2** based on the kinetic analysis of UV-Visible absorption data [97], their earlier ^{18}O–labeling study [98], and spectrophotometric identification of the (non-isolated) key intermediate of Ru^VORu^V [99]. Meyer et al. proposed an O_2

evolution mechanism by a bimolecular reaction of $Ru^V ORu^V$ species. However, there is no direct kinetic information about the water oxidation rate. The researchers speculated that the intermediate for the bimolecular mechanism may have a related di-μ-oxo bridge (see below) rather than a linear Ru – O – O – Ru structure. However, the bimolecular mechanism cannot quantitatively account for the isotopic distribution of O_2 evolved, nor is it consistent with the results of the kinetic studies on the O_2 evolution reported by other groups [100, 101].

$$\left\{ \begin{array}{c} Ru^V-O-Ru^V \\ \parallel \quad \quad / \quad \backslash \\ O \quad \quad O \quad O \\ \quad \quad \backslash / \\ Ru^V-O-Ru^V \\ \parallel \\ O \end{array} \right\}^{8+}$$

On the other hand, Hurst's group proposed another mechanism of water oxidation catalyzed by 2, based on structural [102], kinetic [101], and isotopic distribution data of O_2 evolution [103]. Their mechanism involves a $Ru^V ORu^V$ intermediate with a terminal ruthenyl O atom on each center that is considered to be either an O_2 evolving species or its intermediate precursor. They proposed two pathways for a catalytic mechanism; the elements of H_2O are added as OH and H onto the adjacent terminal ruthenyl O atoms [as pathway (i), Fig. 14a], and OH and H are added to a bpy ring and one of the ruthenyl O atoms for incorporation of two water molecules [as pathway (ii), Fig. 14b] [103]. Their isotopic distribution data can be quantitatively accounted for by two pathways. It excludes a direct O – O bond formation either by the intramolecular coupling of $Ru^V = O$ groups within a single complex or by the intermolecular mechanism by bimolecular reactions of the complexes.

The catalytic activity of 2 in a Nafion film was examined using a Ce(IV) oxidant [100]. The O_2 evolution rate is first-order kinetics with respect to 2, suggesting a unimolecular mechanism of O_2 evolution by 2 in the film. The maximum rate was 2.4×10^{-3} s^{-1}, which is comparable to that (4.2×10^{-3} s^{-1}) in an aqueous solution reported by our group [100].

Recently it was reported that a dinuclear Ru complex, [(terpy)$_2$(H$_2$O)Ru^{II}(dpp)Ru^{II}(H$_2$O)(terpy)$_2$]$^{3+}$ (3), bridged by a bpp$^-$ (3,5-bis(2-pyridyl)pyrazolate) chelating ligand, is active in water oxidation catalysis [104]. [(terpy)$_2$Ru^{II}(μ – OAc)(dpp)Ru^{II}(H$_2$O)(terpy)$_2$](PF$_6$)$_2$ was isolated as the precursor of nonisolated 3 and characterized by X-ray structure analysis (Fig. 15). The key structural feature of 3 is the absence of a μ – O bridged Ru$_2$ species (Ru – O – Ru) that is contained in most water oxidation catalysts [5, 99]. The reaction of 3 (0.914 mM in 2 ml) with 100-fold excess of a Ce(IV) oxidant in 0.1 M triflic acid evolved 34 μmol (turnover number of 18.6) of O_2 after 48 h. The reaction with Ce(IV) led to the fast oxidation of 3 to

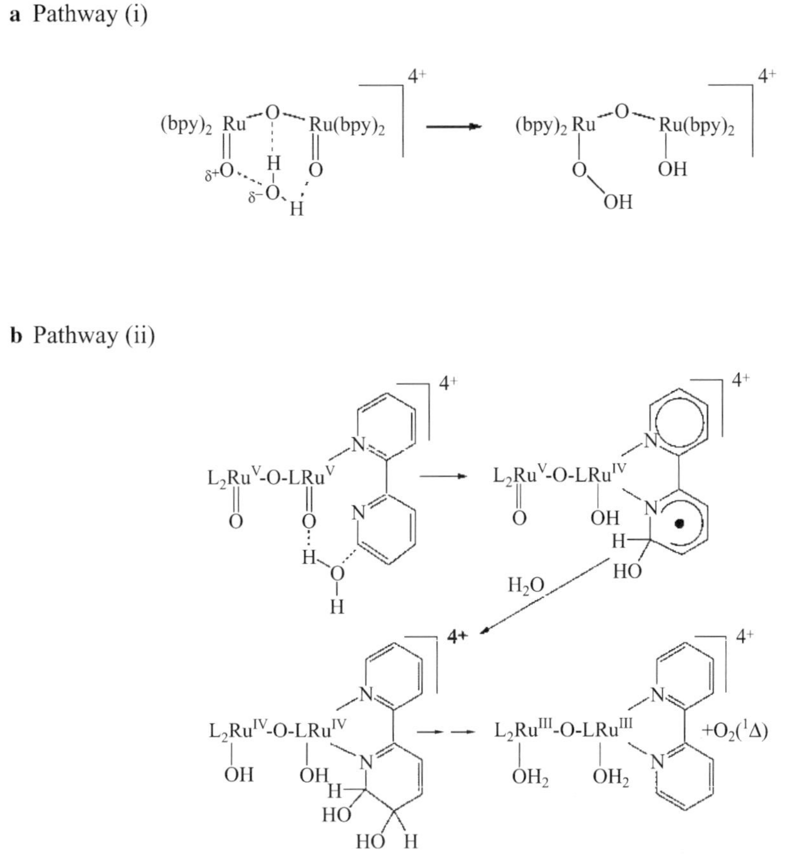

a Pathway (i)

b Pathway (ii)

Fig. 14 Hypothetical mechanisms for O₂ formation. **a** pathway (i) involving one O atom from a solvent water molecule and the other from $Ru^V = O$, **b** pathway (ii) involving both O atoms from solvent water molecules (reprinted with permission from American Chemical Society [103])

$Ru^{IV} - Ru^{IV}$ with ruthenyl oxo groups on each center, which is hypothesized as the active species for the catalysis, followed by a slow pseudo-first-order O₂ evolution with respect to **3** with a rate constant of 1.4×10^{-2} s^{-1}. This value is 3 times higher than that (4.2×10^{-3} s^{-1}) [84] of **2**, but 4 times lower than that (5.6×10^{-2} s^{-1}) [88] of $[(NH_3)_3Ru^{II}(\mu-Cl)_3Ru^{III}(NH_3)_3]^{2+}$ under similar conditions. Hurst et al. argued that the higher rate of O₂ evolution compared with **2** can be attributed to a more favorable disposition of the $Ru^{IV} = O$ groups in **3** rigidly facing each other, in contrast to $Ru^V = O$ in **2** rotating along the Ru–O–Ru bond (O₂ evolution mechanism was presumed by intramolecular coupling of $Ru^V = O$ groups on **2**, but this possibility was excluded by recent works (see above). However, the argument by Hurst et al.

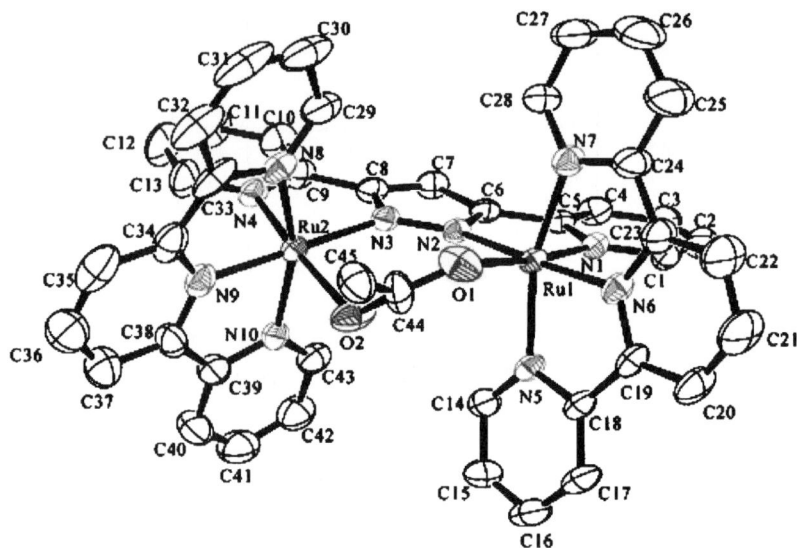

Fig. 15 X-ray structure of [(terpy)$_2$Ru^{II}(μ – OAc)(dpp)Ru^{II}(H$_2$O)(terpy)$_2$]$^{2+}$, which is a precursor of **3** (reprinted with permission from American Chemical Society [104])

includes some speculation about the mechanism; detailed studies with various spectroscopic techniques and ^{18}O-labeling experiments are expected to establish water oxidation catalysis by **3** when compared with the catalytic mechanism by **2**. Additionally, although the catalysis by **3** in solid materials is interesting, it has not been tested yet. The hybrid catalyst of polymeric materials and **3** is anticipated for the development of a heterogeneous catalyst.

Wada et al. synthesized another dinuclear Ru complex, [(*t*-Bu$_2$qui)(OH)Ru^{II}(btpyan)Ru^{II}(OH)(*t*-Bu$_2$qui)]$^{2+}$ (**4**), with 3,6-di-*tert*-butyl-1,2-benzoquinone (*t*-Bu$_2$qui) ligands on each metal center and a bridging ligand of 1,8-bis(2,2′:6′,2″-terpyridyl)anthracene (btpyan) (Fig. 16) [105, 106]. The addition of 2 eq of *t*-BuOK to the **4** solution in methanol caused the reduction of *t*-Bu$_2$qui ligands by intramolecular electron transfer from the hydroxo ligands to quinone ligands coupled with proton dissociation of hydroxo ligands, resulting in [(*t*-Bu$_2$sq)(O)Ru^{II}(btpyan)Ru^{II}(O)(*t*-Bu$_2$sq)]^0 (*t*-Bu$_2$sq = 3,6-di-*tert*-butyl-1,2-semiquinone) (Eq. 8). The formed oxo ligand on each center was considered to be a radical character, probably forming a peroxo O – O bond by intramolecular coupling of the oxo ligand. However, the character of the oxo ligands is not yet clear:

$$[(t\text{-Bu}_2\text{qui})(\text{OH})\text{Ru}^{II}(\text{btpyan})\text{Ru}^{II}(\text{OH})(t\text{-Bu}_2\text{qui})]^{2+}$$

$$\underset{2\text{H}^+}{\overset{-2\text{H}^+}{\longleftrightarrow}} [(t\text{-Bu}_2\text{sq})(\text{O})\text{Ru}^{II}(\text{btpyan})\text{Ru}^{II}(\text{O})(t\text{-Bu}_2\text{sq})]^0 . \qquad (8)$$

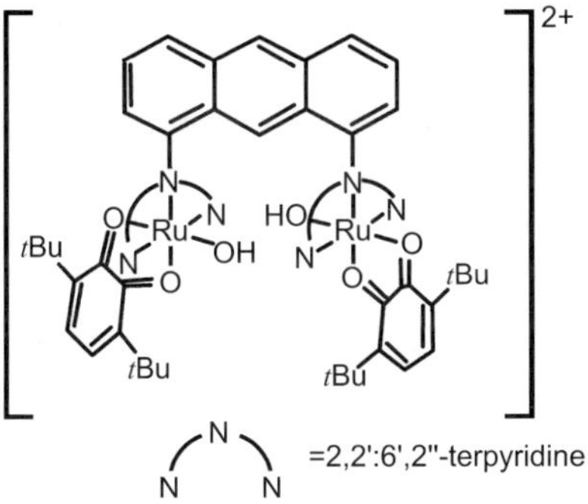

Fig. 16 Chemical structure of [(t-Bu$_2$qui)(OH)Ru^{II}(btpyan)Ru^{II}(OH)(t-Bu$_2$qui)]$^{2+}$ (**4**) (reprinted with permission from WILEY-VCH [105])

The resultant complex [(t-Bu$_2$sq)(O)Ru^{II}(btpyan)Ru^{II}(O)(t-Bu$_2$sq)]^0 underwent ligand-localized oxidation at 0.4 V (vs. Ag/AgCl) in methanol to give [(t-Bu$_2$qui)(O)Ru^{II}(btpyan)Ru^{II}(O)(t-Bu$_2$qui)]$^{2+}$. The latter further underwent metal-localized oxidation at 1.2 V in CF$_3$CH$_2$OH/ether to yield [(t-Bu$_2$qui)(O)Ru^{III}(btpyan)Ru^{III}(O)(t-Bu$_2$qui)]$^{4+}$, which presumably catalyzes water oxidation (Eq. 9):

$$[(t\text{-Bu}_2\text{sq})(\text{O})\text{Ru}^{II}(\text{btpyan})\text{Ru}^{II}(\text{O})(t\text{-Bu}_2\text{sq})]^0$$
$$\underset{\underset{0.4\text{ V}}{2e^-}}{\overset{-2e^-}{\rightleftarrows}} [(t\text{-Bu}_2\text{qui})(\text{O})\text{Ru}^{II}(\text{btpyan})\text{Ru}^{II}(\text{O})(t\text{-Bu}_2\text{qui})]^{2+}$$
$$\underset{\underset{1.2\text{ V}}{2e^-}}{\overset{-2e^-}{\rightleftarrows}} [(t\text{-Bu}_2\text{qui})(\text{O})\text{Ru}^{III}(\text{btpyan})\text{Ru}^{III}(\text{O})(t\text{-Bu}_2\text{qui})]^{4+}. \quad (9)$$

The potentiostatic electrolysis at 1.7 V vs. Ag/AgCl evolved 15.2 ml (turnover number of **4**: 33 500) of O$_2$ for 40 h using a **4**-coated ITO electrode (2.0 × 10^{-8} mol per 10.0 cm^2) dipped in an aqueous buffer solution (pH 4.0, H$_3$PO$_4$/KOH, 1.0 M), although a high overpotential (0.99 V) was applied in the electrocatalysis.

3.2
Proton Reduction Catalysts

Proton reduction (Eq. 10) is one of the simplest chemical reactions but necessary to establish a water-splitting reaction by an efficient combination with water oxidation (Eq. 7):

$$2\,\text{H}^+ + 2\,e^- \xrightarrow[]{\text{proton reduction catalyst}} \text{H}_2. \quad (10)$$

The development of a molecular catalyst for proton reduction is important for the construction of an artificial photosynthetic device. In this section, the reported synthetic molecular catalyst for proton reduction will be reviewed.

Proton reduction enzymatically proceeds by hydrogenase [107]. The enzymatic mechanism would provide a clue to the design of a synthetic catalyst for proton reduction. Hydrogenase represents structurally unique cores that rely on two metal centers of Fe_2 or FeNi [107–110]. The crystallographic structures of Fe-only and FeNi hydrogenases are known at high resolution [108–110]. The active site of Fe-only hydrogenase consists of an $Fe_2(\mu-S_2R)(CN)_2(CO)_3L_n$ core (L = H^-/H and a thiolate-linked Fe_4S_4 cluster) in which two iron(I) cations are bridged by azadithiolate, $SCH_2NHCH_2S(S_2R)$.

Rauchfuss et al. reported that a hydrogenase model complex, $[Fe_2(\mu-S_2C_2H_6)(CN)_2(CO)_4]^{2-}$ (5), reacts with proton to give a substoichiometric amount of dihydrogen (H_2) but is catalytically inactive [111]. They tuned the reducing character of 5 by replacing one of two CN^- ligands with a PMe_3 ligand to synthesize $[Fe_2(\mu-S_2C_2H_6)(CN)(CO)_4(PMe_3)]^-$ (6), which is less reducing than 5. 6 was found to work as a catalyst for proton reduction, and a proposed mechanism is shown in Fig. 17 [112]. The addition of excess aqueous H_2SO_4 to the 6 solution in CH_3CN produced $[HFe_2(\mu-S_2C_2H_6)(CN)(CO)_4(PMe_3)]$ species with irreversible protonation of the Fe–Fe bond, and further treatment with toluenesulfonic acid gave $[HFe_2(\mu-S_2C_2H_6)(CNH)(CO)_4(PMe_3)]^+$ species with protonation of CN^- ligand. The CV of 6 with 1 eq of toluenesulfonic acid in CH_3CN showed reduction peaks of monoprotonated and diprotonated species at 1.13 and – 1.03 V vs. Ag/AgCl. In the potentiostatic electrolysis of the 6 solution

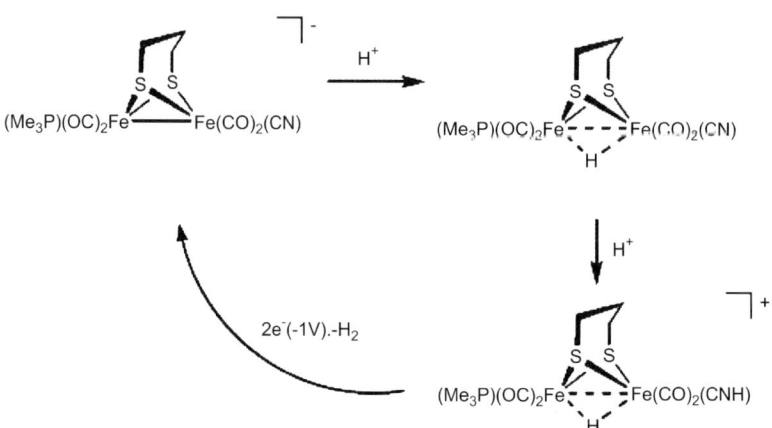

Fig. 17 Proposed catalytic cycle of proton reduction by $[Fe_2(\mu-S_2C_2H_6)(CN)_2(CO)_4]^{2-}$ (5) (reprinted with permission from American Chemical Society [112])

(1×10^{-3} M) with 50 eq of H_2SO_4 at -1.2 V vs. Ag/AgCl, 12 F per mol of **6** was passed over in the course of 15 min, and stoichiometric amounts of H_2 are detected. By contrast, monocyanide $[Fe_2(\mu-S_2C_2H_6)(CN)(CO)_5]^-$ was protonated only at the N of cyanide (not at the Fe – Fe bond) and catalytically inactive, suggesting that the PMe_3 ligand in **6** enhances the basicity of the Fe – Fe bond to be active in catalysis.

Ott et al. synthesized the $[Fe_2(\mu-SCH_2NRCH_2S)(CO)_6]$ (R = p-brombenzyl) [113]. The protonation at N of the azadithiolate bridging ligand on the complex occurred in a CH_3CN solution on the addition of perchloric acid. In the CV of the complex solution with excess perchloric acid, the reduction peak of the complex was shifted by 0.4 V to positive potential by the protonation. The potentiostatic electrolysis of the complex solution (1 mM) with 50 mM perchloric acid at -1.48 V vs. Fc/Fc^+ (-0.97 V vs. Ag/AgCl) showed a 10 times higher initial slope of a charge amount vs. time curve than that without perchloric acid, and 9 F (4.5 turnovers) for the complex amount was passed during electrolysis in 1 min. The catalytic cycle was proposed in Fig. 18, speculating the insertion of a second H^+ into the Fe – Fe bond. However, no evidence for the protonation into the Fe – Fe bond has been obtained.

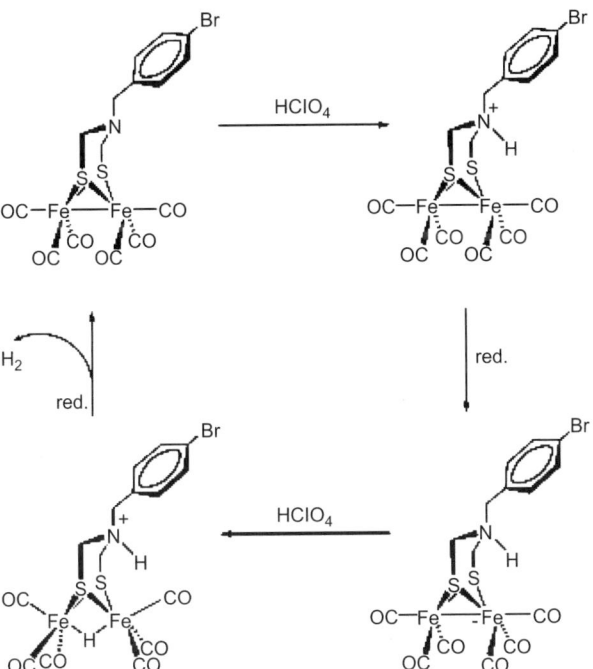

Fig. 18 Proposed catalytic cycle of proton reduction by $[Fe_2(\mu-SCH_2NRCH_2S)(CO)_6]$ (R = p-brombenzyl) (reprinted with permission from WILEY-VCH [113])

The electrocatalytic activity of metal tetraphenyporphyrin (MTPP; M = Mn, Fe, or Co) for proton reduction was studied using a Nafion-film-coated Pt electrode incorporating MTTP [114, 115]. 208 µl of H_2 was evolved using a FeTTP/Nafion film at pH = 1.01 and -0.3 V vs. Ag/AgCl, which is very close to the theoretical H_2 evolution potential (-0.25 V vs. Ag/AgCl) at this pH. When using a neat FeTTP-coated Pt electrode, only 5.31 µl of H_2 was evolved under the same conditions, showing that a polymer matrix is very important for establishing efficient catalysis for proton reduction. Related to this, the electrocatalytic reduction by phthalocyanine cobalt(II) (CoPc) and its derivatives of octcyanophthalocyanine ($CoPc(CN)_8$) and tetrasulfonatophthalocyanine ($CoPc(SO_3H)_4$) incorporated in a poly(4vinylpyridine-co-styrene) film was studied. The turnover rate of CoPc in the film was 2×10^5 h^{-1}, and the order of the turnover rate was $CoPc > CoPc(CN)_8 > CoPc(SO_3H)_4$.

3.3
CO_2 Reduction Catalysts

The reduction of carbon dioxide (CO_2) to produce useful compounds is attracting a great deal of attention because of the greenhouse effect by the increasing concentration of CO_2 in the atmosphere, but catalytic activation of CO_2 is currently not an easy task [116]. Because the use of energy resources such as fossil fuels is meaningless for conversion of CO_2 to useful compounds when considering the environmental problem, the focus of research is on using solar energy (visible light energy) for the CO_2 reduction. Semiconductor catalysts have been investigated as a photocatalyst for this reason to achieve CO_2 reduction by solar irradiation, but this is not included in the present review on molecular catalysis. In the present section electrochemical CO_2 reduction by molecular catalysts confined in a polymer membrane coated on an electrode will be reviewed briefly.

CO_2 reduction reactions are shown in Fig. 19 classified by the number of electrons involved in the reaction. In Fig. 19, proton reduction is also shown as a reference. When the number of electrons involved in the reactions shown in Fig. 19 increases, the redox potential for the reaction tends to shift positively, suggesting that CO_2 reduction becomes thermodynamically easier with an increase in the number of electrons. By contrast, the reduction becomes difficult with an increase in the number of electrons due to a mechanistic difficulty in pooling the electrons for a multielectron reaction. In most CO_2 reductions catalyzed by molecules, the reaction proceeds with two electron reductions producing CO or HCOOH. CO_2 reduction with four, six, or eight electrons is usually a difficult process.

Photochemical CO_2 reduction has been investigated using molecular catalysts such as $[Ru(bpy)_2(CO)_2]^{2+}$ and $[Re(bpy)(CO)_3]^{2+}$, where the complex works both as a sensitizer and a multielectron (usually two) catalyst, but the photocatalysis has mostly been carried out in a homogeneous so-

$$2e + 2H^+ \rightarrow H_2 \qquad -0.41$$
$$CO_2 + 2e + 2H^+ \rightarrow CO + H_2O \qquad -0.52$$
$$CO_2 + 2e + 2H^+ \rightarrow HCOOH \qquad -0.61$$
$$CO_2 + 4e + 4H^+ \rightarrow C + 2H_2O \qquad -0.20$$
$$CO_2 + 4e + 4H^+ \rightarrow HCHO + H_2O \qquad -0.48$$
$$CO_2 + 6e + 6H^+ \rightarrow CH_3OH + H_2O \qquad -0.38$$
$$CO_2 + 8e + 8H^+ \rightarrow CH_4 + 2H_2O \qquad -0.24$$

red.Pot. (V vs.NHE, pH7)

Fig. 19 Redox potential for CO_2 reduction depending on the number of electrons involved

lution [6]. Polymeric $[Ru(bpy)(CO)_2]_n$ was prepared by electroreduction of its original complex and catalyzed CO_2 reduction, forming selectively CO at almost 100% current efficiency [117]. Electrocatalytic CO_2 reduction by metal-polypyridyl complexes and metal-phthalocyanine (MPc) derivatives were studied in a solid polymer matrix [6]. Photocatalytic CO_2 reduction by CoPc or ZnPc embedded in a Nafion membrane was reported to produce formic acid [118].

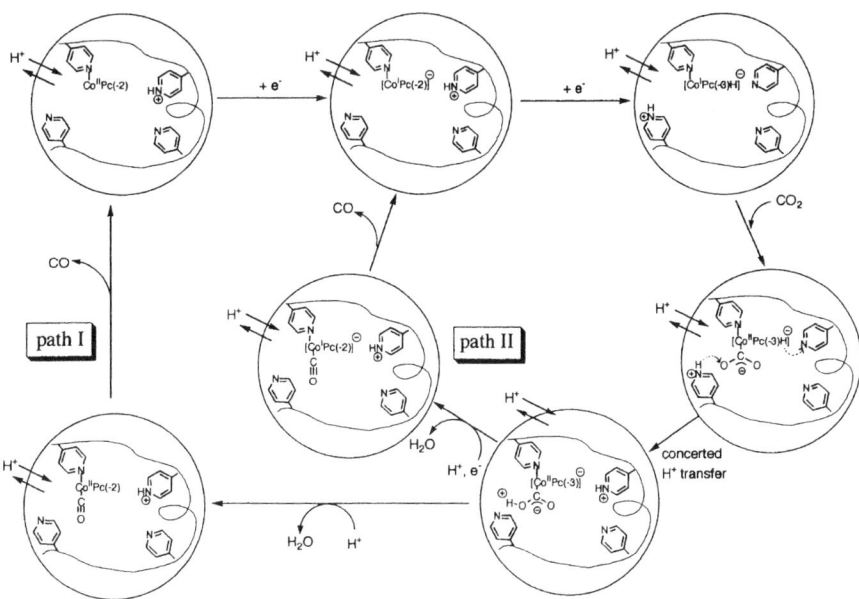

Fig. 20 Mechanism of water phase electrocatalytic CO_2 reduction by CoPc confined in a poly(vinylpyridine) membrane coated on an electrode (reprinted with permission from Elsevier [119])

When water is used as a medium for CO_2 reduction, it is important to suppress proton reduction to produce H_2, which is more favorable in water than CO_2 reduction. A hydrophobic polymer environment can provide such conditions by suppressing proton reduction to selectively carry out CO_2 reduction. In an electrocatalytic CO_2 reduction by CoPc confined in a poly(vinylpyridine) membrane coated on an electrode, selective CO_2 reduction takes place, producing CO [119]. It is of further interest that in the electrocatalytic system, a third electron is injected into the $Co(I)Pc(-3) - CO_2$ intermediate to produce CO and $Co(I)Pc(-2)$ complex after the starting $Co(II)Pc(-2)$ complex is reduced by two electrons to $Co(I)Pc(-3)$ to form $Co(I)Pc(-3) - CO_2$ with CO_2. Such a mechanistic scheme can be represented by Fig. 20 [119]. Path I is the previously accepted scheme, and path II is the proposal of a new two-electron reduction pathway. Note also that for the CO_2 reduction, a proton is also involved in an equilibrium process. It was inferred that in the polyvinylpyridine matrix, protonation and deprotonation takes place easily with the help of the pendant pyridine groups in a concerted fashion, resulting in favorable CO production (see also Fig. 20) [119].

4
Application to Future Energy Conversion Devices

CT and catalysis in a polymer membrane coated on an electrode are important processes to create various sensors and devices for photochemical energy conversion. Among such devices, the application to solar cells and future artificial photosynthetic systems to create energy from sunshine is attracting a great deal of attention. In this section, applications of CT and catalysis in a polymer membrane coated on an electrode to a dye-sensitized solar cell and an artificial photosynthesis are reviewed.

4.1
Dye-sensitized Solar Cell

The photosensitized solar cell composed of nanoparticulated TiO_2 porous film adsorbing dye reported by Graetzel's group (Fig. 21) [52, 120] has shown great success in the relevant research area and is attracting a great deal of attention as a near future solar cell with high-cost performance instead of a conventional amorphous-silicon-semiconductor-based solar cell.

This so-called Graetzel's cell has been reported to give nearly 10% conversion efficiency under AM 1.5 solar irradiation, and so is attracting a great deal of attention as a future commercial solar cell. The most important part of this cell is a monolayered dye covering the nanoporous TiO_2 film coated on a transparent electrode. The dye is attached on the TiO_2 by chemical bonds

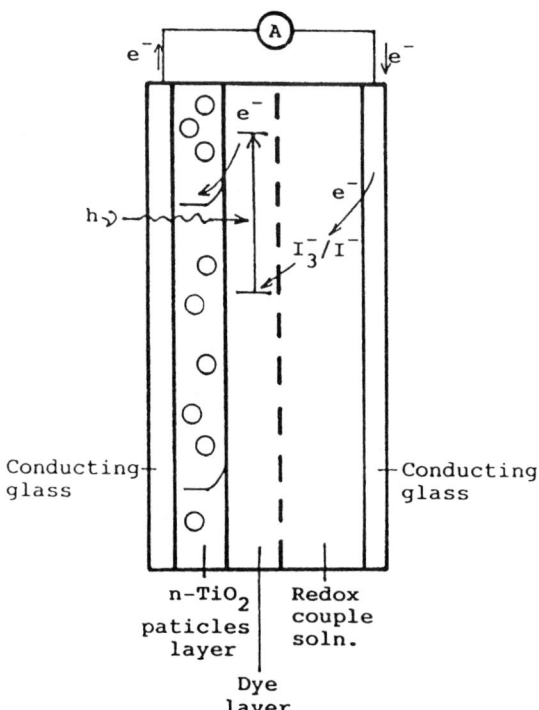

Fig. 21 Dye-sensitized solar cell with the configuration, F-doped ITO/TiO$_2$ nanoparticulated film/dye/(I$_3^-$/I$^-$)redox electrolyte solution/Pt-coated ITO

with –OH groups on the TiO$_2$ surface, so that it forms a kind of polymeric monolayer film on the TiO$_2$. One of the problems of this cell for practical use is that it uses organic liquid. To overcome this problem, solidification of the organic redox electrolyte solution by molten salts and gelator [53] or by polymer film [54] has been achieved. Another approach was to use water instead of an unstable organic liquid [121–123], but the efficiency and stability of the cell in an aqueous phase are low. In order to use a polymeric film as a medium of redox electrolyte, the transport of redox electrolyte through the polymer phase is important, but the transport rate is usually low when a solid-state polymer film is used. Organogels composed of crosslinked polymer and involving excess organic liquid is a good candidate to incorporate redox electrolytes with the liquid medium into the gel, but investigation on the transport of the redox electrolyte through the gel matrix has not been easy due to the lack of a suitable methodology.

As shown in Sect. 2.2, one of the authors of this article has reported that polysaccahrides such as agarose and κ-carrageenan (Sect. 2.2) can form a tight and elastic solid containing excess water, and that electrochemical and photochemical reactions can take place in the solid in the same way as

N3 dye: cis-bis(thiocyanato)bis(4,4'-dicarboxyl-2,2'-bipyridine)ruthenium(II), Ru(dcbpy)$_2$(NCS)$_2$

in pure water. The hardness of the solid is, for instance, almost the same as brick cheese and one third of a conventional rubber eraser for a 2 wt % κ-carrageenan solid containing excess water. Such an interesting solid containing a large amount of excess liquid could offer a solid-state medium for photosensitized cells. We have succeeded in substituting the water in the solid with organic liquid and found that this solid containing organic solvent and I^-/I_3^- redox electrolyte works well as a medium for the TiO$_2$ cell with a well-known N3 dye (see inserted figure) sensitizer [52, 120]. In the next part of this section the solid-state cell using this polysaccharide solid will be reviewed.

The I-V characteristics of a dye-sensitized cell fabricated by ITO/TiO$_2$/N3/solid (containing organic liquid with (C$_3$H$_7$)$_4$NI and I$_2$) working electrodes and the ITO/Pt counter electrodes are shown in Fig. 22, and the results are summarized in Table 4, where the performance of the cell composed of the corresponding liquid medium is also given with the same conditions as the solid-type cell [56].

In Table 4 short-circuit photocurrent (J_{sc}), open-circuit photovoltage (V_{oc}), fill factor (ff), and light-to-electricity conversion efficiency (η) are

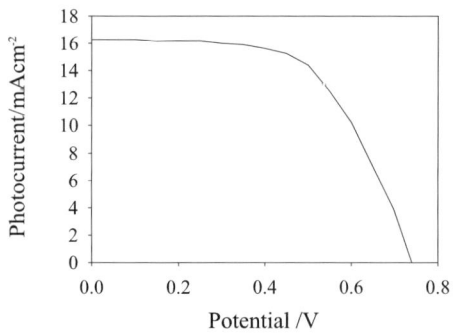

Fig. 22 I-V curve of the solid-state dye-sensitized solar cell, TiO$_2$/N3/-carrageenan solid containing aceonitrile/3-methyl-2-oxazolidinone (1/1) and (C$_3$H$_7$)$_4$NI/I$_2$ (0.3 M/0.03 M)/Pt. The data correspond to run 4 of Table 4 (reprinted with permission from Chemical Society of Japan [56])

Table 4 Performances of solid-state dye-sensitized solar cell, $TiO_2/N3$/Carrageenan solid containing aceonitrile/3-methyl-2-oxazolidinone (1/1) and $(C_3H_7)_4NI/I_2$(0.3 M/0.03 M) /Transparent Pt under 98 mW cm^{-2} irradiation from a 500 W xenon lamp (reprinted with permission from Chemical Society of Japan [56])

Run	Electroyte medium	J_{sc}/mA cm^{-2}	V_{oc}/V	ff	η/%
1	Liquid	13.75	0.68	0.46	4.39
2	Solid	13.50	0.68	0.51	4.82
3*	Solid	17.50	0.71	0.41	5.22
4**	Solid	16.25	0.72	0.61	7.23
5**	Liquid	13.25	0.76	0.56	5.42

* TiO_2/N3 was treated with 2 mol% *t*-butylpyridine/acetonitrile solution before cell fabrication. ** 0.5 M *t*-butylpyridine was added in the solution; fluorine-doped SnO_2 and Pt plate were used for working and counter electrodes. Runs 1/2 and runs 4/5 are comparable, respectively

given. In run 3 the TiO_2/dye electrode was treated with a 2 mol% *t*-butylpyridine/acetonitrile solution before cell fabrication in order to suppress back electron transfer from the injected electron in the TiO_2 layer to the oxidized dye. In runs 4 and 5 a – 0.5 M *t*-butylpyridine solution was added to the cell medium. The solid-state cell showed even better performance than the corresponding conventional (reference) liquid-type cell. In a recent investigation it was shown that the difference between the solid-type and liquid-type cell is attributable to the slight difference in the liquid phase thickness for both cells. A careful comparison between the two types of cells showed that both exhibit an almost identical performance. In run 4 (solid-type cell) the conversion efficiency (7.23%) was calculated for the irradiation conditions with AM (air mass) 1.5 and 100 mW cm^{-2} light intensity to give 5.8%.

The diffusion coefficient of holes via the I^-/I_3^- redox couple in the solid containing 2.5 wt % carrageenan, 0.3 M KI, 0.03 M I_2, and excess water was investigated by an impedance spectroscopy; it was ca. 1.7×10^{-5} cm^2 s^{-1}, which is almost the same as that in liquid water, showing that the hole transport in the solid is not a problem in comparison with the liquid medium system. Similar results have been obtained also for a solid containing redox electrolyte and organic liquid [57]. In the present solid material, transport of small ions and molecules takes place in the same way as in a liquid, showing that the liquid contained in this solid behaves as if it were a pure liquid.

In order to further investigate the effect of a solid-type electrolyte solution on each CT step (Fig. 23), alternating current impedance spectrum was measured under irradiation (Fig. 24).

In Fig. 24 the resistances for CT from ω_1 to ω_4 correspond to the following processes:

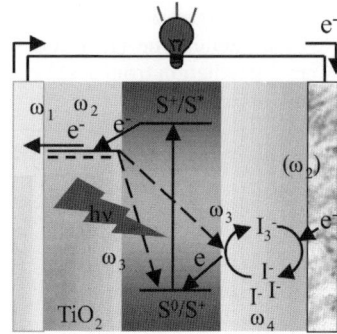

Fig. 23 Charge transport (CT) and resistances (ω_1 to ω_4) of the dye-sensitized solar cell

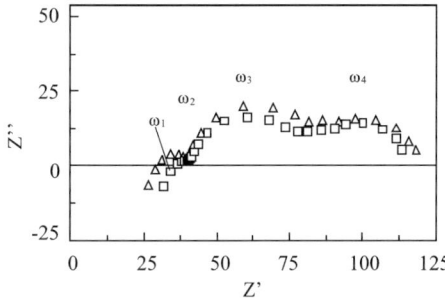

Fig. 24 Impedance spectra of solid-type (*open circle*, κ-carrageenan) and liquid-type cells (*open triangle*) (reprinted with permission from Elsevier [57])

- ω_1 = interface at TiO$_2$/conducting layer of the substrate
- ω_2 = interface between nanoparticles of TiO$_2$
- ω_3 = interface at TiO$_2$/electrolyte or TiO$_2$/dye
- ω_4 = diffusion of I$_3^-$/I$^-$

As shown in Table 5, the CT resistance of the solid-type cell is not very different from that of the liquid-type cell, indicating that the polysaccharide solid can work as effectively as the redox electrolyte medium [57].

Table 5 Results of impedance analysis for the 1 wt % κ-carrageenan solid and liquid type cells with acetonitrile/2-methoxyethanol (1 : 1) containing LiI/Pr$_4$NI/I$_2$ (0.1 M/0.5 M/0.05 M) and TBP (0.5 M). An ITO base electrode was used instead of fluorine-doped SnO$_2$ (reprinted with permission from Elsevier [57])

electrolyte medium	wt %	J_{sc}/mA cm^{-2}	V_{oc}/V	ff	η/%	ω_3	ω_4
Liquid		8.49	0.73	0.54	3.35	41.0	40.3
κ-carrageenan	1.0	8.29	0.75	0.53	3.28	39.9	37.0

A solid-type dye-sensitized solar cell was thus easily fabricated using an inexpensive and commercially available polysaccharide solid containing organic medium and I^-/I^3 redox electrolyte. Polymer gels are usually soft and fragile. However, the present solid is tight and elastic. Although long-term performance has not yet been tested, the photocurrent did not decrease in a 3 h irradiation session without any sealing of the solid-type cell, showing that it is much more stable than a conventional liquid-type cell against evaporation of the organic liquid.

4.2
Artificial Photosynthesis

Artificial photosynthesis is an attractive device for converting solar energy to chemical energy that is stored by production of highly reduced compounds derived from protons or carbon dioxide. It is crucial to use water molecules as the systems electron source, and consequently water oxidation by solar light to evolve dioxygen is an essential process in artificial photosynthesis. A necessary task for designing an artificial photosynthesis system is to efficiently couple the photochemical reaction of the sensitizer with both a water oxidation catalyst and a reduction catalyst for protons or carbon dioxide at donor and accepter terminal reactions, respectively. However, to date this task has been difficult to achieve. Recent examples of strategies for constructing artificial photosynthesis systems will be reviewed here.

Sun et al. reported tris(bipyridine) ruthenium(II) complex covalently linked to a manganese(II) complex (without catalytic activity for water oxidation) and showed that after initial electron transfer from the photoexcited Ru(II) to an external electron acceptor, methylviologen (MV^{2+}), intramolecular electron transfer took place from the Mn(II) to the photogenerated Ru(III) in an acetonitrile solution [124, 125].

Similar strategies for constructing an artificial photosynthesis system were reported by Burdinski et al., in which mono-, di-, and trinuclear manganese complexes (without catalytic activity for water oxidation) with phenolate ligands covalently linked to tris(bipyridine) ruthenium(II) were synthesized and characterized [126]. They showed that intramolecular electron transfer takes place from a Mn^{II}-trimer moiety to the photochemically generated Ru(III), as well as from a phenolate ligand to the Ru(III) in a Mn^{IV}-monomer derivative [127]. Although these works by Sun et al. and Burdinski et al. may offer mechanistic insight into the elucidation of electron transfer in photosynthetic water oxidation, O_2 is not produced from water in the present stage of their projects. These types of strategies, i.e., the use of a water oxidation catalyst covalently linked to a photosensitizer such as tris(bipyridine) ruthenium(II), might be possible to establish a photochemical water oxidation if a highly active catalyst could be developed.

The project by Sun et al. has been extended to photochemical production of H_2 by a Ru(II) photosensitizer covalently linked to a proton reduction catalyst of dinuclear Fe carbonyl complex [128, 129] (Fig. 25, left). The researchers synthesized (but did not isolate) a Ru(II) terpy complex connected to an $[Fe_2(\mu-SCH_2N(Ph)CH_2S)(CO)_6]$ complex by an acetylene spacer [129] (Fig. 25, right). However, the photoproduction of H_2 has not yet been reported.

Aida and coworkers synthesized water-soluble poly(phenyleneethynylene) wrapped with poly(benzyl ether) dendrimer frameworks bearing negatively exterior surfaces, as a photosensitizer [130]. The wrapping of the conjugated backbone suppressed self-quenching of the excited state. The negatively charged surface incorporated MV^{2+} to form a spatially separated donor–acceptor supramolecular complex, and the quenching rate constant by MV^{2+} was 1.5×10^{15} M^{-1} s^{-1}. Upon excitation of the poly(phenyleneethynylene) wrapped in dendrimer frameworks in the presence of MV^{2+} and triethanolamine (sacrificial electron donor) and Pt colloids, H_2 evolution took place with an overall efficiency of 13%, which is one order of magnitude better than precedent examples. The spatial isolation of the conjugated backbone and its long-range π-electronic conjugation, along with electrostatic interactions on the exterior surface, are considered to play important roles in achieving efficient photosensitized proton reduction.

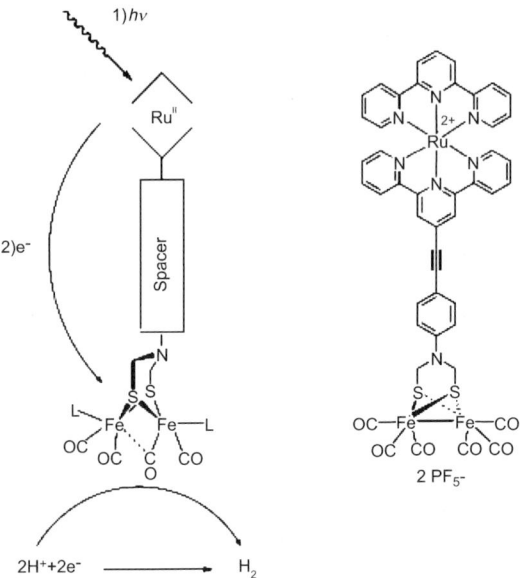

Fig. 25 Schematic representation of the project for photochemical reduction of proton (reprinted with permission from WILEY-VCH [129])

5
Conclusion and Future Directions

CT and catalysis by molecules confined in polymeric materials and their application to future nanodevices for energy conversion have been reviewed, including rather fundamental aspects of the role of polymer matrixes in the functioning of molecules. Functional molecules have been attracting a great deal of attention as key materials for future nanotechnologies. The design and development of superior functional molecules will become more significant in the future. Basic and applied research related to molecular catalysts for fundamental reaction toward energy conversion, water oxidation, reduction of protons, and carbon dioxide should be promoted as urgent and crucial scientific topics. Using the same techniques that are used to assemble and organize molecules, many functional molecules will be explored to create efficient nanodevices. To create nanodevices in the future, the use of polymeric materials is undoubtedly one of the most important research directions. The mechanistic insight of CT in a polymer film confining functional molecules could allow for the design of efficient functional polymeric films. This review has also introduced polymeric solid materials with physicochemical properties that are very similar to physicochemical properties relevant to charge and mass transport as in solution. The materials are expected to be applied to various electronic and photoelectronic devices, and a successful application to a dye-sensitized solar cell has been described here. The significant development of new functional molecules and its nanotechnology based on polymeric solid materials could open the door to innovative devices in the 21st century.

References

1. Bard AJ, Fox MA (1995) Acc Chem Res 28:141
2. Graetzel M, Kalyanasundaram K (1994) Curr Sci 66:706
3. Meyer GJ (ed) (1997) Molecular Level Artificial Photosynthetic Materials Progress in Inorganic Chemistry. Wiley-Interscience, New York
4. Kaneko M (1997) Photoelectric conversion by polymeric and organic materials. In: Nalwa HS (ed) Handbook of Organic Conductive Molecules and Polymers. Wiley, Chichester, p 661
5. Yagi M, Kaneko M (2001) Chem Rev 101:21
6. Abe T, Kaneko M (2003) Prog Polym Sci 28:1441
7. Murray RW (ed) (1992) Techniques of Chemistry: Molecular Design of Electrode Surfaces. Wiley, New York
8. Bard AJ (ed) (1996) Electroanalytical Chemistry. Marcel Dekker, New York
9. Zen J-M, Kumar AS (2001) Acc Chem Res 34:772
10. Yagi M, Kinoshita K, Kaneko M (1996) J Phys Chem 100:11098
11. Yagi M, Kinoshita K, Kaneko M (1997) J Phys Chem B 101:3957
12. Torres GR, Dupart E, Mingotaud C, Ravaine S (2000) J Phys Chem B 104:9487

13. Kaneko M, Woehrle D (1988) Adv Polym Sci 84:141
14. Li W, Osora H, Otero L, Duncan DC, Fox MA (1998) J Phys Chem A 102:5333
15. Kaneko M (2001) Prog Polym Sci 26:1101
16. Majda M, Faulkner LR (1984) J Electroanal Chem 169:77
17. Buttry DA, Anson FC (1981) J Electroanal Chem 130:333
18. Blauch DN, Saveant JM (1992) J Am Chem Soc 114:3323
19. Blauch DN, Saveant JM (1993) J Phys Chem 97:6444
20. White HS, Leddy J, Bard AJ (1982) J Am Chem Soc 104:4811
21. Leddy J, Bard AJ (1985) J Electroanal Chem 189:203
22. Oyama N, Ohsaka T, Kaneko M, Sato K, Matsuda H (1983) J Am Chem Soc 105:6003
23. Martin CR, Rubinstein I, Bard AJ (1982) J Am Chem Soc 104:4817
24. Anson FC, Saveant JM, Shigehara K (1983) J Phys Chem 87:214
25. Anson FC, Saveant JM, Shigehara K (1983) J Am Chem Soc 105:1096
26. He P, Chen X (1988) J Electroanal Chem 256:353
27. Yagi M, Mitsumoto T, Kaneko M (1997) J Electroanal Chem 437:219
28. Yagi M, Mitsumoto T, Kaneko M (1998) J Electroanal Chem 448:131
29. Inzelt G, Day RW, Kinstle JF, Chambers JQ (1983) J Phys Chem 87:4592
30. Inzelt G, Chambers JQ, Kinstle JF, Day RW (1984) J Am Chem Soc 106:3396
31. Majda M, Faulkner LR (1984) J Electroanal Chem 169:97
32. Buttry DA, Anson FC (1982) J Am Chem Soc 104:4824
33. Yagi M, Sato T (2003) J Phys Chem B 107:4975
34. Yagi M, Fukiya H, Kaneko T, Aoki T, Oikawa E, Kaneko M (2000) J Electroanal Chem 481:69
35. Yagi M, Yamase K, Kaneko M (1999) J Electroanal Chem 476:159
36. Yagi M, Yamase K, Kaneko M (2002) Electrochim Acta 47:2019
37. Zhang J, Abe T, Kaneko M (1997) J Electroanal Chem 438:133
38. Zhao F, Zhang J, Abe T, Kaneko M (1999) J Porphyrins Phthalocyanines 3:238
39. Yagi M, Takahashi M, Teraguchi M, Kaneko T, Aoki T (2003) J Phys Chem B 107:12662
40. Yagi M, Nagai K, Kira A, Kaneko M (1995) J Electroanal Chem 394:169
41. Zhang J, Zhao F, Abe T, Kaneko M (1999) Electrochim Acta 45:399
42. Yagi M, Nagai K, Onikubo T, Kaneko M (1995) J Electroanal Chem 383:61
43. Zhang J, Yagi M, Hou XH, Kaneko M (1996) J Electroanal Chem 412:159
44. Kinoshita K, Yagi M, Kaneko M (1999) Electrochim Acta 44:1771
45. Yagi M, Kinoshita K, Nagoshi K, Kaneko M (1998) Electrochim Acta 43:3277
46. Osada Y, Kajiwara K (eds) (1997) Gel Handbook. NTS, Tokyo
47. Kaneko M, Mochizuki N, Suzuki K, Shiroishi H, Ishikawa K (2002) Chem Lett 31:530
48. Ueno H, Kaneko M (2004) J Electroanal Chem 568:87
49. Mochizuki N, Ueno H, Kaneko M (2004) Electrochim Acta 49:4143
50. Ueno H, Endo Y, Kaburagi Y, Kaneko M (2004) J Electroanal Chem 570:95
51. Kaneko M, Gokan N, Takato K (2004) Chem Lett 33:686
52. O'Regan B, Graetzel M (1991) Nature 353:737
53. Kubo W, Kitamura T, Hanabusa K, Wada Y, Yanagida S (2002) Chem Commun 374
54. Mikoshiba S, Sumino H, Yonetsu M, Hayase S (2000) 13th International Conference on Photochemical Conversion and Storage of Solar Energy. Snowmass,CO, 2000:W6-70
55. Murakami TN, Kijitori Y, Kawashima N, Miyasaka T (2004) J Photochem Photobio A Chem 164:187
56. Kaneko M, Hoshi T (2003) Chem Lett 32:872
57. Kaneko M, Hoshi T, Kaburagi Y, Ueno H (2004) J Electroanal Chem 572:21

58. Arnott S, Fulmer A, ScottI WE, Dea CM, Moorhouse R, Rees DA (1974) J Mol Biol 90:269
59. Uzuhashi Y, Nishinari K (2003) FFI J 208:791
60. Ikeda S (2003) FFI J 208:801
61. Crumbliss AL, Perine SC, Edwards AK, Rillema DP (1992) J Phys Chem 96:1388
62. Oyama N, Anson FC (1979) J Am Chem Soc 101:3450
63. Sende JAR, Arana CR, Hernandez L, Potts KF, Keshevarz-K M, Abruna HD (1995) Inorg Chem 34:3339
64. Zhang J, Zhao F, Kaneko M (2000) J Porphyrins Phthalocyanines 4:65
65. Watanabe M, Nagasaka H, Ogata N (1995) J Phys Chem 99:12294
66. Williams ME, Murray RW (1999) J Phys Chem 103:10221
67. Zamponi S, Kijak AM, Sommer AJ, Marassi R, Kulesza PJ, Cox JA (2002) J Solid State Electrochem 6:528
68. Lev O, Wu Z, Bharathi S, Glezer V, Modestov A, Gun J, Rabinovich L, Sampath S (1997) Chem Mater 9:2354
69. Alber KS, Cox JA, Kulesza PJ (1997) Electroanalysis 9:97
70. Lee M-H, Kim Y-T (1999) Electrochem Solid State Lett 2:72
71. Zouni A, Jordan R, Schlodder E, Fromme P, Witt HT (2000) Biochim Biophys Acta Bioenerg 1457:103
72. Ferreira KN, Iverson TM, Maghlaoui K, Barber J, Iwata S (2004) Science 303:1831
73. Limburg J, Vrettos JS, Liable-Sands LM, Rheingold AL, Crabtree RH, Brudvig GW (1999) Science 283:1524
74. Collomb MN, Deronzier A, Richardot A, Pecaut J (1999) New J Chem 23:351
75. Limburg J, Vrettos JS, Chen HY, de Paula JC, Crabtree RH, Brudvig GW (2001) J Am Chem Soc 123:423
76. Lister MW, Petterson RC (1962) Can J Chem 40:729
77. Yagi M, Narita K (2004) J Am Chem Soc 126:8084
78. Narita K (2004) Studies on catalytic activity for water oxidation induced by adsorption of di μ-O dinuclear manganese complex onto heterogeneous matrixes. Master's thesis, Niigata University
79. Naruta Y, Sasayama M, Sasaki T (1994) Angew Chem Int Ed Engl 33:1839
80. Shimazaki Y, Nagano T, Takesue H, Ye B-H, Tani F, Naruta Y (2004) Angew Chem Int Edit 43:98
81. Yagi M, Tokita S, Nagoshi K, Ogino I, Kaneko M (1996) J Chem Soc Faraday Trans 92:2457
82. Yagi M, Nagoshi K, Kaneko M (1997) J Phys Chem B 101:5143
83. Kinoshita K, Yagi M, Kaneko M (1998) Macromolecules 31:6042
84. Nagoshi K, Yagi M, Kaneko M (2000) Bull Chem Soc Jpn 73:2193
85. Yagi M, Sukegawa N, Kasamastu M, Kaneko M (1999) J Phys Chem B 103:2151
86. Yagi M, Sukegawa N, Kaneko M (2000) J Phys Chem B 104:4111
87. Kinoshita K, Yagi M, Kaneko M (1999) J Mol Catal A Chem 142:1
88. Yagi M, Osawa Y, Sukegawa N, Kaneko M (1999) Langmuir 15:7406
89. Ogino I, Nagoshi K, Yagi M, Kaneko M (1996) J Chem Soc Faraday Trans 92:3431
90. Yagi M, Ogino I, Miura A, Kurimura Y, Kaneko M (1995) Chem Lett 863
91. Gilbert JA, Eggleston DS, Murphy WR, Geselowitz DA, Gersten SW, Hodgson DJ, Meyer TJ (1985) J Am Chem Soc 107:3855
92. Gersten SW, Samuels GJ, Meyer TJ (1982) J Am Chem Soc 104:4029
93. Rotzinger FP, Munavalli S, Comte P, Hurst JK, Graetzel M, Pern F-J, Frank AJ (1987) J Am Chem Soc 109:6619
94. Comte P, Nazeeruddin MK, Rotzinger FP, Frank AJ, Graetzel M (1989) J Mol Catal 52:63

95. Meyer TJ, Huynh MHV (2003) Inorg Chem 42:8140
96. Ruettinger W, Dismukes GC (1997) Chem Rev 97:1
97. Chronister CW, Binstead RA, Ni JF, Meyer TJ (1997) Inorg Chem 36:3814
98. Geselowitz D, Meyer TJ (1990) Inorg Chem 29:3892
99. Binstead RA, Chronister CW, Ni JF, Hartshorn CM, Meyer TJ (2000) J Am Chem Soc 122:8464
100. Nagoshi K, Yamashita S, Yagi M, Kaneko M (1999) J Mol Catal A Chem 144:71
101. Lei Y, Hurst JK (1994) Inorg Chim Acta 226:179
102. Yamada H, Hurst JK (2000) J Am Chem Soc 122:5303
103. Yamada H, Siems WF, Koike T, Hurst JK (2004) J Am Chem Soc 126:9786
104. Sens C, Romero I, Rodriguez M, Llobet A, Parella T, Benet-Buchholz J (2004) J Am Chem Soc 126:7798
105. Wada T, Tsuge K, Tanaka K (2000) Angew Chem Int Edit 39:1479
106. Wada T, Tsuge K, Tanaka K (2001) Inorg Chem 40:329
107. Cammack R (1999) Nature 397:214
108. Peters JW, Lanzilotta WN, Lemon BJ, Seefeldt LC (1998) Science 282:1853
109. Nicolet Y, Piras C, Legrand P, Hatchikian CE, Fontecilla-Camps JC (1999) Structure 7:13
110. Volbeda A, Garcin E, Piras C, Lacey ALd, Fernandez VM, Hatchikian EC, Frey M, Fontecilla-Camps JC (1996) J Am Chem Soc 118:12989
111. Schmidt M, Contakes SM, Rauchfuss TB (1999) J Am Chem Soc 121:9736
112. Gloaguen F, Lawrence JD, Rauchfuss TB (2001) J Am Chem Soc 123:9476
113. Ott S, Kritikos M, Akermark B, Sun L, Lomoth R (2004) Angew Chem Int Edit 43:1006
114. Abe T, Taguchi F, Imaya H, Zhao F, Zhang J, Kaneko M (1998) Polym Adv Technol 9:559
115. Taguchi F, Abe T, Kaneko M (1999) J Mol Catal A Chem 140:41
116. Ayers WM (ed) (1988) Catalytic Activation of Carbon Dioxide. American Chemical Society, New York
117. Collomb-Dunand-Sauthier M-N, Deronzier A, Ziessel R (1994) Inorg Chem 33:2961
118. Premkumar J, Ramaraj R (1997) J Photochem Photobio A Chem 110:53
119. Abe T, Yoshida T, Tokita S, Taguchi F, Imaya H, Kaneko M (1996) J Electroanal Chem 412:125
120. Kaneko M, Okura I (eds) (2002) Photocatalysis – Science and Technology. Kodansha-Springer, Tokyo
121. Dai Q, Rabani J (2001) Chem Commun 2142
122. Kaneko M, Nomura T, Sasaki C (2003) Macromol Rapid Commun 24:444
123. Hoshi T, Nomura T, Shiroishi H, Kaneko M (2004) J Appl Electrochem 33:1239
124. Sun LC, Berglund H, Davydov R, Norrby T, Hammarstrom L, Korall P, Borje A, Philouze C, Berg K, Tran A, Andersson M, Stenhagen G, Martensson J, Almgren M, Styring S, Akermark B (1997) J Am Chem Soc 119:6996
125. Sun LC, Hammarstrom L, Norrby T, Berglund H, Davydov R, Andersson M, Borje A, Korall P, Philouze C, Almgren M, Styring S, Akermark B (1997) Chem Commun 607
126. Burdinski D, Bothe E, Wieghardt K (2000) Inorg Chem 39:105
127. Burdinski D, Wieghardt K, Steenken S (1999) J Am Chem Soc 121:10781
128. Wolpher H, Borgstrom M, Hammarstrom L, Bergquist J, Sundstrom V, Styring S, Sun L, Akermark B (2003) Inorg Chem Commun 6:989
129. Ott S, Kritikos M, Akermark B, Sun L (2003) Angew Chem Int Edit 42:3285
130. Jiang D-L, Choi C-K, Honda K, Li W-S, Yuzawa T, Aida T (2004) J Am Chem Soc 126:12084

Editor: Akihiro Abe

Conducting Polymer Nanomaterials and Their Applications

Jyongsik Jang

Hyperstructured Organic Materials Research Center
and School of Chemical and Biological Engineering, Seoul National University,
Shinlimdong 56-1, 151-742 Seoul, Korea
jsjang@plaza.snu.ac.kr

1	Introduction	192
2	Generic Synthesis of Conducting Polymers	195
2.1	Introduction	195
2.2	Chemical Oxidation Polymerization	195
2.3	Electrochemical Polymerization	198
3	Fabrication of Conducting Polymer Nanomaterials	199
3.1	Introduction	199
3.2	Soft Template Method	199
3.3	Hard Template Method	203
3.4	Template-Free Method	205
4	Polypyrrole Nanomaterials	207
4.1	Nanoparticle	207
4.2	Core-Shell Nanomaterial	209
4.3	Hollow Nanosphere	212
4.4	Nanofiber	213
4.5	Nanotube	215
4.6	Thin Film and Nanopattern	216
4.7	Nanocomposite	217
5	Polyaniline Nanomaterials	218
5.1	Nanoparticle	218
5.2	Core-Shell Nanomaterial	219
5.3	Hollow Nanosphere	220
5.4	Nanofiber and Nanorod	221
5.5	Nanotube	222
5.6	Thin Film and Nanopattern	223
5.7	Nanocomposite	224
6	Polythiophene Nanomaterials	225
6.1	Nanoparticle	226
6.2	Hollow Nanosphere	226
6.3	Nanofiber and Nanowire	227
6.4	Nanotube	227
6.5	Thin Film	228
6.6	Nanocomposite and Nanohybrid	229

7	**Poly(3,4-ethylenedioxythiophene) Nanomaterials**	230
7.1	Nanoparticle	230
7.2	Nanofiber, Nanorod, and Nanotube	232
8	**Poly(*p*-Phenylene Vinylene)**	234
8.1	Nanoparticle	235
8.2	Nanofiber, Nanoribbon, and Nanotube	235
9	**Applications**	236
9.1	Chemical Sensor and Biosensor	236
9.2	Transistor and Switch	240
9.3	Data Storage	241
9.4	Supercapacitor	242
9.5	Photovoltaic Cell	243
9.6	Electrochromic Device	243
9.7	Field Emission Display	244
9.8	Actuator	244
9.9	Optically Transparent Conducting Material	245
9.10	Surface Protection	245
9.11	Substituent for Carbon Nanomaterial	246
10	**Conclusions**	247
References		248

Abstract A paradigm shift takes place in the fabrication of conducting polymers from bulky features with microsize to ultrafine features with nanometer range. Novel conducting polymer nanomaterials require the potential to control synthetic approaches of conducting polymer on molecular and atomic levels. In this article, the synthetic methodology of conducting polymer has been briefly considered with chemical oxidation polymerization and electrochemical polymerization. The recent achievements in the fabrication of conducting polymer nanomaterials have been extensively reviewed with respect to soft template method, hard template method and template-free method. It also details the morphological spectrum of conducting polymer nanomaterials such as nanoparticle, core-shell nanomaterial, hollow nanosphere, nanofiber/nanorod, nanotube, thin film and nanopattern and nanocomposite. In addition, their applications are discussed under nanometer-sized dimension.

Keywords Conducting polymer · Nanomaterial · Poly(3,4-ethylenedioxythiophene) · Polyaniline · Polypyrrole

Abbreviations

AAO	Anodic aluminum oxide
AFM	Atomic force microscopy
AOT	Sodium bis(2-ethylhexyl) sulfosuccinate
APS	Ammonium persulfate
CNT	Carbon nanotube
C-PPV	Poly(*p*-phenylene vinylene) with crown ether substituents
CSA	Camphorsulfonic acid
CVD	Chemical vapor deposition

1-D	One-dimensional
2-D	Two-dimensional
3-D	Three-dimensional
DBSA	Dodecylbenzenesulfonic acid
DeTAB	Decyltrimethylammonium bromide
DTAB	Dodecyltrimethylammonium bromide
E^0	Oxidation potential
EDOT	3,4-ethylenedioxythiophene
EMI	Electromagnetic interference
h	Hour(s)
ICP	Inherently conducting polymer
ITO	Indium tin oxide
LB	Langmuir–Blodgett
LBL	Layer-by-layer
LED	Light emitting diode
min	Minute(s)
mm	Millimeter(s)
μm	Micrometer(s)
MEH-PPV	Poly[2-methoxy-5-(2-ethylhexyloxy)-*p*-phenylenevinylene]
MH-PPV	Poly[2-methoxy-5-(*n*-hexadecyloxy)-*p*-phenylenevinylene]
MIMIC	Micromolding in capillaries
nm	Nanometer(s)
NMP	*N*-methylpyrrolidone
NSA	Naphthalene sulphonic acid
OTAB	Octyltrimethylammonium bromide
p	Para
PADPA	*p*-amino-diphenylamine
PANI	Polyaniline
PC	Polycarbonate
PEDOT	Poly(3,4-ethylenedioxythiophene)
PEO	Poly(ethylene oxide)
P3HT	Poly(3-hexylthiophene)
PL	Photoluminescence
PMMA	Poly(methyl methacrylate)
PPV	Poly(*p*-phenylene vinylene)
PPy	Polypyrrole
PS	Polystyrene
PS-PIAT	Diblock copolymer consisting of 40 styrene and 50 3-(isocyano-1-alanyl-aminoethyl)thiophene
PSS	Poly(sodium 4-styrenesulfonate)
PT	Polythiophene
PVA	Poly(vinyl alcohol)
PVD	Physical vapor deposition
PVG	Polyaniline-porous Vycor glass
PVK	Poly(*N*-vinylcarbazole)
PVP	Poly(vinyl pyrrolidone)
RT	Room temperature
s	Second(s)
SDS	Sodium dodecyl sulfate
SEM	Scanning electron microscopy

SPIAD	Surface polymerization by ion-assisted deposition
T_g	Glass transition temperature
TEM	Transmission electron microscopy
THF	Tetrahydrofuran
TPPS	5-, 10-, 15-, 20-tetrakis(4-sulfonatophenyl)porphyrin
p-TSA	p-toluenesulfonic acid
UV-vis	Ultraviolet-visible
VDP	Vapor deposition polymerization

1
Introduction

From a material viewpoint, the advancement of science and technology provides the smaller and smaller dimensions with higher precision and enhanced performance. Currently, nanotechnology is concerned with fabrication and various applications of functional materials and structures in the range of 1 nm to 100 nm using chemical and physical methods [1–3]. Nanoscale size control of material leads to superior physical and chemical properties with molecular and supermolecular structures. Assembling the nanostructures into the ordered array is often necessary to render them functional and operational. Novel nanostructured materials and devices with the enhanced capabilities can be generated by a combination of nanobuilding units and strategies for assembling them. Nanomaterials include nanoparticle, core-shell nanostructure, hollow nanosphere, nanofiber, nanotube, nanopattern, nanocomposite, and so forth. They are divided into nano-sized metal, inorganic material, semiconductor, biomaterial, oligomer and polymer, etc. The widespread interest in nanostructured materials mainly originates from the fact that their properties (optical, electrical, mechanical and chemical performance) are usually different from those of the bulk materials. These phenomena arise from the quantum chemical effects including quantum confinement and finite size effect as well as the nano-sized filler effect. The ability to selectively tune defects, electronic states, and surface chemistry has motivated the development of a variety of methods to fabricate metallic, inorganic, and polymeric nanomaterials. In the last decades, nanoparticles of inorganic material have been produced on a large scale by employing both physical and chemical methods [4–12]. In addition, there has been considerable progress in the preparation of nanocrystals of metals, semiconductors and magnetic materials using colloid chemical method [13–18]. While metallic and inorganic semiconductor nanomaterials are routinely made, the preparation of polymeric nanomaterials has been unexploited relatively. However, polymer nanomaterials provide a variety of advantages over other materials because they have a wide range of source materials and tunable surface functionalities. Among the synthetic polymers, conducting polymers have attracted considerable attention as important polymer materials since the initial discovery of polyacetylene [$(CH)_2$] in the late 1970s [19–26]. Inherently conducting polymers (ICP) have the conjugated double-bonded backbone that provides the electronic conductivity after doping with suitable dopants. In general, ICPs are semiconductors with versatile properties and various applications. These peculiar characteristics have an impact on the discovery of vast conducting polymers by means of different synthetic methods. Numerous research papers have been published concerning a variety of the synthesis, physical and chemical properties and application of conducting polymers [27–63]. On the other hand, this review has been focused on recent achievements of the fabrication and application of these conducting nanomaterials from the earlier part of this century due to the limited space. Previous reviews are helpful to get the initial information for the reader [64–66].

Herein five conducting polymers such as polypyrrole (PPy), polyaniline (PANI), polythiophene (PT), poly(3,4-ethyelenedioxythiophene) (PEDOT) and poly(p-phenylene vinylene) (PPV) are selected for this review. Although other conducting polymers are also noteworthy, I do not describe them here because I would like to concentrate on the fabrication methods and applications of common conducting polymers that are prepared from simple monomers. Molecular structures of these typical conducting polymers are shown by Table 1.

Table 1 Typical conducting polymer structures (undoped form)

Name	Structure
Polypyrrole	
Polyaniline	
Polythiophene	
Poly(3,4-ethylenedioxythiophene)	
Poly(para-phenylene vinylene)	

Section 2 mainly focuses on generic synthesis of conducting polymers: chemical oxidation and electrochemical polymerization. Although a large amount of work has been devoted to characterize physical and chemical properties of newly developed conducting polymers, this part only describes the information of synthetic methodology of conducting polymers.

Next section covers extensive discussions of various fabrication methods for conducting polymer nanomaterials in detail. This section is divided by the soft template method, hard template method, and template-free method.

In terms of the nanomaterial morphology, specific fabrication methods for five typical conducting polymers (PPy, PANI, PT, PEDOT, PPV) have been reviewed in Sects. 4–8. It deals with nanoparticle, core-shell nanomaterials, hollow nanospheres, nanofibers, nanotubes, nanopatterns, and nanocomposites of each conducting polymer.

Based on conducting polymer nanomaterials, various applications are reviewed in the final section. These applications include chemical sensor and biosensor, transistor and switch, data storage, supercapacitor, photovoltaic cell, electrochromic device, field emission display, actuator, optically transparent conducting material, surface protection, and substituent for carbon nanomaterials (Fig. 1). Because large amounts of research have been dedicated to this field, it is very difficult to cover whole application fields of conducting polymers. Some comprehensive review articles related to applications of conducting polymers are available [67–73].

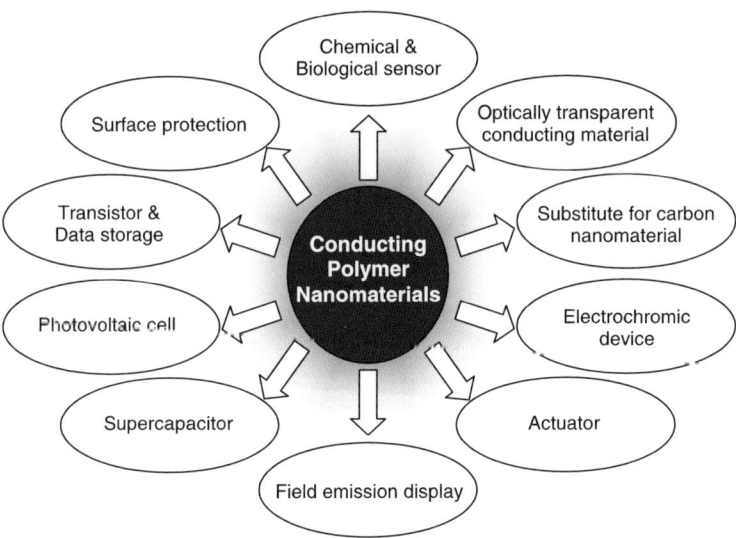

Fig. 1 Application fields of conducting polymer nanomaterials

2
Generic Synthesis of Conducting Polymers

2.1
Introduction

Conducting polymers are categorized as the cationic and anionic salts of highly conjugated polymers. Whereas the cation salts are obtained by chemical oxidation and electrochemical polymerization, it is also possible to produce the anion salts of the highly conjugated polymers using electrochemical reduction or chemical reduction with reagents such as sodium naphthalide. An oxidized conducting polymer has electrons removed from the backbone, resulting in a cationic radical [74]. On the other hand, a reduced conducting polymer has electrons added to the backbone, resulting in an anionic radical. However, this reduced conducting polymer is much less stable than the cation counterpart. In general, there are two synthetic methods of conducting polymers: chemical polymerization and electrochemical polymerization.

Chemical oxidation polymerization has been desirable for mass production. However, it is necessary to restrict by-products and want materials to keep the global environment clean and safe [75]. This paper mainly describes the chemical oxidation polymerization of simple monomers using oxidizing agents to produce large quantities of conducting polymers. Electrochemical polymerization is useful particularly when thin films are desired. Both methods should be considered separately.

2.2
Chemical Oxidation Polymerization

Typically, major chemical polymerization of conducting polymers represents the addition polymerization and differs from electrochemical polymerization in a basic mechanism [76–81].

Scheme 1 describes a typical mechanism for chemical oxidation polymerization of pyrrole. In the initiation step, radical cations ($C_4NH_5^+$) are generated by the oxidation of pyrrole monomer. A radical-radical coupling occurs between two radical cations, and forms a dimer with deprotonation, leading to a bipyrrole. The bipyrrole is reoxidized and couples with other radical cations. This process is repeated consecutively during the propagation step. The termination takes place due to the nucleophilic attack of water molecules or impurities in the polymer chains.

Scheme 2 represents the possible chemical structure of PPy molecules during chemical oxidation polymerization. In the case of monomer-monomer coupling reaction, protons in the α-position are more readily eliminated than those in the β-position. The α,α coupling makes a major contribution to coupling reaction and leads to a linear PPy. This α,α coupling of alternat-

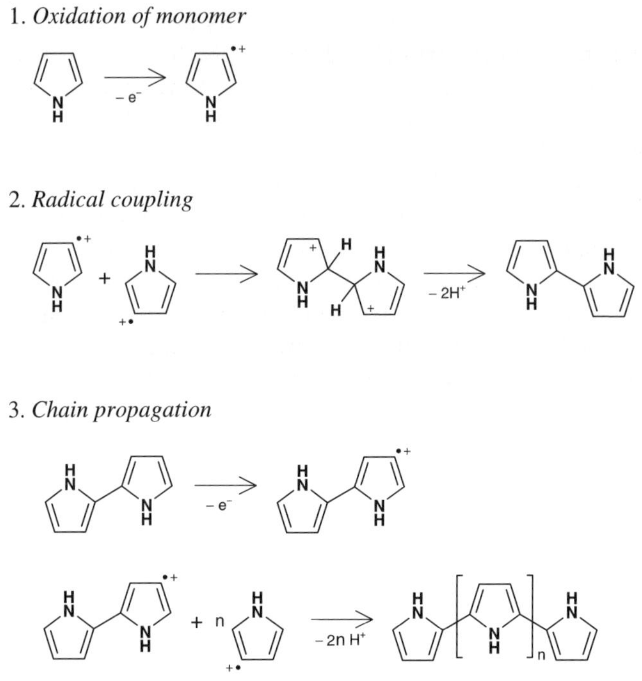

1. *Oxidation of monomer*

2. *Radical coupling*

3. *Chain propagation*

Scheme 1 Polymerization mechanism of pyrrole

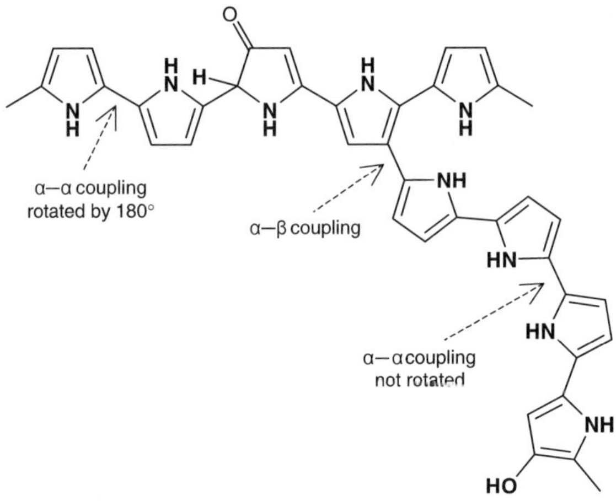

Scheme 2 Possible chemical structures in PPy chains

ing pyrrole units provides a high degree of conjugation. On the other hand, α,β coupling occurs to the same extent and forms a crosslinked PPy chain, breaking the linearity and planarity of PPy chains. Owing to these, PPy is insoluble in common solvents [88]. It is known that PPy has four different electronic band structures with changing oxidation levels. In the neutral state,

PPy shows a large $\pi - \pi^*$ band gap of 3.2 eV. As one electron is removed to form a polaron (a radical cation, 1/2 spin), two localized electronic levels (bonding and antibonding cation levels) appear in the band gap. Further electron removal generates a double charged spinless bipolaron (a dication). At a higher oxidation level (a doping level of ca. 33%), the overlap between bipolarons can be occurred, leading the formation of two wide bipolaron bands in the gap [89]. In aqueous polymerization, over-oxidation produces a small portion of carbonyl and hydroxyl groups into the PPy chains [90].

Chemical polymerization of aniline has been carried out in an acidic solution [91–101]. This acidic condition provides the solubilization of the monomers as well as the formation of emeraldine salt as a conducting PANI. Scheme 3 depicts a chemical polymerization mechanism of aniline in acidic

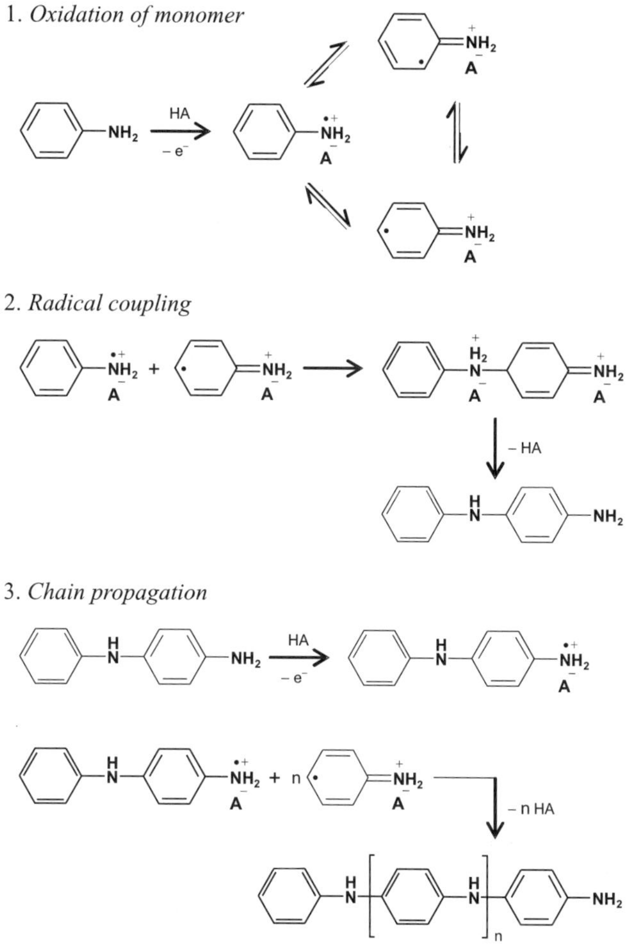

Scheme 3 Polymerization mechanism of aniline

solution using an oxidizing agent. Aniline monomer forms the anilinium ion in acidic medium and chemical polymerization results in the formation of protonated, partially oxidized form of PANI. The initial step involves formation of the aniline radical cation. The next step is followed by the coupling of N- and *para*-radical cations with consecutive rearomatization of the dication of *p*-amino-diphenylamine (PADPA). The oxidation process of the diradical dication makes the fully oxidized pernigraniline salt form of PANI due to the high oxidizing power of the oxidant such as ammonium persulfate (APS) (($NH_4)_2S_2O_8$). Whereas "head-to-tail" coupling is predominant, some coupling in the *ortho*-position also occurs, leading to conjugation defects in the final product. After consuming all the oxidant, the unreacted aniline monomer reduces the pernigraniline to produce the green emeraldine salt as the resultant polymer. From the viewpoint of color change in solution, the formation of PADPA reflects pink, and the formation of protonated pernigraniline becomes deep green. Green emeraldine salt precipitates after reduction of pernigraniline in the final step. In the termination step, green emeraldine salt forms as a conductive PANI form and it can be converted to emeraldine base with an alkaline solution or an excess of water. The emeraldine base can be transformed to two non-conductive PANI such as the completely oxidized pernigraniline ($m = 0$) and reduced leucoemeraldine ($m = 1$) depending on oxidation states. Imine sites of the emeraldine base are easily protonated in acid condition, which results in the formation of green emeraldine salt as a conducting PANI [96]. The conductivity of PANI was affected by degree of protonation and oxidation. In addition, the structural and conformational factors derived from polymerization condition also affect the conductivity of PANI.

2.3
Electrochemical Polymerization

Compared with chemical oxidation polymerization, electrochemical polymerization is performed at an electrode (conductive substrate) using the positive potential [97–104]. Whereas the powder forms are obtained by chemical polymerization, the electrochemical method leads to films deposited on the anode. When a positive potential is applied at the electrode, pyrrole monomer such as a heterocyclic compound is oxidized to form a delocalized radical cation, which includes the possible resonance forms. Radical-radical coupling reaction produces the dimerization of the monomer radicals at the α-position. Removal of $2H^+$ ions consequently forms the neutral dimer. Next step is chain propagation which includes the oxidation of the neutral dimer to form the dimer radical. The resultant radical can react with other monomer or dimer and this radical coupling and the electrochemical oxidation processes repeat in order to extend the polymer chain. The final step involves the termination of chain growth and the resultant PPy film is formed on the anodic electrode.

In the case of aniline electropolymerization, the radical cation of aniline monomer is formed on the electrode surface by oxidation of the monomer [100, 105–114]. This process is considered to be the rate-determining step. Radical coupling and elimination of two protons make mainly *para*-formed dimers. Chain propagation proceeds with oxidation of the dimer and aniline monomer on the electrode surface. In this step, the radical cation of the oligomer couples with a radical cation of aniline monomer. In the final step, PANI is doped by the acid (HA) present in solution. The growth of PANI has been considered to be self-catalyzed. This means that the polymers are formed at the higher rate as the more monomers are deposited onto the polymer surface. It involves the adsorption of the anilinium ion onto the oxidized form of PANI, followed by electron transfer to form the radical cation and subsequent reoxidation of the polymer to its most oxidized state.

3
Fabrication of Conducting Polymer Nanomaterials

3.1
Introduction

A variety of fabrication methods have been developed for conducting polymer nanomaterials. Among the various synthetic strategies, template method is a very promising and powerful tool to fabricate conducting polymer nanomaterials. Template method involves the inclusion of guests such as inorganic or organic constituents inside the void spaces of a host material. These voids act as the template, deforming the shape, size, and orientation of the compound produced.

In general, template method is classified by soft and hard templates. Whereas anodic aluminum oxide (AAO) membrane, track-etched polycarbonate (PC) and zeolite can be used as hard templates, soft templates include surfactant, cyclodextrin, liquid crystal, etc. Compared with soft and hard templates, template-free method represents the fabrication technique of conducting polymer nanomaterials without the template, which is discussed in this section [115].

3.2
Soft Template Method

Recently, soft template method has been used for the fabrication of various morphologies of polymer nanomaterials. There are several soft templates such as surfactant, liquid crystalline polymer, cyclodextrin, and functionalized polymer [100, 101, 116–122]. Among them, surfactants, which imply cationic, anionic and non-ionic amphiphiles, are mostly used for the formation of micelle as a nanoreactor [123–129]. Microemulsions are macroscopically homogeneous mixtures of oil, water and surfactant, which on the microscopic level consist of individual domains of oil and water separated by a monolayer of amphiphile [130]. Micelle formations in microemulsion are represented in Scheme 4. Microemulsions should not be regarded as emulsions with very small droplet size; micro- and macroemulsions are fundamentally different. Macroemulsions mean conventional emulsions. Whereas emulsions are inherently unstable systems in which the droplets eventually will undergo coalescence, microemulsions are thermodynamically stable with a very high degree of dynamics with regard to the internal structure. In emulsion, phase separation is rapid unless the system is well mixed. Droplets continuously collide and coalesce, and are broken by the shear exerted on the system. The droplet size is dependent on the system components (oil, stabilizer, phase ratio) and the mixing characteristics. On the other hand, microemulsions are thermodynamically stable (i.e., indefinitely stable) with droplet sizes varying from 10 to 100 nm. Relatively large quantities of mixed emulsifiers typically consisting of an ionic surfactant (e.g., sodium dodecyl sulfate (SDS)) and a short chain alcohol are usually used to prepare these emulsions [131]. During the polymerization, in a conventional emulsion polymerization, the monomer is located in the following four locations: (1) monomer droplets; (2) inactive monomer swollen micelles; (3) active micelles that become monomer-swollen polymer particles where the polymerization occurs; (4) solute monomer in an aqueous phase. Two characteristics of oil-in-water microemulsion polymerization are different from those of conventional emulsion polymerization: (1) no monomer droplets and no inactive micelles exist; (2) the system is optically transparent.

Microemulsions act as attractive media for polymerization reactions. Microemulsion polymerization is a novel fabrication technique which allows the preparation of ultrafine latex particles within the size range from 10 nm

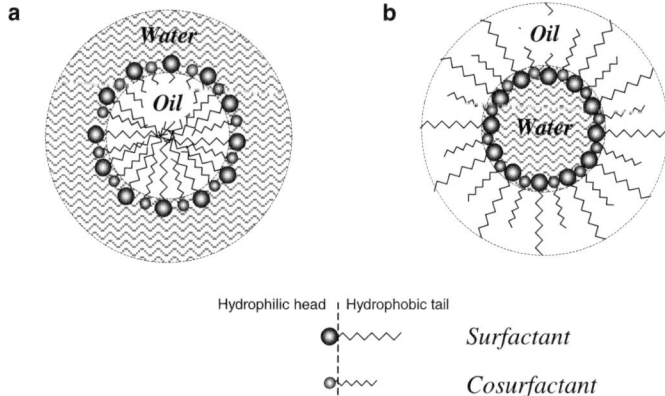

Scheme 4 Schematic illustrations of **a** oil-in-water and **b** water-in-oil micelles

to 100 nm and with narrow size distribution. However, the formulation of microemulsions is subject to severe constraints due to high emulsifier level (over 10 wt %) needed for achieving their thermodynamic stability [132, 133].

Miniemulsion systems are somewhere in between macro- and microsystems. They contain both micelles and monomer droplets, but the monomer droplets are smaller than in macrosystems [134–137]. For both micro- and minireaction systems in which the initiator is soluble in the continuous phase, the mechanism for polymerization is determined by the relative surface areas of micelles versus monomer droplets. Compared with the miniemulsion (5–10 wt % of surfactant used), high concentration (15–30 wt %) of surfactant forms robust and compact micelle, and the inner space of micelle can be used as a nanoreactor. Besides sphere and layer morphologies, a wide range of morphological spectra could be obtained by carefully controlling the synthetic conditions. The surfactant templates for sphere, rod, and layer nanomaterials are schematically represented in Scheme 5.

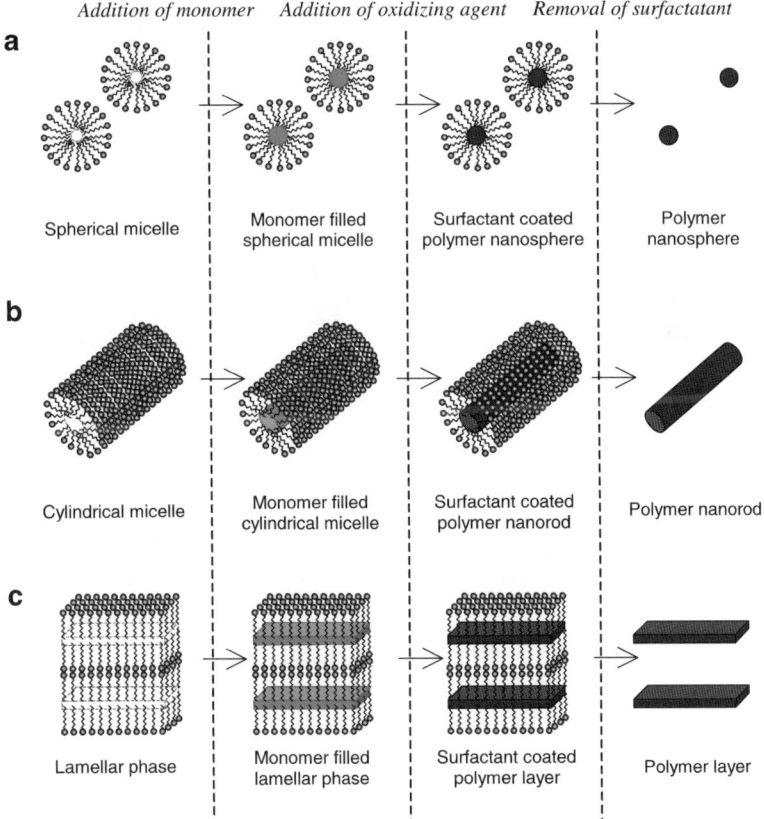

Scheme 5 Schematic illustration for the surfactant templating of **a** PPy nanosphere, **b** nanorod, and **c** layerd assemblies

In general, microemulsion polymerization has been widely accepted for synthesizing conducting polymer nanoparticle, hollow nanosphere, core-shell nanostructures, and nanofibers [138–153]. As shown by Fig. 2, various morphologies of PPy nanomaterials, such as ellipse, hexagon, tetrahedron, rod, needle, and comb shapes were observed in a specific condition. The driving force in determining the morphogenesis is not clear, but it is obvious that the soft templates played an important role in the structural development of PPy nanomaterials.

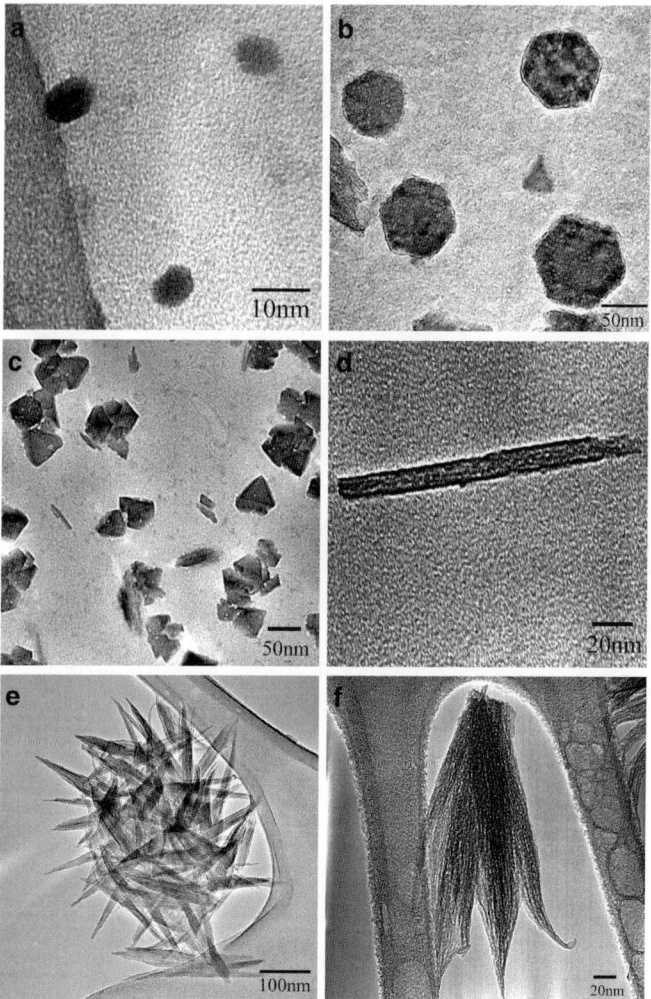

Fig. 2 Various morphologies of PPy nanomaterials synthesized using the template of cationic surfactants: **a** ellipse, **b** hexagon, **c** tetrahedron, **d** rod, **e** needle, and **f** comb-like PPy nanomaterials

However, it is difficult to control the micelle formation during microemulsion polymerization. In general, polymerization process is kinetically and thermodynamically unstable because of Ostwald ripening, the growth by collision between monomer droplets and monomer consumption during polymerization [154, 155]. It is noteworthy that precise control of the micelle is essential to produce monodisperse and nano-sized conducting polymer nanomaterials.

3.3
Hard Template Method

Hard template method has been used for the 1-D nanostructures such as nanotubes, nanorods and nanofibers of conducting polymers. The commonly used templates are AAO membrane, and track-etched PC membrane, whose pore size ranges from 10 nm to 100 µm. Hard template methods for synthesizing conducting polymer nanomaterials have been extensively reviewed in recent years [156–160].

In general, the polymerization of a conducting monomer has been performed at nanochannel as a nanoreactor and hard templates are removed after polymerization in order to fabricate 1-D conducting polymer nanomaterials. When the pore is filled with appropriate material, it generates a self-assembled nanofiber. The membranes are commercially available in fixed sizes with specific pore diameters, and also prepared in the laboratory using electrochemical means. Desired pore length and diameter can be controlled by synthetic parameters. Martin et al. have used hard template method for preparing polymer nanomaterials [161–165]. Especially, nanotubes and nanofibers composed of conducting polymers were fabricated within the pores of nanoporous membranes. They synthesized the nanotubes of PPy, PANI, and poly(3-methylthiophene) with hard templates using chemical oxidation and electrochemical polymerization. During the polymerization process, the conducting polymer preferentially nucleates and grows on the pore walls of membranes. Resultant polymer tubular structures are tuned by polymerization time. Whereas short polymerization time provides the thin wall of conducting polymer nanotube, long polymerization time produced thick walls. In addition, PPy nanotubes and nanofibers were selectively fabricated depending on the polymerization time. Several researchers also focused on hard template method in order to synthesize various conducting polymer nanomaterials [166–171].

Most template methods can be accomplished by simply immersing the hard template into a monomer/oxidant solution. Recently, Jang et al. produced PPy nanotube and carbon nanotube (CNT) using vapor deposition polymerization (VDP) mediated AAO membrane method [172]. An experimental scheme of this method is represented by Scheme 6.

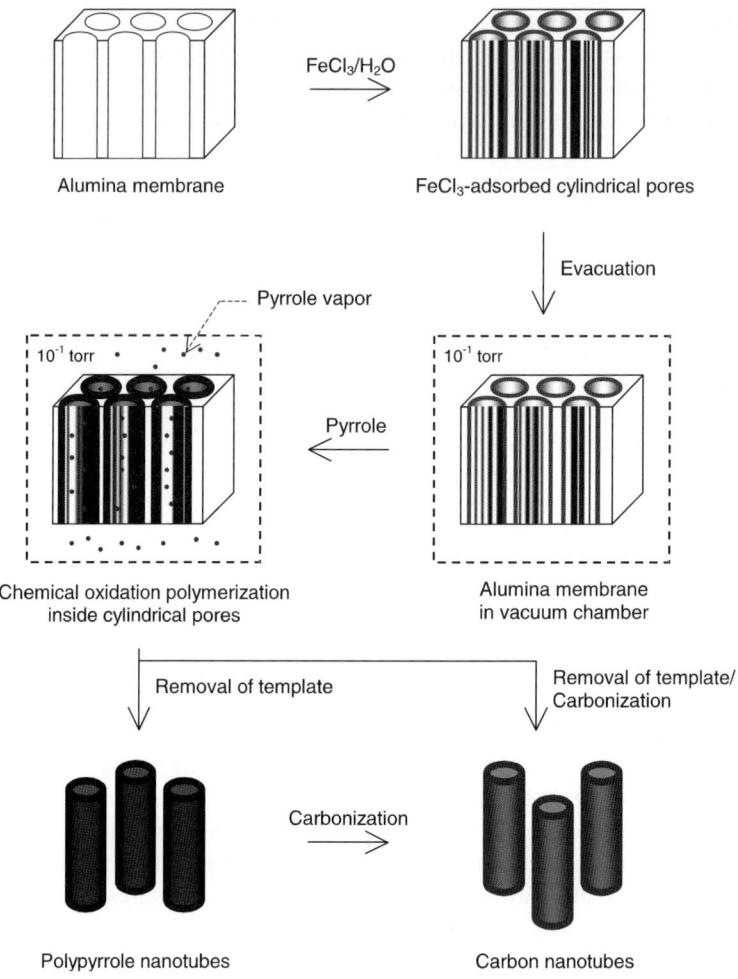

Scheme 6 VDP of PPy nanotubes using AAO membrane

Vapor deposition of inorganic thin films has a vital role in the semiconductor and coating industries. In addition, vapor deposition process has been developed to obtain high quality metal, semiconductor, or ceramic films. The deposition process can be broadly categorized into either physical vapor deposition (PVD) or chemical vapor deposition (CVD). Chemical and physical vapor phase syntheses are well-established technologies for large-scale production of metal, metal oxides, and inorganic nanoparticles. While a chemical reaction occurs in CVD, PVD includes evaporation and sputtering.

In VDP, a monomer or pre-polymer is first deposited, and then polymerization is carried out on the surface via thermal, photochemical, or other initiation process. VDP has distinct advantages over simple deposition of polymer to produce uniform, defect-free films and has the advantage over CVD in that various starting monomers can be employed. Since the thickness of monomer films can be precisely and reproducibly controlled, polymer films from a few nanometers to submicrometers are formed. This approach eliminates the use of organic solvents, so the undesired pinhole defects and cloudiness are no longer issues. There is also a growing interest in eliminating solvent waste from processes due to environmental concerns. Since all processing has been conducted in a well-controlled vacuum environment, any residual monomer and solvent can be precisely managed and recovered. Although the VDP process is experimentally convenient, the nature of polymerization has not been fully explored. Additionally, in this configuration, it is difficult to distinguish the polymers that have been formed in the gas phase with subsequent deposition onto the surface, from those polymerized by monomers adsorbed on the surface.

3.4
Template-Free Method

Template-free techniques have been extensively studied for the fabrication of conducting polymer nanomaterials fabrication. Compared with hard and soft template methods, these methodologies provide a facile and practical route to produce pure, uniform, and high quality nanofibers. Template-free methods encompass various methods such as electrochemical synthesis, chemical polymerization, aqueous/organic interfacial polymerization, radiolytic synthesis, and dispersion polymerization.

First of all, the forerunning work was performed with electrochemical polymerization of PANI micro- and nanorod fabrication [173, 174]. Iridium and platinum needles were used as electrodes, and a PANI microrod was grown on the electrode surface. Another approach was related to the fullerene derivatives doping with electrochemical polymerization in order to fabricate PANI microfibers and nanofibers [175, 176]. In addition, PPy microtubes have been synthesized in the presence of β-naphthalene sulphonic acid (β-NSA) by electrochemical polymerization. This microtube has the diameter of 0.8–2.0 μm and the length of 15–30 μm. Electrochemical deposition of PANI has been also carried out within multi-walled CNT. In addition, PPy microtube (a few mm in length and hundreds μm in diameter) is formed in the presence of Na-toluene sulphonic acid along platinum wire [177].

PANI micro- and nanotubes have been prepared by chemical polymerization using fullerene, NSA derivative, and azobenzene sulphonic acid [178, 179]. PPy microfiber and microtube were also fabricated in the presence of β-NSA and the morphology of conducting polymer could be tuned by controlling the concentration of β-NSA [101].

PANI nanofibers were fabricated under ambient environment using aqueous/organic interfacial polymerization. Despite large-scale production, the diameter of resultant PANI nanofibers ranged from 30 to 50 nm and fiber length varied between 500 nm and several micrometers [115]. Two solutions were prepared before interfacial polymerization of aniline. Aniline monomer was dissolved in an organic solvent (e.g., CCl_4, benzene, toluene, CS_2). The other solution consisted of ammonium peroxydisulfate and camphorsulfonic acid (CSA) in water. After mixing two solutions, green PANI formed at the interface between two layers and consecutively diffused into aqueous phase. Dark-green PANI nanofiber was collected in water phase. Polymerization yields ranged from 6 to 10 wt % and approximately 95 vol % of the sample are PANI nanofiber with the diameter of 30–50 nm.

Bertino and coworkers fabricated PANI nanofibers using radiolytic synthesis as a new template-free pathway. PANI nanofibers were formed in aqueous solutions of aniline, ammonium peroxydisulfate, and hydrochloric acid with gamma ray irradiation. Typical PANI nanofibers had the diameters of 50–100 nm and the length of 1–3 μm [180].

Tremendous research works have been performed on the synthesis of conducting polymer nanomaterials using dispersion polymerization method [181–188]. There are two categories of dispersion polymerization in order to fabricate the conducting polymer colloids. The first approach forms polymer stabilizer coated conducting polymer nanoparticles. In this case, the monomer and oxidant are dissolved in a stabilized liquid medium and the formation of insoluble conducting polymer nanoparticles occurs as the polymerization proceeds.

The other approach involves inorganic oxide conducting polymer core-shell nanostructures. In-situ polymerization of monomer is performed in the presence of silica nanoparticles and the conducting polymer is adsorbed onto the silica nanoparticles as a seed. The resultant silica-conducting polymer core-shell nanomaterial is obtained in the absence of additional polymers or surfactant. The particle size and morphology of silica-conducting polymer can be readily varied by adjusting size of silica, concentration of silica and monomer, and type of oxidant and monomer. In order to synthesize conducting polymer nanoparticle as a core, various water soluble polymers such as poly(vinyl alcohol) (PVA), poly(vinyl methyl ether), cationic and anionic polyelectrolytes have been used as polymeric stabilizers. This water soluble polymer is physisorbed onto the surface of conducting polymer core due to hydrogen bonding in most cases. Many research groups have mainly fabricated PPy, PANI coated silica nanoparticle using dispersion polymerzation. Different inorganic oxides (haematite, ceria and titania) and polymer latex particles such as polyurethane, poly(vinyl acetate) , and alkyl resins are utilized as the core part [189–196].

In recent years, dispersion polymerization method has been developed for the fabrication of conducting polymer nanorods [197]. Compared to conventional polymerization using hydrochloric acid (HCl) as a dopant, 35 wt % of hydrochloric acid was utilized in order to form amphiphilic monomer ions such as anilinium ions. PANI nanorods with average diameter of 21–53 nm and length of 0.5–1.0 μm were produced depending on the type of salt, stirring speed, and reaction temperature.

Dispersion polymerization with APS oxidant produced shorter PANI nanorods compared with $FeCl_3$ oxidant. In addition, magnetic stirring decreased the length of PANI nanorods and also the average diameter of nanorods decreased with increasing polymerization temperature.

Currently, electrospinning method has been recognized as a simple, efficient and versatile technique for the fabrication of conducting polymer nanofibers [198–200]. This electrostatic process involves the application of a high voltage electrostatic field to charge the surface of a conducting polymer solution droplet and the ejection of a liquid jet through a spinneret.

It has been reported that the diameter and shape of conducting polymer nanofibers can be affected by experimental parameters such as solution viscosity and conductivity, surface tension, the kind of polymer and solvent, applied electrical potential and distance between the capillary and collection screen [201, 202].

Some researchers studied the effect of solution viscosity on the formation of PANI-CSA fibers [203, 204]. They found that a stable polymer droplet was formed with adding 2 wt % poly(ethylene oxide) (PEO) into solution, whereas no fiber formation of PANI/CSA occurred in chloroform without PEO addition.

It was also reported that polystyrene (PS) and polyacrylonitrile fibers were coated with PPy or Au from aqueous solution by electrostatic deposition [205].

4
Polypyrrole Nanomaterials

PPy consisting of five-membered heterocylclic rings is one of the most promising conducting polymers. PPy has been extensively explored because of their easy synthesis, tunable conductivity, reversible redox property, and environmental stability. PPy can be easily prepared by chemical or electrochemical polymerization via the oxidation of pyrrole monomers [206]. In general, chemical polymerization leads to intractable powder, whereas electrochemical polymerization results in film deposited on the electrode.

4.1
Nanoparticle

Spherical PPy nanoparticles have been prepared by chemical oxidation polymerization with the aid of surfactant or stabilizer in an aqueous solution. Above all, microemulsion polymerization has been extensively utilized to synthesize various nanometer-sized conducting polymer particles. A typical example is the synthesis of PPy nanoparticles with diameter of several nanometers using low temperature polymerization [146]. Low temperature polymerization is appropriate for reducing the inner space of micelles by virtue of deactivating the chain mobility of the surfactant. Thus PPy nanoparticles as small as 2 nm in diameter could be prepared through chemical oxidation polymerization inside the micelles made of catioinic surfactants at low temperature (Fig. 3). As the polymerization temperature increased, PPy nanoparticle grew as a result of the enhanced chain mobility of the surfactant. In addition, the size of PPy nanoparticle decreased with shortening the chain length of the surfactant. The micelle aggregation number, which is defined as the number of surfactant molecules required to form a micelle, becomes smaller as the chain length of surfactant decreases. The reduced micelle aggregation number gives rise to the formation of smaller particles. On the other hand, the longer surfactant chains provide more free volume inside micelle, which lead to the increment of particle size. Importantly, the thermodynamically stable micelle successfully acted as a nanoreactor in synthesizing PPy nanoparticles.

Another approach to synthesize PPy nanoparticles was performed in a microemulsion system; ferric chloride (FeCl$_3$) was employed as an oxidizing agent, and dodecylbenzenesulfonic acid (DBSA) and butanol were used as a surfactant and a cosurfactant, respectively [207]. It was reported that microemulsion polymerization system increased not only the yield of the resultant PPy but the extent of conjugation length in the polymer as compared with solution and conventional emulsion polymerization.

Surfactant-mediated methodology allows the morphological transition of spherical PPy nanoparticles into structured aggregates. Intriguingly, amorph-

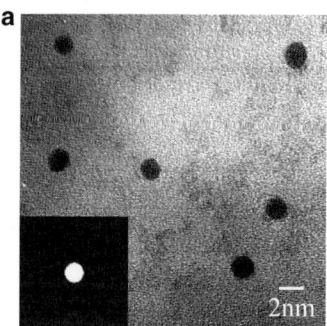 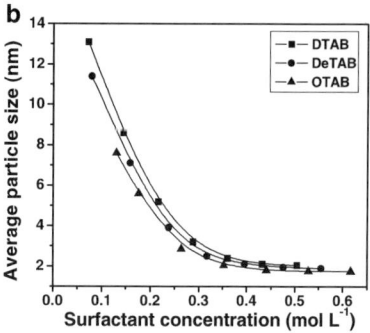

Fig. 3 a TEM image and nanobeam electron diffraction pattern of the PPy nanoparticles prepared using decyltrimethylammonium bromide (DeTAB) (0.4 M) at 3 °C. **b** Average change in nanoparticle size as a function of surfactant concentration. The average nanoparticle size was determined by TEM (50 particles counted) (reproduced with permission from [146])

ous ultrafine PPy nanoparticles associated with one another to form crystalline lamellar structures, and the regularly linked PPy nanoparticles could play the role of crystalline lattices in the supramolecular assembly [208].

Similarly, PPy nanoparticles could form self-assemblies during the evaporation process of the liquid medium [209]. Evaporation-induced self-assemblies with diverse morphologies were selectively fabricated by drying a solution containing PPy nanoparticles. The evaporation rate and nanoparticle concentration are strongly affected on the formation of the PPy assemblies. When the liquid medium rapidly evaporated, a 2-D dendrite or pseudo 1-D dendrite was observed. In the case of slow evaporation, ring and spherulitic assemblies took place on the carbon film grid.

Dispersion polymerization has been also carried out to prepare PPy nanoparticle by several research groups. The synthesis of sterically stabilized PPy colloid was carried out using a tailor-made reactive polymeric stabilizer [210]. In this study, a copolymer stabilizer was formed by free radical copolymerization of thiophene-based vinylic monomer with various hydrophilic vinyl monomers such as 2-(dimethylamino)ethyl methacrylate, 2-vinylpyridine, and N-vinylpyrrolidone. These copolymer stabilizers were grafted onto the surface of PPy nanoparticle during dispersion polymerization and contributed to effective steric stabilization of the nanoparticle. The resulting nanoparticle had the size distribution in the range of 50–100 nm.

Zelenev et al. described the preparation of PPy nanoparticles with the diameter of 20–60 nm by the oxidation of pyrrole with sodium persulfate (oxidizing agent) and 4-ethylbenzenesulfonic acid (doping agent) in the presence of Rhodasurf TB970 (polymeric stabilizer) [211]. Larger nanoparticles were obtained when the concentrations of all components including monomer, stabilizer, doping agent, and oxidizing agent increased.

Recently, PPy nanoparticles with the diameter of 60–90 nm were polymerized with $FeCl_3$ in aqueous solutions containing PVA as a stabilizer [212]. At room temperature (RT), the polymerization of pyrrole occurred at a high rate. When the concentration of pyrrole increased, the resultant PPy nanoparticles became coarser with broadening the particle size distribution. Furthermore, the increase in concentration of PVA resulted in faster polymerization and finer PPy nanoparticles. Such a phenomenon was due to the reinforcement of the structural-mechanical barrier formed by the stabilizer at the surface of the nanoparticle, preventing the growth of PPy nanoparticles during the polymerization process.

4.2
Core-Shell Nanomaterial

To date, versatile PPy-based core-shell nanomaterials have been reported in the literature [213–217]. PPy has been selectively employed as either core or shell materials to realize various interesting properties.

In the case of PPy as the conductive core material, a PPy-poly(methyl methacrylate) (PMMA) core-shell nanosphere with an average diameter of several tens of nanometers was synthesized via a two-step microemulsion polymerization procedure (Fig. 4) [213]. The size of core and shell parts was easily tuned by adjusting the amount of surfactant and monomer. The shell thickness was maintained in nanometer-scale to minimize the loss of electrical conductivity.

Very recently, PPy-poly(N-vinylcarbazole) (PVK) core-shell nanoparticle was synthesized by nanoparticle-seeded dispersion polymerization [214]. PVK has been applied to various optoelectronic devices because of its characteristic optical properties (e.g., fluorescence and electrochromic property). In particular, PVK has a wide bandgap, which can be decorated with dyes, and thus it has been considered as a promising material for advanced electroluminescent devices. N-vinylcarbazole can be polymerized through a vinyl group as well as carbazole ring, and PPy nanoseed composed of α,α'-linked or α,β-linked pyrrole rings provides the initiation sites for the growth of

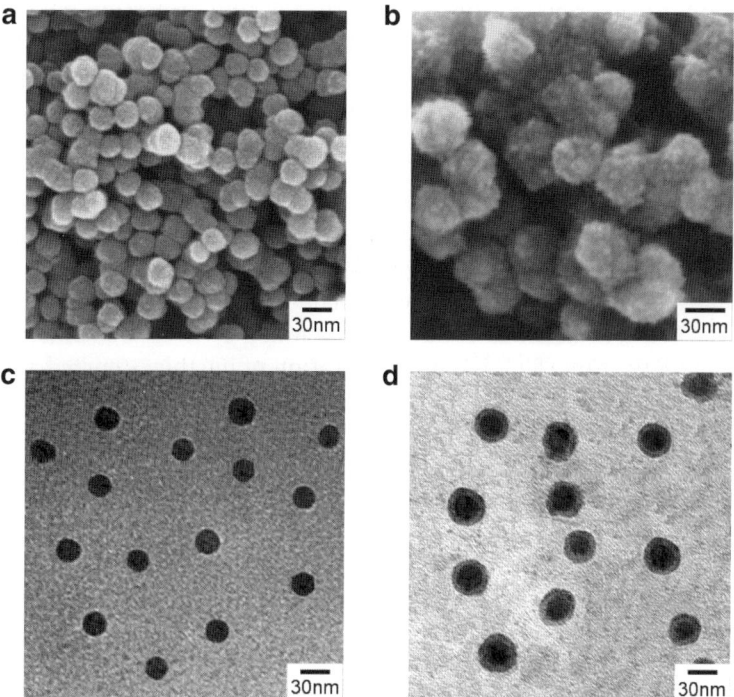

Fig. 4 SEM and TEM images of PPy nanoparticles and PPy/PMMA core/shell nanospheres synthesized using the feeding ratio of pyrrole/MMA of 1/2.5 (by wt%) at 0.21 M of dodecyltrimethylammonium bromide (DTAB): **a,b** SEM images; **c,d** TEM images (reproduced with permission from [213])

PVK branches. As a result, ultrathin and robust a PVK shell was coated onto the surface of a PPy nanoseed. The shell thickness was controlled within the range of a few nanometers. Compared with pure PVK (conductivity, ca. 10^{-10} S cm^{-1}), the core-shell nanoparticle showed superior conductivity (ca. 10^{-1} S cm^{-1}) and similar fluorescence, and revealed only 2–3% decrease in conductivity after 20 days because the shell protected the PPy core from environmental perturbation. Accordingly, this methodology opens up a route to the fabrication of similar core-shell nanocomposites through a broad selection of solvents, oxidizing agents, monomer types, and polymerization temperatures.

Another way to PPy-based core-shell nanomaterial is the coating of a PPy nanolayer onto the surface of various nanomaterials. Pyrene-PPy core-shell nanoparticles were fabricated using micelle as a nanoreactor [215]. During the polymerization of pyrrole, pyrene molecules were phase-separated and gathered together toward the interior of micelle. Therefore, PPy nanoparticles embedded with a pyrene core were successfully synthesized. The adsorption state between pyrene molecules in the core part was tunable over a wide range with a small amount of pyrene, due to the nanometer-sized reaction site of micelle and the packing constraint of pyrene crystal. The emission color of the core-shell nanoparticles was controllable from violet to blue by changing the amount of loaded pyrene. The synthetic approach presented a facile methodology for regulating the adsorption state of organic dye as well as a new concept for building the hole-transporting layer/emitting layer nanohybrid in an electroluminescence device.

PPy-coated ceramic nanomaterial (zeolite and titanium silicate) has been fabricated with microscopic structural homogeneity [216]. The core-shell nanoparticle was synthesized via a self-assembled array of cetylpyridinium chloride on the surface of core material. Cetylpyridinium chloride played a critical role for sustaining the colloidal stability of resulting product and providing a nanoscopically confined environment for the growth of ordered PPy film. An ultrathin PPy layer (thickness: 10–30 nm) was successfully deposited on the ceramic nanoparticle (diameter: 100 nm). Even with a fairly low amount of PPy in the core-shell nanoparticle (8%), a high conductivity (5 S cm^{-1}) was obtained. The result was attributed to the enhanced molecular order of PPy chains compared with conventional PPy.

Gold-PPy encapsulated nanoparticles were also prepared with cetylpyridinium chloride (cationic surfactant) in a colloidal gold solution [217]. The relaxation dynamics of bare gold nanoparticle and PPy encapsulated gold nanoparticle were compared with various photoexcitation energies. It was found that the PPy shell could provide efficient channels for the thermal energy relaxation to the core gold nanoparticle in the PPy encapsulated gold nanoparticle.

Magnetite-PPy nanoparticles with core-shell structure were formed by means of surfactant templating [218]. The synthetic process included the preparation of magnetite nanoparticles using the precipitation-oxidation method, followed by the emulsion polymerization of pyrrole in the presence of magnetite nanoparticles. Sodium dodecylbenzenesulfonate played the role of both surfactant and dopant. The resulting product was a spherical nanoparticle with the diameter of 30–40 nm. The conductivity and magnetization of the core-shell nanoparticle depended on the degree of doping and the magnetite content. The conductivity decreased at RT with increasing magnetite content, whereas the saturated magnetization and coercivity increased.

4.3
Hollow Nanosphere

Polymer hollow nanospheres have been attracting intense attention because of their diverse potential applications such as drug delivery, heterogeneous catalysis, dye encapsulation, contaminated waste removal, and enzyme/protein protection. Polymer hollow nanospheres are also of particular interest in respect of their feasibility to encapsulate a wide range of guest molecules and to modify the surface functionalities of nanoparticles. Therefore, these polymeric nanocapsules have been fabricated with various methods including colloidal templating, layer-by-layer (LBL) adsorption, encapsulation of a non-solvent, crosslinking of micellar coronas, and self-assembly approach using covalent bond and hydrogen bond.

PPy hollow nanoparticles have been commonly obtained using various template approaches [84, 219–224]. A gold nanoparticle was utilized as a template for synthesis of a PPy hollow nanocapsule [219, 220]. The nucleation and growth of PPy around the gold nanoparticle, followed by dilute ferricyanide etching, led to the formation of a PPy hollow nanocapsule. The inner diameter (5–200 nm) of the PPy nanocapsule was governed by the size of the gold nanoparticle, and the shell thickness (5–100 nm) was determined by the polymerization time.

A PPy-chitosan hollow nanosphere (core diameter: 20 ± 3 nm, shell thickness: 15 ± 4 nm) has been fabricated by using AgCl nanoparticle as a sacrificial core at 2 °C [221, 222]. The core and shell were sequentially formed in the same reaction medium. During the synthetic process, chitosan stabilized the AgCl nanoparticle and prevented the aggregation of PPy. In addition, the PPy hollow nanosphere was stable in acidic aqueous media and insoluble in basic media due to the presence of chitosan in the shell part.

Recently, a novel approach to PPy hollow nanospheres has been developed using core-shell nanomaterial composed of an identical polymer [84, 223]. The core-shell nanoparticle composed of only PPy was generated via the microemulsion polymerization using two oxidizing agents with different chemical oxidation potentials, and used as a precursor for the fabrication of the PPy hollow nanoparticle (Scheme 7). In the synthetic procedure of PPy

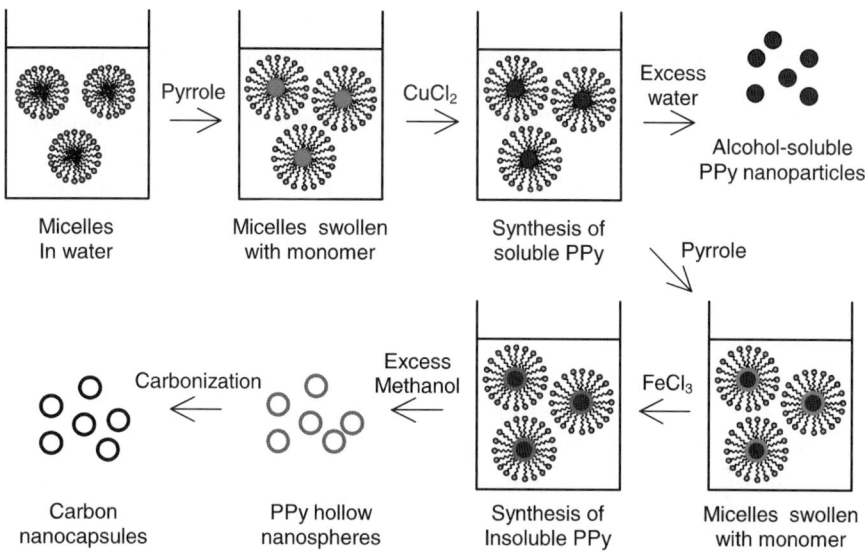

Scheme 7 Schematic representation of the fabrication of PPy and carbon hollow nanospheres (reproduced with permission from [84])

core-shell nanoparticle, pyrrole and two different kinds of oxidizing agents, i.e., cupric chloride ($E^\circ = +0.16$ V) and ferric chloride ($E^\circ = +0.77$ V), were stepwise added into the micellar solution. Cupric chloride, with lower oxidation potential, created a linear PPy core, which was soluble in alcohol. On the other hand, ferric chloride, with higher oxidation potential, yielded a crosslinked PPy shell. By adding excess methyl alcohol into the reaction solution, the linear PPy core was etched out, and the intractable and insoluble PPy nanocapsule could be obtained. The inner diameter (19–33 nm) and shell thickness (5–13 nm) were controlled by changing the amount of monomer in polymerization step. The diameter of the PPy hollow nanoparticle decreased with increasing the concentration of surfactant and decreasing the spacer length of surfactant.

4.4
Nanofiber

Most approaches to 1-D PPy nanomaterials have been dependent on the chemical and electrochemical polymerization methods using various templates such as AAO and polymer membranes [165, 224–230], mesoporous silica [168, 231], inorganic nanofibers [232], surfactants [233], and biomaterials [234, 235]. The fibrillar and tubular PPy nanomaterials with controlled diameters were mainly fabricated within the cylindrical pores of AAO or PC membranes [165, 225, 226]. The 1-D PPy nanomaterials have shown the electrical conductivites that were orders of magnitude higher than those of conventional PPys (e.g., powder and thin film), which was attributed to alignment of polymer chains along the pore axis. The 1-D nanocomposites such as Au-PPy and Au-PPy-Au nanorods could be also fabricated using AAO membrane, and their self-organization into flat 2-D or curved 3-D structures was explored [227]. The mesoscopic amphiphiles consisting of hard hydrophilic and soft hydrophobic blocks were assembled into different architectures by controlling the composition and the block ratio of the nanorods.

PPy-PMMA coaxial nanocable was fabricated through sequential polymerization of methyl methacrylate and pyrrole inside the channel of mesoporous SBA-15 silica, followed by removal of the silica template [168]. In addition, a PPy nanofiber insulated within 1-D silicate channel was produced using the pyrrole-containing surfactant monomer. During the sol-gel process, pyrrole domain was segregated and insulated by 1-D silicate nanochannel, and then the chemical polymerization of pyrrole resulted in the formation of PPy nanofiber [231].

A bulk synthesis of PPy nanofiber (diameter: 60–90 nm) was performed using nanofiber seeds as a template [232]. V_2O_5 nanofiber (diameter: 15 nm), which was chemically treated with pyrrole monomer, was used as a reactive seed template, and pre-polymerization reaction on the surface of fibrillar template provided the evolution of bulk fibrillar morphology with subsequent addition of the oxidizing agent.

The preparation method of 1-D PPy nanomaterials (nanowires and nanoribbons) was developed using a lamellar mesostructure formed in-situ between surfactant cation and oxidizing anion during polymerization [233]. The diameter of as-prepared PPy nanowires was in the range of 20–65 nm, and the length was up to several micrometers.

The helical superstructures of PPy and PEDOT could be prepared via the electrochemical polymerization using anionic synthetic lipid assemblies [234]. Since the oxidative polymerization of corresponding monomers produces cationic intermediates, the anionic assemblies acted as an excellent template owing to the mutual electrostatic attractive force.

Another approach to fabricate PPy nanofiber has entailed the electrochemical polymerization process. PPy nanowire was formed through electropolymerization at the electrode surface modified by a grafted polyethylacrylate film. This nanowire had the diameter of 100–300 nm and its length could be controlled by varying the polymerization time [236, 237].

Several research groups have studied the fabrication of diverse 1-D nanomaterials sheathed with PPy. In particular, CNT-PPy nanomaterials have been made by in-situ chemical polymerization [238–241] and electrochemical deposition [242, 243]. For example, Zhang and coworkers described a facile approach to the synthesis of size-controllable PPy-CNT nanocables by in-situ chemical oxidation polymerization using a cationic surfactant or a non-ionic surfactant [238]. On the other hand, PPy-CNT coaxial nanowire was fabricated by electrochemically depositing a PPy layer onto each of the aligned CNTs [243].

Besides a CNT, different kinds of 1-D nanomaterials have been employed as the inner part in functional nanomaterials. A one-step process was developed to produce a silver-PPy coaxial nanocable in aqueous solution at RT [244]. Pyrrole and silver nitrate were used as a reducing agent and an oxidizing agent, respectively, and poly(vinyl pyrrolidone) (PVP) was acted as an structure-directing agent to form the silver-PPy nanocable. Silver grew into nanowire with the aid of PVP, and simultaneously PPy was polymerized along the surface of the silver nanowire. This synthetic technique might be extended to the preparation of versatile metal-conducting polymer nanocables under mild condition.

The nanocoating of PPy on natural cellulose fiber was performed without disrupting the hierarchical network structure of individual fibers [245]. Since pyrrole was hardly adsorbed onto a cellulose surface, this approach was based on the adsorption of the growing polymer chain from the solution and on the subsequent immobilization to form thin film. The conformation of the PPy chain was parallel to the surface of the cellulose fiber, because the corresponding oligomer is adsorbed parallel to the surface and further polymerized in the lateral direction.

The well-aligned array of the core-shell nanostructure was formed using copper sulfide nanorods coated with PPy nanolayers [246]. PPy was homogeneously deposited on a copper sulfide nanorod via in-situ polymerization at the interfacial layer between chloroform and water. The uniform PPy coating was achieved by two factors of confinement: (1) it was vertically confined at the interfacial layer of chloroform and water; (2) it was horizontally confined in the interrod spaces of the nanorod arrays. The growth of PPy film was tunable by adjusting polymerization time, monomer concentration, and monomer-to-oxidant weight ratio.

4.5
Nanotube

In general, PPy nanotubes have been mainly produced by the hard template method [165, 172, 225, 226, 247, 248]. For example, PPy nanotube with highly uniform surface and controlled wall thickness was fabricated by one-step VDP using AAO membrane [172]. A template-mediated VDP was found to be a facile and effective method to fabricate polymer nanotubes. The vapor phase polymerization provides highly uniform tubular walls as well as easy control over the wall thickness.

Recently, a facile soft template synthesis was developed for fabricating PPy nanotubes against the hard template synthesis [153, 249]. PPy nanotubes could be readily produced through a cylindrical micelle templating in reverse (water-in-oil) emulsion system. Sodium bis(2-ethylhexyl) sulfosuccinate (AOT) was selected as a surfactant to form a micelle. When aqueous $FeCl_3$ solution was added into AOT/apolar solvent mixture, a reverse cylindrical micelle was generated through a cooperative interaction between $FeCl_3$ and AOT. The anionic head-group of AOT extracted iron cation from the aqueous phase due to the electrostatic attraction. Since iron cation acted as an oxidizing agent, the introduction of pyrrole into reverse cylindrical micelle phase gave rise to the chemical polymerization of pyrrole along the exterior of the reverse cylindrical micelle, which resulted in the formation of a PPy nanotube. Consequently, this soft-template approach may be easily scaled up or extended to the preparation of other 1-D nanomaterials. It was noteworthy that PPy nanotubes had higher conductivities than bulky PPy material. The structure-property relationships of PPy nanotube have been also studied by several research groups [250–253].

4.6
Thin Film and Nanopattern

Despite the intractability of conducting polymers, a variety of methods have been developed for the fabrication of conducting polymer thin films and nanopatterns [254, 255]. Nanoscale PPy thin films or patterns were fabricated through several techniques such as surfactant and block copolymer templating [256, 257], microphase separation and molecular mask [258], dip-pen lithography [259], and so forth [260].

The array of PPy dots (diameter: 80–180 nm) was patterned in a parallel fashion using block copolymer micelle as a soft template [256]. The Langmuir–Blodgett (LB) film composed of PS-b-poly(2-vinylpyridine) was deposited onto the solid substrate, and the aligned PPy dots could be formed through the chemical oxidation polymerization of pyrrole in the presence of the block copolymer micelle template. The chemical differentiation (hydrophilic vs. hydrophobic) between the core and the corona in block copolymer micelle led to spatially-limited PPy growth.

PPy and PANI nanostructures (sphere, wire, and planar film) were prepared with sub-100-nm features on flat surface using adsorbed surfactant molecules as the nanoreactor [257]. The morphology of the polymer nanostructures could be controlled by the addition of coadsorbing molecules, which induced the phase transition of surfactant aggregates. Furthermore, the hydrophobic chain length of surfactant and the surface property of the adsorbing substrate were also important factors affecting the formation of resulting products.

A new technique was developed to fabricate the pattern of PPy thin film with micrometer to nanometer scales by adsorbed surfactant bilayers as a molecular mask [258]. Surfactant molecules were regioselectively adsorbed on a substrate due to a microphase-separated polymer blend or block copolymer. As a result, PPy could be selectively grown up to 50 nm along the hydrophobic domain.

In recent years, dip-pen nanolithography has been used for conducting polymer nanopatterning [259]. The electrostatic interaction between water-soluble PPy ink and negatively charged substrate could provide a primary driving force for the formation of stable nanopatterns (the smallest feature size: 130 nm).

Another approach was developed using two-step deposition polymerization to make the direct formation of patterned PPy thin film [260]. The surface roughness of PPy films was much lower than that of conventional PPy film. The film surface was formed with an apparent granular structure of the dimension less than 300 nm. The conductivity of iodine-doped PPy was also comparable to that of PPy film formed by plasma polymerization or electrochemical deposition.

4.7
Nanocomposite

PPy nanocomposites have been extensively reported in the literature [261-267]. In the case of inorganic nanoparticles/conducting polymer nanocomposites, various inorganic nanomaterials including silica, palladium, platinum, and maghemite have been formed via inclusion techniques using both chemical and electrochemical approaches [261]. Multifunctional nanocomposites could be fabricated by the judicious choice of synthetic techniques and inorganic materials.

A titania-PPy hybrid nanocomposite was built from in-situ generated organically functionalized anatase building blocks [262]. A pyrrole-covered nanocrystalline (2.5–5.0 nm) anatase nanoparticle was made by a simple one-pot synthesis. At the same time, a bifunctional ligand, 3-[6-(pyrrol-1-yl)hexyl]penta-2,4-dione, was used to control the particle size and the chemical properties at the surface. This ligand was composed of a complexing acetylacetone function linked to a pyrrole moiety through a hexamethylene chain. These monomer-capped anatase nanoparticles could be polymerized by electrochemical or chemical methods, leading to conductive nanocomposite film.

PPy-Fe_3O_4 nanocomposite was also prepared using common ion adsorption effect [263]. When the Fe_3O_4 nanoparticle was treated with aqueous $FeCl_3$ solution, the iron cation from $FeCl_3$ was absorbed on the surface of the Fe_3O_4 nanoparticle. The subsequent addition of the monomer induced the chemical oxidation polymerization on the surface of the Fe_3O_4 nanoparticles, which resulted in the formation of PPy-Fe_3O_4 nanocomposites.

The conductive and biodegradable nanocomposite was made of PPy and polylactide [264, 265]. PPy nanoparticles were incorporated into the polylactide matrix via the emulsion polymerization of pyrrole in the aqueous solution containing surfactant and polylactide. The surface resistivity of resultant nanocomposite ranged from 2×10^7 to $15\,\Omega\,\text{square}^{-1}$ with various PPy content (1–17 wt %).

5
Polyaniline Nanomaterials

PANI is one of the most promising conducting polymers with enhanced conductivity, good environmental stability, and diverse color change corresponding to different redox states. It has been known that PANI nanomaterials can be applied for many practical fields such as chemical sensor, supercapacitor, corrosion protection, battery and energy storage, and antistatic coating. Therefore, there have been prodigious research papers concerning the synthesis and application of PANI nanomaterials.

5.1
Nanoparticle

The processability of PANI is relatively poor because it is infusible and insoluble in common solvent. In order to improve its processability, the alternative approach is achieved by preparing PANI nanoparticles and dispersing them uniformly in aqueous/organic solvent. The PANI nanoparticle has been synthesized using polymer surfactant, and the diameter of the PANI nanoparticle is controlled by different surfactants [268]. The amphiphilic polymer molecule such as hydrophobically end-capped poly(ethylene oxide) was used to form micelle, and the micelle dimensions could be tuned by changing the spacer group of the hydrophilic block. The size of the PANI nanoparticle was distributed in the range of 5–40 nm. Chemical oxidation polymerization of aniline has been performed in DBSA–isooctane–water reverse micelle system [269]. It was reported that the PANI nanoparticle exhibited needle-like shape (diameter: 10 nm, length: 20–30 nm) and had a uniform dispersion in solvent. It is a feasible and attractive method to readily prepare DBSA-PANI nanoparticle in the reverse micelle system.

Another approach to fabricate PANI nanoparticles was reported with aqueous/ionic liquid interfacial polymerization [270]. Interfacial polymerization involves the step polymerization of two reactive monomers or agents, which are respectively dissolved in two immiscible phases. The polymerization reaction takes place at the interface of the two liquids. Green PANI nanoparticle was formed at the interface and gradually diffused into the aqueous phase. The entire water phase was homogeneously filled with dark-green PANI, while the ionic liquid layer showed a color of orange stemming from the formation of aniline oligomers. The yield of PANI nanoparticles was ca.

25 wt %. PANI nanoparticles with the diameter of ca. 30–80 nm could be obtained in environmentally benign solvent.

Chemical oxidation polymerization in dilute and semi-dilute solutions of poly(sodium 4-styrenesulfonate) (PSS) was demonstrated for the synthesis of PANI nanocolloids (particle size: ca. 2–3 nm) [271]. A similar approach was performed to synthesize PANI nanoparticles using ammonium peroxydisulfate in aqueous medium comprising poly(ε-caprolactone)-PEO-poly(ε-caprolactone) amphiphilic triblock copolymer micelle [272]. Micelle size conspicuously affects the morphology of the PANI nanoparticle, and the PANI nanoparticle size is strongly dependent on PEO molecular weight. The diameter of the nanoparticle increased from 30 nm to 100 nm as the PEO molecular weights decreased.

Electropolymerization was performed using constant potential of 0.7 V and obtained a PANI nanoparticle with diameter of ca. 80 nm [273]. In this report, the effect of erbium chloride and magnetic field on the property of PANI deposited onto two different platinum electrodes was investigated. The PANI nanoparticle was prepared on a highly oriented pyrolytic graphite surface from dilute PANI acidic solution (1×10^{-3} M aniline/1 M $HClO_4$) using a pulsed potentiostatic method [274]. This nanoparticle was disk-shaped with the height of 1–3 nm and the diameter of 20–30 nm.

5.2
Core-Shell Nanomaterial

To date core-shell nanomaterials based on PANI have been synthesized using PS, titanium oxide, gold, and vanadium oxide as seeds [275–283]. Various ways to nanometer-size PANI core-shell materials have been continuously developed, including ultrasonic irradiation polymerization, oxidative polymerization and electrochemical polymerization.

Ultrasonic irradiation was employed to produce the aggregation of TiO_2 in nanometer size, and the resulting PANI-TiO_2 core-shell nanomaterial was found to be spherical [275]. In Fourier transform infrared analysis, the characteristic peaks of benzoid and quinoid rings originating from PANI shifted to higher energies with increasing TiO_2 content due to the hydrogen bonding interaction between PANI and TiO_2. Moreover, X-ray photoelectron spectroscopy showed that the ratio of Ti and N atoms on the core-shell nanomaterial surface was lower than that in the bulk. These facts strongly supported the formation of a PANI-encapsulated TiO_2 nanostructure. The introduction of aniline before or after the sol formation was the critical factor to form the PANI/TiO_2 core-shell nanomaterial in sol-gel method [276].

Since the adsorption of monomer on the oxide surface was a dominant step in sol-gel process, PANI/TiO_2 nanocomposite was affected by the amount of monomer prior to polymerization. This new finding helped to produce a novel oxide-polymer core-shell material or higher polymer content hybrid. PANI-encapsulated TiO_2 nanoparticle was also fabricated using oxidative polymerization with ammonium peroxodisulfate [277]. TiO_2 nanoparticle with the diameter of 80–100 nm was used as the seed and HCl was applied as the doping agent. The TiO_2 surface was modified with silane coupling agent [278]. High resolution X-ray photoemission spectroscopy confirmed the formation of PANI on the surface of TiO_2 from the similar ratio of imine nitrogen and amine nitrogen in TiO_2/PANI nanocomposite compared with that of PANI. The conductivity of PANI/TiO_2 composite film in m-cresol was higher than that of the film cast from chloroform owing to the unfolded conjugation chains of PANI [279]. The combination of surfactant-coated TiO_2 and PANI solution afforded TiO_2/PANI nanocomposite, which contained well-dispersed TiO_2 up to 50 wt % of filler content. Gold nanoparticle-doped PANI nanocomposite was also fabricated using hydrogen tetrachloroaurate as an oxidizing agent [280], and an extended work was performed using in-situ intercalation polymerization [281].

Another synthetic method for PANI core-shell nanomaterial has been developed using 25 nm silica, PS, and PANI [282]. Nanometer-sized SiO_2 particle was synthesized by a sol-gel process and PS was coated with SiO_2 nanoparticle using soapless seeded emulsion polymerization. Consecutively, SiO_2-PS core-shell nanomaterial was dispersed in SDS solution to adsorb aniline over the entire surface of SiO_2/PS core-shell nanomaterial. Then SiO_2/PS/PANI conductive nanocomposite was fabricated by chemical oxidation polymerization. The conductivity of SiO_2/PS/PANI nanocomposite was ca. 10^5 S cm^{-1}. In addition, PANI surface coverage and conductivity of PANI nanocomposite have been also monitored [283]. PANI surface coverage was not consistent with the conductivity of SiO_2/PS/PANI nanocomposite, because the conductivity was highly correlated with the thickness of PANI coated layer onto the PS nanoparticle. Therefore, uniform coverage of the core nanoparticle is not required to improve the conductivity of the SiO_2/PS/PANI nanocomposite.

5.3
Hollow Nanosphere

Hollow microsphere (diameter: 360–1200 nm) of PANI-NSA has been fabricated using an emulsion template method at low temperature [284]. In this template method, the target material is precipitated or polymerized on the surface of the template, which results in a core-shell structure. On removing the template, hollow microsphere can be obtained. However, the removal of the template often affects the spherical structure, especially for hollow polymer microsphere. Therefore, they select the emulsion template method as the emulsion can be readily removed through dissolution or evaporation after polymerization.

Uniform PANI thin shells and hollow capsules have been formed using polyelectrolyte-coated microspheres as a template [285]. Multilayers of poly (diallyldimethylammonium chloride) and PSS pre-coated onto melamine formaldehyde particle via the LBL self-assembly procedure was used as a template for the subsequent deposition of PANI.

5.4
Nanofiber and Nanorod

Recently, PANI nanofiber has been synthesized using interfacial polymerization without templates or functional dopants [116, 286]. Interfacial polymerization is performed in an aqueous/organic biphasic system. Therefore, this method provides PANI nanofibers in a large scale [287]. The scanning electron microscopy (SEM) images of PANI nanofiber are displayed in Fig. 5 [288]. The fabrication of PANI nanofiber has been performed with a broad selection of solvents, doping agents, monomer concentrations, and reaction temperatures. The average diameter of the nanofiber could be tuned from 30 nm using hydrochloric acid to 120 nm using perchloric acid.

PANI nanofiber has been also prepared using oligomer-assisted synthesis [289]. In this method, the polymerization has been carried out in concentrated CSA solution, and aniline oligomers are used to accelerate the polymerization reaction. The resultant PANI nanofiber showed the diameter in ranges from 20 to 40 nm with several micrometer lengths.

A nanofiber seeding method has been developed in order to fabricate PANI nanofiber without organic dopants, surfactants, and insoluble templates [290]. With even very small amounts of inorganic or organic nanofibers, the morphology of the resulting doped PANI powder was changed

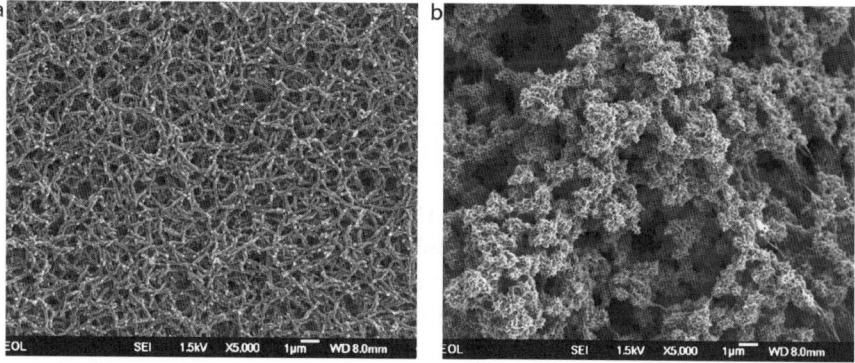

Fig. 5 SEM images showing the morphology of PANI synthesized from **a** a rapidly mixed reaction and **b** a slowly mixed reaction. High-quality nanofibers are obtained in the rapidly mixed reaction, while irregular agglomerates form in the slowly mixed reactions (reprinted with permission [288])

from particulate to nanofiber. These findings provide the groundwork for widespread applications with respect to shape control.

In addition, one-phase surfactant-assisted chemical method has been utilized to synthesize PANI nanofiber, which was doped with CSA and 2-acrylamido-2-methyl-1-propanesulfonic acid, in large quantities [291]. A chemical oxidative polymerization of aniline has been carried out using ammonium peroxydisulfate as an oxidizing agent in the presence of non-ionic surfactant. A precipitate of doped emeraldine salt is composed of PANI nanofiber, which has the diameter of 30–50 nm and exhibits the conductivity of 1–5 S cm^{-1} at RT. Another piece of research has been done through chemical oxidation polymerization of aniline in a surfactant gel, which was formed by a mixture of hexadecyltrimethylammonium chloride, acetic acid, aniline, and water at $-7\,°C$ [292]. The dendritic PANI nanofiber has the diameter of 60–90 nm and the length of 1–2 μm. Extended works have been performed by the electrospinning method [293]. It should be taken into account that PANI-CSA fiber shape could be influenced by the synthetic variables such as solvent, surface tension, viscosity, and solution conductivity.

Recently, Jang et al. have demonstrated a dispersion polymerization for the mass production of PANI nanorod [197]. PANI nanorod (diameter: 20–50 nm, length: 0.2–1.0 μm) was exclusively fabricated in multi-gram scale quantity (ca. 10 g) and high yield (95%). Dispersion polymerization has been performed using hydrochloric acid as a dopant and ammonium peroxydisulfate as an oxidant. The experimental parameters such as oxidant, stirring, and temperature were thoroughly examined during the polymerization process.

5.5
Nanotube

In general, PANI nanotubes have been mainly prepared using a nanoporous membrane [170, 294, 295] and polymer fiber template [296], which is similar to the case of the PPy nanotube. The chemical and electrochemical polymerization of aniline inside the cylindrical nanopores (pore diameter: 20–100 nm) generated the tubular structures with maximum conductivities of ca. 130 and 90 S cm^{-1}, respectively [170]. PANI nanotubes filled with cobalt nanowire have been also fabricated using an alumina membrane template. It exhibited the enhanced coercivities in the range of 180–209 Oe, and the conductive PANI shell protected the metal nanowires from oxidation and corrosion [294]. A fiber templating approach has been also developed to yield the tubular structures of PANI [296]. The electrospun poly(L-lactide) fiber was utilized as template, and the PANI layers (thickness: 70 nm) were coated onto the fiber. The thermal decomposition has been performed at 310 °C to remove the fiber core. The resulting PANI tube has a conductivity of ca. 0.28 S cm^{-1}.

One of the alternative approaches is the soft template technique. The extensive works have advanced with respect to the generation of PANI nanotubes via a self-assembly process [121, 179, 297–299]. PANI nanotubes were prepared through chemical and electrochemical routes in the presence of organic functionalized acid ($-SO_3H$) and inorganic acids (HCl, H_2SO_4, H_3PO_4, and HBF_4) as the dopant. The shell thickness of the nanotubes could be tuned by changing the synthetic variables such as the molar ratio of monomer to dopant, the concentration of dopant, and the polymerization temperature.

Physical performances of the PANI nanotube have been explored in terms of electrical [300, 301] and magnetic properties [302]. For example, the electrical property of a single PANI nanotube has been measured by a standard four-terminal technique. While the bulk conductivity of pelletized PANI nanotubes was as low as 3.5×10^{-2} S cm^{-1}, the conductivity of a single PANI nanotube showed a relatively high conductivity of ca. 30.0 S cm^{-1} [300].

5.6
Thin Film and Nanopattern

Over the last decades, there have been a lot of efforts to fabricate polymer thin film. However, it is difficult to obtain a conducting polymer in thin film or monolayer owing to lack of processability and solubility. Among the conducting polymers, PANI thin film has a great potential to apply for chemical sensor due to its sensing ability and conductivity. Similar to PPy, there have been several methods to fabricate PANI thin film: LB technique, self-assembly, electropolymerization, evaporation, and plasma-mediated polymerization.

The LB technique has been a well-established technique for the preparation of organic thin films utilizing the amphiphilic property of materials. In the case of PANI, thin film has been successfully fabricated by derivatizing the polymer with long alkyl chains, doping with an amphiphilic dopant, and LB deposition along with specific film forming materials [303]. For example, soluble PANI was synthesized in N-methylpyrrolidone (NMP) or NMP-$CHCl_3$ mixed solvent, and PANI nanofilm was formed with this mixed solution by the LB technique [304].

Self-assembled methods have been also used for preparing and patterning PANI thin films. A monolayer of PANI has been assembled on hydroxyl-terminated surfaces, based on chemical or electrochemical polymerization of electrostatically bound anilinium [305]. The hydroxyl terminated silicon substrate was initially treated with silane coupling agent and sulfonated. The negative charge of the sulfonated surface plays a role of the active site where the anilinum is attached onto the silicon surface. With similar conceptual approach, complete alignment of PANI monolayers has been achieved epitaxially on a lattice-matched substrate [306]. The oxygen group of mica surface was used as the active site to deposit PANI chains. Such ordered films provided the enhanced performance of the PANI nanofilm.

With development of polymerization technique, an in-situ polymerization has been adapted for preparing PANI thin film. In general, polymer thin films have been produced by spin coating, thermal evaporation, and cyclic voltammetry. Since that discovery, there have been significant efforts to fabricate PANI thin film using several synthetic methods such as anodic polymerization, and plasma-mediated polymerization. Anodic polymerization has offered a simple method to obtain thin film on indium tin oxide (ITO) or gold substrate [307]. Using plasma-mediated polymerization, PANI thin film was deposited on the specific substrate by in-situ polymerization and consecutively doped by injecting dopant gas (e.g., iodine) into the reaction chamber [308, 309].

Thermal evaporation of PANI has been also developed to fabricate PANI thin film by vacuum deposition of PANI powder on the reference electrode [310]. The chemically synthesized PANI powder was formed in pellet type and the pelletized PANI was evaporated on glass substrates at a pressure of 10^{-6} mm Hg to form PANI thin film. The pre-cleaned glass substrate was covered uniformly with PANI thin film and this thin film was utilized as a carbon monoxide sensor. Thus, thermal evaporation method could be used for thin film formation of conducting polymer nanomaterials [311–313].

5.7
Nanocomposite

PANI nanocomposite has been electrosynthesized using PC medium in the presence of trichloroacetic acid [314]. In this battery study, PANI matrix exhibited high discharge capacity as well as high electrochemical stability during multiple charge-discharge cycles. The thickness of PANI films strongly affected the electrochemical properties of conducting polymer nanocomposites and the values of discharge capacity and COULOMBIC efficiency were 107 A h kg^{-1} and 97.5% for thin film and 82 A h kg^{-1} and 95.0% for thick film, respectively. These nanocomposites showed excellent stability after charge-discharge cycles compared with that of bulk PANI films. The electrical properties of a cathode based on PANI-silica composite were studied for the lithium battery. In this case, PANI nanocomposite with silica host matrix has been prepared by a template method using silica sol-gel as a template [315]. The PANI network in the porous silica host provided a large electroactive surface area and formed parallel ion and electronic conducting pathways to improve the process of charge transfer and mass transport. This synergistic mechanism between PANI and silica is responsible for the improved performance as the cathode electrode of the lithium battery. When the COULOMBIC efficiency of the PANI nanocomposite was ca. 95%, the discharge reached a stable value of 223 A h kg^{-1}. Another approach for sensor application has been developed based on the electrical conductivity, high stability, and color change of the PANI nanocomposite by variation of pH value. The PANI-porous Vycor glass (PVG) nanocomposite was fabricated via the infiltration of PANI into the pores of a PVG. The optical sensor was assembled by fixing a PVG slide onto a bifurcated optical fiber bundle with a cyanoacrylic resin [316]. As the PANI exists inside porous PVG, well-oriented PANI molecules were formed without crosslinking and structural defects, and endowed the good reproducibility and long life time for sensor application. In addition, PANI nanofiber-gold nanoparticle composites have been prepared by redox reaction of metal ions (Au^{3+}) in the dedoped PANI nanofiber [317] and the combination of PANI with gold nanoparticles provided an application for non-volatile memory. For the application on electrochemical supercapacitor devices, the nanocomposite based on PANI and polyoxometalates has been synthesized by electrochemical method and further deposition of the PANI nanocomposite on graphite provided a supercapacitor cell to measure the electrochemical property [318]. This nanocomposite displayed a balanced performance of PANI matrix and inorganic component to electrochemical charge reaction. The self-assembly method has also been developed to fabricate PANI nanotube-titanium dioxide nanocomposite using β-NSA as the dopant [319]. The morphology, size, conductivity, and hydrophobicity of the PANI nanotube-titanium dioxide nanocomposite have been tuned by changing the content of TiO_2 nanoparticles in the nanotube.

6
Polythiophene Nanomaterials

Polythiophene (PT) is an important conducting polymer that constructs the environmentally and thermally stable materials. It has been applied as electrical supercapacitor, non-linear optics, polymer light emitting diodes, electrochromics, photoresists, antistatic coatings, sensors, batteries, electromagnetic shielding materials, solar cells, memory devices, imaging materials, and transistors [320, 321]. In PT, the rings coupled in the 2 and 5 positions allow the conjugation of the π-orbitals along the polymer chain. The large band gap of neutral PT provides a relatively low electrical conductivity of about 10^{-8} S cm^{-1}. The enhanced electrical conductivity has been achieved by oxidizing (or doping) PT. Consequently, the oxidation of PT affords a drastic change in the electronic band structure as new mid-gap states are created and resonance structure is formed. The first chemical preparations of PT were reported in 1980 by two scientists, Yamamoto and Dudek. Both groups synthesized PT by a metal-catalyzed polycondensation polymerization of 2,5-dibromothiophene [322, 323]. Several PT derivatives and substituted PTs have been developed and exploited in diverse afore-mentioned applications.

Owing to the recent push toward conducting polymer nanomaterials, the creative design and development of new PT nanomaterials have provided novel and fascinating materials and led to enhanced performance in certain devices. Therefore, the synthesis of PT nanomaterials has become a critical research subject. This leads to the intriguing prospect that chemical and physical properties of PT nanomaterials can be selectively tuned through specific polymerization and assembly. Herein, the recent works in the syntheses of PT nanomaterials will be highlighted comprehensively.

6.1
Nanoparticle

In contrast with studies on PPy and PANI, limited information has been available concerning the fabrication of the PT nanostructure due to the relatively low reactivity of the monomer. While much effort has been devoted to the synthesis of PT nanoparticles, no remarkable progress has been achieved yet.

Ong and coworkers reported the molecular self-assembly behavior of poly(3,3'-polydialkylquarterthiophene) in solution and the unique molecular property of the polymer for the preparation of structurally ordered conducting polymer nanoparticles [324]. They demonstrated that molecular organization in the semiconductor materials could be achieved before deposition to enhance the charge transport capability. The atomic force microscopy (AFM) image exhibited an array of small crystal domains dispersed in an amorphous polymer matrix. On the other hand, highly crystalline film with extensive nanocrystal domains was formed after annealing at 145 °C and cooling down to RT. From this result, it has been demonstrated that structurally ordered nanoparticles can be generated in suitable solvents under proper experimental conditions.

6.2
Hollow Nanosphere

PT nanovesicle structure has been reported by Rowan and Nolte [325]. They described the aggregation behavior of a rod-coil diblock copolymer derived from an isocyanoamino acid containing a thiophene group. This diblock copolymer consists of styrene and 3-(isocyano-l-alanylaminoethyl)thiophene (PS-PIAT) units and is an amphiphile polymer in both organic and aqueous solvents. The diblock copolymer molecules were aggregated by the poor solubility of the polyisocyanide block in $CHCl_3$ and the formed vesicles had a high polydispersity. The aggregation behavior of PS-PIAT has been studied in water system by injecting a PS-PIAT solution in tetrahydrofuran (THF). After the solution was allowed to equilibrate, the morphology of the aggregates was examined by cryogenic SEM and normal SEM. The cryogenic SEM image represented PT hollow nanosphere and the aggregates were vesicular in nature.

6.3
Nanofiber and Nanowire

PT nanofiber and nanowire have been realized by various polymerization strategies. One of the facile synthetic methodologies is the template strategy. While PANI and polydiacetylene fibers have been prepared inside MCM-41 by oxidation and free-radical initiation of the monomers [326, 327], the fabrication of PT nanowire from MCM-41 has been addressed recently [328]. The proposed polymerization process is divided into two steps: (1) loading of the monomer into preformed silica channels by CVD; (2) polymerization by irradiation or oxidation of monomer. In the article, it has been illustrated that massive, coiled and individual strands are distributed over large areas. These strands have a tendency to entangle with each other. Similar template method has also been employed to produce PT nanofiber in which nanostencil shadow masks are used as molds for nanofiber formation [329]. Field effect transistor based on regioregular poly(3-hexylthiophene) (P3HT) nanofiber had electrical performance with mobility values of $0.02\,\text{cm}^2\,\text{V}^{-1}\,\text{s}^{-1}$ and on/off current ratios of 106. Current densities of $700\,\text{A}\,\text{cm}^{-2}$ were achieved in a single nanofiber. PT nanofiber from THF fraction exhibited elevated current level and decreased activation energy. On the other hand, P3HT-b-PS copolymer exhibited high electrical conductivity, enhanced mechanical property and processability [330]. Formation of ordered supermolecular assembly in these regioregular materials strongly associates with their excellent electrical conductivity. The nanowire morphologies of regioregular poly(alkylthiophenes) provided the possibility of guiding the intrinsic self-assembly for regular conjugated polymer chains by coupling with incompatible segments chemically.

6.4
Nanotube

Researches on PT nanotubular structures have been undertaken by Shi's research group [331, 332]. The aligned PT nanotubules with diameters of 20 nm and 200 nm could be synthesized by direct electrochemical polymerization of thiophene monomer in boron trifluoride diethyl etherate solution with microporous alumina membranes as the templates. The enhanced tubular alignment has been achieved compared to the conducting polymer tubules reported previously [333, 334]. This is because the alumina membrane templates have much higher pore densities and much better pore uniformity. They also verified that tubules could be grown as long as the template thickness if sufficient charge was passed into the cell. The aligned nanotubules were readily formed into bundles due to their large surface area, which resulted in strong interactions between the tubules. Furthermore, this methodology has been expanded to the fabrication of aligned microtubular heterojunctions of poly(*p*-phenylene) and PT [335]. The length of microtubule was about 40 mm and was controlled by the total charge passed through the cell during the electropolymerization. The skin of PT nanocomposite was flexible and smooth.

In addition, the electronic structure and the charge behavior of PT nanotubular film have been also investigated [336]. It was revealed that there were two new photovoltage responses assigned to the localized electronic states transition appearing under the external electric field. The large surface area of PT nanotubules allowed the tubules to absorb the oxygen atoms in the atmosphere. The interaction between oxygen and PT chains resulted in the formation of charge surface states. These states could be reflected on the photovoltage responses. In addition, PT nanotubular film was characterized by the response of the local electronic state in the near-infrared region.

6.5
Thin Film

Historically, diverse approaches have been performed to fabricate a PT nanolayer and thin film. One of the convenient ways is LBL strategy in order to construct PT assembled structure. McCullough's group reported the first evidence for the fabrication by self-assembly of LBL based on regioregular PT using only polyanion and polycation dilute solutions of PT. This method leads to structurally ordered nanolayered sheets of conducting polymers [337]. Therefore, it furnishes a basic tool for the construction of organized PT nanolayer.

A similar synthetic way to LBL electrodeposition of PT film has been designed by Advincula and coworkers [338]. They described electrodeposition of polysiloxane precursor polymer to form crosslinked conjugated PT ultrathin film. In this study, the precursor polymer contained a polysiloxane backbone with pendant electroactive thiophene monomer, which was electropolymerized by cyclic voltammetry. The morphology of PT film was transformed from a relatively globular to membrane-like shape with increasing the number of cycles.

An alternative process to PT thin film has been developed using direct surface polymerization. It was proved that PT thin film was formed via surface polymerization by ion-assisted deposition (SPIAD), in which 55–200 eV thiophene ions and R-terthiophene neutrals are codeposited on surfaces [339–342]. This PT film displayed unique optical properties in its photoluminescence (PL) and ultraviolet-visible (UV-vis) absorption compared with the film prepared by direct thiophene ion deposition only. This method was clearly applicable to a wide range of different ions and neutral species. However, vaporized and ionized reagent species could not be applied by SPIAD.

Recently, "nanorubbing" surface modification methodology has been published concerning homogeneous rubbing and crystallization for PT thin film [343-345]. On the basis of this study, the usage of an AFM stylus was focused to impart orientation to the PT surface with a nanometric resolution.

6.6
Nanocomposite and Nanohybrid

A nanostructure of PT dendrimers has been synthesized and characterized with their unique supramolecular assembly into 2-D crystals, nanowires, and nanoparticle aggregates [346]. It was elucidated that self-organization of the dendrimers on the solid substrate was dependent on the nature of the substrates, preparation methods, and the molecule-molecule and molecule-substrate interactions. It was remarkable that the PT dendrimer exhibited different nanostructures depending on the property of substrate surface. The results demonstrated the unique potential of thiophene dendrimers to form nanostructures on substrate surfaces.

PT can be used as component for various nanocomposites. Several types of nanohybrids have been fabricated such as polymer/polymer hybrid or metal/polymer nanocomposite. PT with enhanced electrical conductivity was combined with PVK showing excellent optical property and luminescence behavior [347]. A conducting composite comprising PVK and PT was prepared by oxidative crosslinking of pendant carbazole moieties in the presence of ferric chloride and thiophene monomer. It was found that the particle size was in the range of 24-72 nm for the composite. A reaction between thiophene and PVK in presence of anhydrous ferric chloride resulted in the formation of an intractable PVK and PT nanocomposite which showed improved thermal stability relative to PVK and PT. The DC conductivity of the composite was higher than that for PVK and was comparable to that for the PT.

To date, a large number of transition metal-polymers have been fabricated and these materials are of interest because they allow the metal complexes to be incorporated in a processable form. There were several arrangements in metal-containing conducting polymers [348, 349]. The metal may be tethered to the backbone of a conjugated polymer by a linker (Type I). The metal and backbone of conducting polymer are electronically coupled, and this coupling can influence physical and chemical properties of both metal and conjugated polymer in Type II. Finally, the metal group can be located directly in the conjugated backbone (Type III). Therefore strong electronic interactions between the organic bridge and metal group are formed in this arrangement.

Several reports were published on the PT/metal hybrids or PT/inorganic nanocomposites. Novel bithiophene with a pendant fullerene substituent was synthesized by electrochemical polymerization [350]. It was revealed that a photoinduced electron was transferred from the donor cable (PT) to the pendant acceptor cable (fullerene moieties). On the other side, it was demonstrated that a highly conducting cobalt selen-PT hybrid material catalyzed the reduction of oxygen [351]. A selective four-electron reduction process occurred in order to produce H_2O as the sole product. In addition, an enhanced photocurrent was obtained in Al/porphyrin-sensitized regioregular PT Schottky-barrier cell because the top energy level of the valence band for PT solid was more negative than the highly occupied molecular orbital (HOMO) energy level of porphyrin [352]. From these results, a larger thermodynamic driving force for a photoinduced hole has been transferred from porphyrin to PT and a closer distance between two conjugating planes of porphyrin and PT is needed to obtain a larger photocurrent of Al/porphyrin-sensitized PT Schottky-barrier cell.

7
Poly(3,4-ethylenedioxythiophene) Nanomaterials

PEDOT had been developed as one of the PT derivatives by the Bayer AG research laboratories in Germany during the 1980s [353, 354]. PEDOT was prepared using standard oxidative or electrochemical polymerization methods. Initially, PEDOT was found to be an insoluble polymer, yet exhibited some very interesting properties such as high conductivity (ca. $300\,\mathrm{S\,cm^{-1}}$), transparency in oxidized thin film, and excellent stability in an oxidized state. The solubility problem was subsequently overcome by a water-soluble polyelectrolyte such as PSS and this polyelectrolyte was used as the charge-balancing dopant during polymerization to form a PEDOT/PSS solution. The combination of PEDOT and PSS electrolyte resulted in a water-soluble conducting polymer with good film forming properties: conductivity (ca. $10\,\mathrm{S\,cm^{-1}}$), high visible light transmittance, and excellent stability. Thin film of PEDOT/PSS annealed in air at 100 °C for 1000 h showed only a minimal change in conductivity.

7.1
Nanoparticle

While a lot of literatures concerning conducting polymer nanoparticle were published, there were limited reports on the fabrication of PEDOT nanoparticle due to the relatively low solubility of the 3,4-ethylenedioxythiophene (EDOT) monomer in aqueous media. Several synthetic methods for the PEDOT nanoparticle have been reported using seed polymerization, emulsion polymerization and dispersion polymerization.

In the case of the PEDOT nanoparticle, DBSA was used as a doping agent and a surfactant simultaneously to form a micellar solution. Two different oxidants such as $FeCl_3$ and APS were employed in order to polymerize EDOT monomer in aqueous solution. The resultant product was quite spherical and its diameter ranged from 35 to 100 nm. When ferric chloride was applied as oxidant, PEDOT nanoparticle had higher electrical conductivity compared to APS oxidant due to strong attraction of oxidant-anionic surfactant in $FeCl_3$-DBSA. The highest conductivity of PEDOT nanoparticle was measured to be 50 S cm^{-1} [355].

Seed polymerization has been also considered as an alternative synthetic route to prepare the PEDOT nanoparticle. In 1999, Armes and coworkers reported the fabrication of PEDOT-coated PS latex using seed polymerization [356, 357]. PVP-stabilized PS latex with a diameter of 1.8 μm was used as the core seed, since PS particles could be readily prepared with narrow particle size distributions. The polymerization in aqueous media proceeded using ferric tris(p-toluenesulfonate) at 85 °C. Various PEDOT-coated latexes were obtained by varying the loaded amount of PEDOT from 4.9 to 38 wt %. PEDOT residues remaining after extraction of the PS latex revealed a broken eggshell morphology, which was tangible evidence for the core-shell morphology. Another report for the fabrication of the PEDOT/silica colloidal nanocomposite was published using a 20 nm diameter silica sol [358]. Raspberry-shaped PEDOT-silica nanocomposite with submicrometer dimensions was produced with silica contents ranging from 19 wt % to 80 wt % and p-toluenesulfonic acid (p-TSA) and APS were added as dopant and oxidant, respectively. The mean particle diameter varied from 150 to 510 nm and the conductivity of PEDOT-silica nanocomposite was as high as 0.2 S cm^{-1}. PEDOT/silica nanocomposite particle has been also prepared using the methanolic silica sol. In addition, PEDOT bulk powder was synthesized under similar conditions in the absence of any silica sol. PEDOT/silica nanocomposite and PEDOT bulk powder were compared by transmission electron microscopy (TEM) and SEM. The distinctive raspberry shape of PEDOT/silica nanocomposite particles was obtained while this raspberry shape was non-existent in PEDOT bulk powder.

Several reports related to PEDOT-coated particles and PEDOT hollow particles have been pronounced in the literature [359, 360]. Dispersion polymerization has been applied for PEDOT-coated PS particles fabrication. 100 nm PS nanoparticle was used as the core material [359]. In order to improve the stability of the PS particle, DBSA was used as the surfactant. It was presumed that hydrophobic alkyl chains of the surfactant were positioned towards the surface of PS particles and the sulfonic acid group toward the water phase. EDOT monomer was adsorbed on the surface of the PS nanosphere and polymerization was initiated by the addition of the APS oxidant. PS-PEDOT core-shell structure was distinctively visualized by TEM. The doped PEDOT shell had a higher electron density than the PS core and the thickness of the PEDOT shell was ca. 8 nm.

Another approach has been performed for preparation of PEDOT-coated silica core-shell particles and PEDOT hollow particles [360]. Silica particle size of 130 nm was utilized as the core seed and p-TSA was used as a good dopant. p-TSA played a role of improving the solubility of EDOT monomer in water possibly due to the enhanced protonation of EDOT. Hollow PEDOT particles were obtained by removing the silica core using 20 wt % HF solution. Filler content of PEDOT-coated silica particles affected the reflection characteristics of an opalescent sample. The most dilute sample exhibited a reflection peak at ca. 610 nm, while the most concentrated sample represented a peak at ca. 400 nm. With increasing filler content, this shift in peak position was attributed to the decreased interparticle spacing and corresponding stopband wavelength.

7.2
Nanofiber, Nanorod, and Nanotube

Fabrication methods for 1-D PEDOT structures such as nanorod, tube and wire have been proposed via various synthetic routes: microemulsion polymerization, template synthesis, electrochemical polymerization, and so forth. In the case of the template synthesis, various commercial templates were produced and available currently such as track-etched PC and AAO membranes, and mesoporous silica.

PEDOT fiber has been synthesized using PC micro-filtration membranes with a pore size of 10 nm to 10 μm [361]. Electrochemical polymerization was performed using 0.5 M monomer and 0.5 M LiClO$_4$ in acetonitrile with a potential of 1.3 V vs. Ag/AgCl. The electrical conductivity and crystallinity of PEDOT in the pores of the membrane increased and tracked the common behavior of heterocyclic polymers. In this study, PEDOT fiber was embedded in the pore of template as polymeric light-emitting diode (LED) of submicron size and revealed the possibility of polymer nanoLED, although the light output of PEDOT fiber-embedded membrane was lower than that of PEDOT film [362]. Similar results were also presented in the literature [363]. The track-etched PC membrane with a thickness of 20 μm and a pore diameter of 75–150 nm was employed. Electrochemical polymerization was carried out using sodium dodecylsulfate, LiClO$_4$ and EDOT solution. The diameter of the PEDOT nanofiber was measured to be 150 nm by SEM. Raman shift from 1419 cm^{-1} to 1424 cm^{-1} corresponded to the difference in conjugation length because the 1064 nm excitation wavelength was in resonance with the longer chains of PEDOT nanofiber.

The fabrication of PEDOT nanofiber, nanotube and nanowire has been mainly focused on the AAO membrane, because AAO has advantages such as rigid shape, uniform diameter, and various pore sizes [364–368]. Electrochemical polymerization was performed in the pores of AAO membrane using SDS, LiClO$_4$ and EDOT solution and measured the resistance of the PEDOT nanofiber with a diameter of 35 nm and 150 nm, respectively. The resistance of PEDOT nanofiber was between 1.5 K and 300 K and the resistance ratio of R(T)/R (300 K) was the relevant term to investigate the intrinsic electrical properties of the conducting polymer. The resistance ratio increased drastically when the diameter of the PEDOT nanofiber was reduced from 150 nm to 35 nm.

Recently, PEDOT nanomaterials have been fabricated in the shape of rods, tubes, thimbles, and belts through chemical polymerization in the pore of AAO membrane [368]. EDOT-filled AAO membrane was employed to overcome the poor solubility of EDOT monomer in water and the difficulty in controlling the reaction time. Chemical polymerization has been accomplished by transferring EDOT-filled AAO membrane into an aqueous oxidant solution. EDOT monomer was retained in the pore of the AAO due to the extremely low solubility of EDOT in the aqueous solution. The elevated polymerization temperature and high concentrated $FeCl_3$ solution increased the rate of polymerization, which resulted in the augment of wall thickness. Under these conditions, different nanostructures of PEDOT such as belt-like structure, thimble-like structure and nanorod were realized by different $FeCl_3$ concentration and polymerization temperatures.

Soft templates with the aids of surfactants or emulsifiers have several advantages compared with hard templates: facile removal of soft template by washing, readily controllable diameter of soft template (micelle), and mild polymerization condition. Recently, Jang et al. focused on the fabrication of PEDOT nanorods using micellar soft templates [369]. Ferric cations adsorbed reverse cylindrical micelles were prepared in hexane via a coordinative interaction between an aqueous $FeCl_3$ solution and AOT. AOT was dissolved in hexane, and then the addition of aqueous $FeCl_3$ solution generated the reverse cylindrical micelles. One-dimensional water-oil interfaces compartmentalized by reverse cylindrical micelles were constructed and all requisites for carrying out the interfacial polymerization were fulfilled at the nanometer scale. Formation of the PEDOT nanorod could be achieved by chemical polymerization of EDOT owing to relatively higher preference to the organic phase, high oxidation potential of EDOT and steric hindrance of AOT tail groups. Consequently the polymerization of EDOT monomer took place locally on the reverse cylindrical micelle surface, where the monomer came into contact with the oxidizing agent, and finally PEDOT nanorod (diameter: 40 nm, length: 200 nm) was fabricated in the 1-D micelle surface.

Another interesting fabrication method has been introduced as MIMIC (micromolding in capillaries) [370]. The elastomer stamp was conformally contacted with a piece of cleaned glass or a Si wafer and then a drop of PEDOT-PSS (Baytron P) solution was applied in front of the capillary openings of the stamp. PEDOT-PSS solution migrated into capillaries in several hours and dried out. Polymer nanowire standing on the surface was left after the stamp was peeled off. Scanning force microscopy displayed that PEDOT-PSS nanowire had 278 nm periods and ca. 25 nm height. The height of molded PEDOT-PSS nanowire could be controlled, as loading weight placed on the top of the stamp was varied and the concentration of PEDOT-PSS solution was diluted. When the compressed stamp was used, the height was reduced dramatically up to 6 nm. Furthermore, the feasibility of patterning in two dimensions was exploited in this report. However, there were several defects between the nanowires due to the roughness of the glass and defects in the stamp, and some PEDOT-PSS nanowires were connected with each other at some points. A different approach has been performed in order to fabricate PEDOT-PSS nanowire with diameters under 10 nm using a molecular combining method [371]. Two microliters of the PEDOT-PSS dispersion was deposited on the nanoelectrode, fabricated on the SiO_2/Si substrate and subsequently sucked at the speed of $0.1\,\mu L\,s^{-1}$. The volume and the speed of suction of PEDOT-PSS droplet were controlled using a video-based contact angle meter. The conductivity of individual PEDOT-PSS nanowire on nanoelectrode was measured using an atomic force microscope. The conductivity of two single nanowires was determined to be 0.6 and $0.09\,S\,cm^{-1}$, which was of the same order as that of the PEDOT-PSS film.

Shinkai et al. used the superstructures of anionic porphyrin (5-, 10-, 15-, 20-tetrakis(4-sulfonatophenyl)porphyrin, TPPS) as the template, because porphyrins tended to aggregate into a 1-D rod-like structure [372]. J-aggregates of TPPS immobilized on ITO substrates were confirmed by UV-vis absorption at 489 nm. They also demonstrated the influence of TPPS on the formation mechanism of PEDOT nanorod. PEDOT films were prepared after 10, 30 and 60 cycles in the presence of TPPS and after 10 cycles in the absence of TPPS. When TPPS was not applied into solution, nanorod-like morphology was not found. Therefore, it could be concluded that TPPS played an important role in forming J-aggregates as the template. The dimension of PEDOT nanorod was 300–500 nm in length and 30–50 nm in diameter. Slow scan rate led the PEDOT nanorod to form less ordered structure in the film layer.

8
Poly(*p*-Phenylene Vinylene)

Synthetic routes for the preparation of PPV are mainly related with precursor-mediated methods. The Wessling–Zimmerman method and the Gilch polymerization technique have been widely used [373, 374]. The former method involves water-soluble sulfonium salt precursor, whereas organic soluble precursor is formed in the second route. Since PPV has been known to exhibit the excellent electroluminescence and non-linear optical properties as well as high electrical conductivity, PPV nanotubes or nanorods have a great advantage for producing optical nanodevices such as light emitting diodes and photonics applications. Therefore, most research related with PPV nanomaterials has been conducted to fabricate nanotubes, nanorods or thin film and elucidate their unique optical properties.

8.1
Nanoparticle

There have been few reports with respect to the fabrication of nanoparticles consisting of only PPV. On the other hand, researches concerning PPV-quantum dot nanocomposites have been extensively carried out. For example, Emrick and coworkers could achieve the uniform dispersion of CdSe nanocrystals into PPV thin film without aggregation [375]. First, they synthesized highly photoluminescent CdSe nanocrystals using functional ligands with excellent thermal stability. Subsequently, PPV was grafted onto the surface of the nanocrystals through functional ligands. As-prepared PPV-CdSe nanocomposite showed unique optical properties which were not found in conventional blends. Namely, the ability to tailor and disperse quantum dots in PPV thin film allowed enormous changes in the photophysical properties of the nanocomposite.

8.2
Nanofiber, Nanoribbon, and Nanotube

There have been a variety of attempts to fabricate 1-D nanomaterials consisting of PPV by a template-free method. The assembly of a conjugated polymer was induced by K^+ in a chloroform solution and nanoribbons of PPV were formed by the self-assembly technique [376]. It was postulated that the formation of 2:1 sandwich complex between K^+ ions and crown ether substituents brought two polymer chains closer together. Interpolymer bridges were created and the formation of PPV with crown ether substituents (C-PPV) aggregates in the solution was induced by the K^+ ions and $\pi - \pi$ stacking interaction of PPV molecules. The average length of the nanoribbon was ca. 400 nm, whereas that of C-PPV was less than ca. 100 nm. Similarly, the fabrication of PPV-based organogels has been published [377]. Hydrogen bonding and π-stacking for PPV played the important role of the formation of supramolecular gel nanostructures at RT. The average thickness of nanogels was approximately several micrometers with high aspect ratio and was nearly monodisperse. A novel approach had been developed to synthesize well-ordered poly[2-methoxy-5-(n-hexadecyloxy)-p-phenylenevinylene] (MH-PPV) nanotubes at the air/water interface using the LB technique [378]. The wall thickness and diameter of MH-PPV nanotube were ca. 20 nm and 130 nm, respectively.

Among the several fabrication methods, hard template method, which was pioneered by Martin et al., has been the most famous route of nanotubes and nanowires. Nanotubes of conducting polymers could be readily prepared by filling the nanopores with polymer or polymer solution using AAO template or track-etched PC membrane. PPV nanotube and nanorod had been fabricated in the pores of alumina or PC filters with pore diameter 10–200 nm by the CVD polymerization method [379]. It was postulated that this approach could open up a new fabrication method for the preparation of insoluble organic polymer nanomaterials.

Recently, there have been some attempts to prepare PPV nanowires and nanotubes by electrospinning [380–382]. Electrospun nanofiber consisted of a binary blend of poly[2-methoxy-5-(2-ethylhexyloxy)-p-phenylenevinylene] (MEH-PPV) with regioregular P3HT or poly(9,9-dioctylfluorene) [380]. The resultant nanofiber was found to have the diameter of ca. 100–500 nm with tunable optical and charge transport properties. In this manner, electrospinning has been a powerful technique to fabricate polymer, ceramic and inorganic nanowire. Nanotubular structure and coaxial nanowire composite could be also fabricated with blend of soluble core and insoluble wall material by the electrospinning method.

Another approach had been demonstrated that nanofiber of conjugated polymers and their blends were conveniently formed by the coelectrospinning method [381]. Electrospun nanofiber consisted of a more extended conformation and better spatial orientation. The coaxial nanowire indicated that MEH-PPV was homogeneously incorporated into PPV nanofiber. MEH-PPV nanowire composite exhibited the combined properties of two or more materials and discriminative properties, which were different from those of each material. Moreover, pristine MEH-PPV nanofiber had also fibrous structures with homogeneous fluorescence emission. Electrospun MEH-PPV/SBA-15 composite nanofiber had been fabricated using a dual syringe method [382]. Similarly, MEH-PPV/SBA-15 nanofiber could be fabricated in the shape of a homogeneous nanofiber.

9
Applications

9.1
Chemical Sensor and Biosensor

Many kinds of sensors using conducting polymers have been described for both chemical and biological uses. The oxidation level of conducting polymers can be easily influenced by their inherent reversible doping-dedoping (oxidation-reduction) mechanisms, causing the variations in conductiviy, color, mass, volume, and so forth. Thus, conducting polymers are capable of exhibiting sensitive responses to specific chemical or biological species. Furthermore, the sensitivity of conducting polymers can be tailored by attaching functional groups to the polymer chains or incorporating appropriate counterions during the polymerization. Owing to these characteristics, much effort has been devoted to the fabrication of sensor devices based on conducting polymers. In recent years, various conducting polymer nanomaterials have been employed to detect a wide range of chemical and biological species. It is noteworthy that the nanostructures including nanorods, nanofibers, and nanotubes not only provide a high surface-to-volume ratio but also allow rapid diffusion of analytes into and out of the material.

The most frequently used conducting polymers for developing new sensor technologies were PPy and PANI. Above all, PPy nanomaterials have been applied to various sensors such as toxic gas noses, non-toxic gas noses, aroma sensors, humidity sensors, and microbial noses [383, 384]. For example, CNT-PPy gas sensors were fabricated through a simple chemical polymerization, followed by spin-casting onto pre-patterned electrodes [240]. PPy was uniformly coated on the single wall CNTs to increase the specific surface area. CNT-PPy sensors showed an n-type behavior attributed to the anion doping in PPy during the chemical polymerization. Upon exposure to NO_2 gas, the sensitivity of the nanocomposites was about ten times higher than that of pristine PPy, and the recovery time could be shortened by taking advantage of the Joule-heating effect in the nanocomposites.

PPy wires and sensor devices with feature sizes as small as 200 nm in width could be successfully fabricated through a lift-off process [385]. Surfactants and surface-coupling agents were employed in order to improve the adhesion of PPy to the silicon substrate. Upon cyclic exposure to ammonia gas, the PPy wire sensor was at least six times more sensitive than the PPy bulk film sensor.

The electrochemical behavior of the electrodes modified with PPy nanowires was described as a chemical sensor [386, 387]. The modified electrodes had good activity towards nitrite reduction. The electroreduction current depended linearly on the concentration of nitrite and increased with increasing the PPy thickness, acidity of electrolyte solution, temperature, and scan rate. In the range of 2.28×10^{-4} to 0.02 M, the nitrite concentrations could be accurately determined from the good linearity between the electroreduction current and the concentration of nitrite.

Surface molecularly imprinted PPy nanowire was produced inside the cylindrical pores of alumina membrane [388]. The imprinted molecule (glutamic acid) was immobilized on the pore wall of the alumina membrane. The cylindrical pores were filled with pyrrole, and then the chemical polymerization proceeded. The subsequent removal of the template provided PPy nanowire with glutamic acid binding sites situated at the surface. The resulting PPy nanowire showed a high selectivity for glutamic acid. A glucose biosensor based on PPy nanomaterial was developed by several research groups [389–391]. The feasibility of a glucose sensor was performed with an aligned CNT array coated with PPy as a substrate electrode [390]. Furthermore, an amperometric enzyme electrode was prepared based on the coimmobilization of CNT and glucose oxidase within an electropolymerized PPy film [391]. CNT retained their electrocatalytic activity towards hydrogen peroxide to impart high sensitivity upon entrapment within a PPy network.

Such a simultaneous incorporation of CNT and glucose oxidase imparted biocatalytic and electrocatalytic properties onto amperometric transducers and represented a facile and effective route for preparing an enzyme electrode.

Individually addressable PPy and PANI nanowire was synthesized with controlled dimensions (width: ca. 100 nm; length: up to ca. 13 μm) using electrodeposition between electrodes as a pH sensor [392]. The nanowire sensor displayed superior changes in resistance compared with micron or submicron wire. This research was further extended to the fabrication of PPy nanowire biosensor for label-free bioaffinity sensing [393]. The one-step incorporation of functional biological molecules into PPy nanowire during its electropolymerization was a critical advantage of a novel fabrication method over the CNT and silicon nanowire biosensors that required post-synthetic modification and positioning.

Recently, PANI nanofiber has been widely applied to diverse sensor performances [115, 394–397]. The nanofiber was deposited on the interdigitated electrodes in the form of fiber networks, and their resistance changes were monitored in real-time. PANI nanofiber sensors presented the improved responses for various analytes (e.g., HCl and NH_3 [115, 395], N_2H_4 [395, 396], $CHCl_3$ and CH_3OH [395], and H_2S [397]), relative to the bulk counterparts. On the other hand, the chemical sensors composed of a single PANI nanowire were fabricated by means of scanned-tip electrospinning deposition [398]. The sensor devices provided reversible responses to NH_3 gas concentration as low as 0.5 ppm. Moreover, the diameter-dependent behavior of the sensors was investigated: PANI wire with smaller diameter had a faster response due to the more rapid diffusion of gas molecules through the wire.

The selective and facile incorporation of sensing materials into sensor devices is an important issue. From this point of view, PANI nanoframework was electrochemically grown between the two platinum microelectrodes [399]. The nanoframework comprised numerous intercrossing nanowires with diameters of 40–80 nm, and was utilized as resistive sensors for detecting HCl, NH_3, C_2H_5OH, and NaCl.

PT has also received attention as an excellent sensing material. The sensing behaviors of PT were extensively examined by Leclerc and coworkers [400, 401]. An interesting methodology was developed in order to allow a simple optical detection of iodide [400]. This method was based on the conformational modification of conjugated backbone of cationic poly(3-alkoxy-4-methylthiophene) upon binding of an iodide anion. It was also reported that the electroactive, water-soluble, and cationic PT was employed in detecting DNA to diagnose various diseases [401]. As the neutral probes were hybridized with a complementary DNA target, the solid-state electrochemical detection occurred due to an attractive electrostatic interaction with a cationic PT.

Compared with other conducting polymers, little research has been done on the development of sensor systems based on PEDOT nanomaterial. Glucose sensors based on the PEDOT nanotube membrane were developed for amperometric biosensor applications [389, 402–404], and the composite film composed of single-walled CNT, negatively charged DNA, and positively charged PEDOT provided a novel electrochemical and photochemical functions [405]. Recently, PEDOT nanomaterials are emerging as the promising sensing material because of their outstanding environmental stability and low bandgap. For example, PEDOT nanorod (diameter: ca. 40 nm, length: ca. 200 nm) was utilized in order to detect HCl and NH_3 vapor [369]. When the PEDOT nanorod was exposed to NH_3 vapor, a steep increase in resistance was observed (Fig. 6a). Upon exposure to NH_3 vapor, NH_3 molecule diffused into the PEDOT nanorod, and thereafter the dedoping of the PEDOT nanorod rapidly proceeded through the interaction between NH_3 vapor and polymer backbone. On the contrary, when PEDOT nanorod was exposed to HCl vapor, a decrease in resistance could be observed through the doping mechanism (Fig. 6b). The PEDOT nanorod chemical sensors showed the reversible and reproducible responses (more than 10 cycles) for HCl and NH_3 analytes (Fig. 6c,d). In addition, the nanorod sensors showed a measurable response to NH_3 and HCl vapor concentration as low as 10 ppm and 5 ppm, respectively.

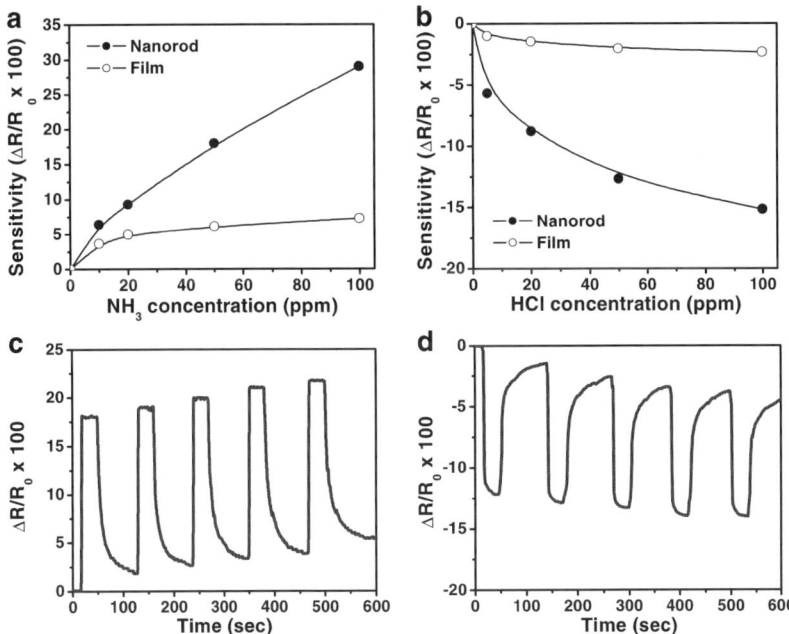

Fig. 6 The sensitivity change of PEDOT nanorods and conventional PEDOT films as a function of **a** NH_3 and **b** HCl vapor concentration, and the reversible and reproducible response of PEDOT nanorods upon periodic exposure to **c** NH_3 and **d** HCl vapor (reproduced with permission from [369])

9.2
Transistor and Switch

The use of conducting polymers as logic or switching elements is of key importance in the development of new electronic devices [68–71]. As a typical example, organic field-effect transistor contains a gate, drain, source, dielectric layer, and semiconducting layer [349]. The current flow between the drain and source electrodes is modulated by the applied gate voltage. An "off" state of a transistor is when no voltage is applied between the gate and the source electrodes. When a voltage is applied to the gate, charges can be introduced into the semiconducting layer at the interface between the semiconductor and dielectric layers. Consequently, the drain-source current increases due to increased charge carriers, which corresponds to the "on" state of a transistor. On the basis of such a concept, n-type PT has been synthesized [351, 352]. It was reported that the field-effect mobility in PT field-effect transistors increased with reduced channel lengths at high source/drain voltages. This was contradictory to the decrease in mobility caused by short-channel effects in amorphous Si thin film transistors [406]. The effect of the physical channel length on the current-voltage characteristics of thin film transistors was examined with regioregular PT [407]. In addition, the field-effect transistor based on P3HT was investigated to determine the influence of moisture on device characteristics and to understand the mechanism underlying the susceptibility to air [408]. The fundamental output characteristics remained almost the same for every current-voltage profile in a vacuum, and N_2 and O_2 atmospheres. However, the operation in N_2 humidified with water caused enlarged off-state conduction and deterioration in the saturation behavior.

Cobalt-PPy-cobalt nanowire was electrochemically synthesized inside alumina membrane and the field-effect transistors were fabricated by patterning a gate on one side of the cobalt-PPy-cobalt nanowire [409]. The measured output and transfer characteristics are as good as or better than PPy film field-effect transistors. The gain of the nanowire field-effect transistors could be controlled with successive electrochemical doping of the PPy segment.

A self-assembled nanojunction (diameter: ca. 20 nm, length: ca. 40 nm) was fabricated between gold microelectrodes on the SiO_2/Si substrate using an end-functionalized poly(p-phenylene ethynylene) derivative [410]. The nanojunction operated as a nanometer-scale photoswitch. With light on or off, the nanojunction was capable of switching between low or high resistance states. The switching of two states was reversible and fast. As the back silicon substrate of the nanojunction was connected ohmically as a gate electrode, the output characteristics suggested that the nanojunction could operate as a p-type transistor.

The photoconducting nanotube device composed of graphitic carbon and PPV layers showed high photoconductiviy efficiencies under illumination from a xenon lamp [411]. This result suggested that the nanotube device could be used in photoswitching applications.

9.3
Data Storage

The storage mechanism of conducting polymer memory is perfectly different from that of silicon memory. Rather than encoding "0" and "1" as the amount of charges accumulated in a memory cell, conducting polymer memory generally records data using its bistable electrical response to an applied voltage [317, 412–417].

Recently, a non-volatile digital memory device was fabricated using PANI nanofiber (diameter: ca. 30 nm) decorated with gold nanoparticles (diameter: below 5 nm) [317]. As shown in Fig. 7a, the polymer memory device displayed bistable electrical behavior (ON/OFF states). The transition from the OFF state to the ON state was ascribed to an electric field-induced charge transfer between the PANI nanofibers and the gold nanoparticles. The write-read-erase cycle tests could also be performed more than seven times (Fig. 7b).

In a similar manner, PT has also been applied to data storage devices [417]. The memory devices were fabricated with the oriented thin film of PT derivatives. Polymer thin film exhibited the presence of two different conducting states (ON/OFF states), which was dependent on the voltage-sweep

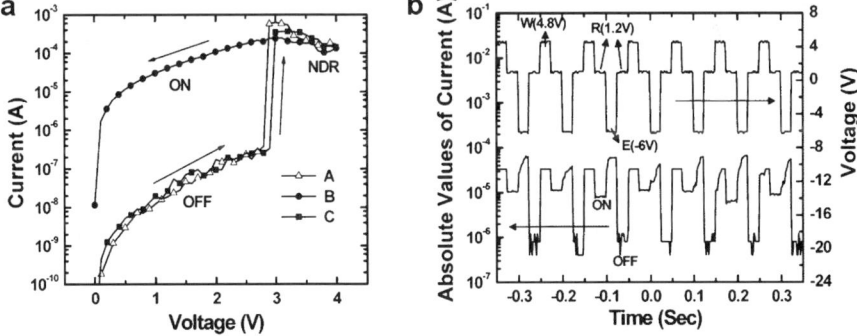

Fig. 7 a Current-voltage characteristics of the polyaniline nanofiber/gold nanoparticle device. The potential is scanned from (A) 0 to +4 V, (B) +4 V to 0 V, and (C) 0 to +4 V. Between +3 and +4 V, a region of negative differential resistance is observed. The *inset* shows the retention time test of the ON state (*top*) and OFF state (*bottom*) currents when biased at +1 V every 5 s. **b** Current response (*left axis*) of the polyaniline nanofiber/gold nanoparticle device to applied potentials (*right axis*) during write-read-erase testing cycles. A potential of +4.8 V is used to write, –6 V is applied to erase, and +1.2 V is used to read. W = write, R = read, and E = erase. The duration of each cycle pulse is 0.1 s, during which time the current response is recorded using an oscilloscope (reprinted with permission from [317]; copyright 2005, American Chemical Society)

direction. A change in the conducting state induced a switching between high and low values in the PL intensity. Consequently, the PL from the polymer thin film could be used as a tool to read the ON/OFF state of the device.

9.4
Supercapacitor

Supercapcitors based on conducting polymers have advantages such as high capacitive energy density and low material cost. On the other hand, they also have disadvantages such as lower life cycle and slow ion transport. For this reason, the nanocomposite electrodes based on CNT coated with conducting polymers have been widely employed for supercapacitors [242, 243, 418–423]. The high surface area and conductivity of CNT enhanced the redox property of conducting polymers. However, the structure and capacitance of the nanocomposites are inherently susceptible to the synthetic route. Electrochemical growth methods are limited by the amount of polymer that can be deposited on the CNT without blocking electrolyte channels, and chemical oxidation polymerization methods can suffer from aggregation of the polymer deposited on the CNTs. To solve these problems, an electrochemical approach to grow the CNT-PPy nanocomposite was created by the simultaneous deposition of multi-walled CNT and PPy during the film growth in a single step [242, 243]. The specific capacitance per mass and geometric area of the nanocomposite was as high as $192\,F\,g^{-1}$ and $1.0\,F\,cm^{-2}$, respectively. These values were at least two times higher than those of conventional PPy film. It was reported that maximum specific capacitance of $265\,F\,g^{-1}$ was measured from the single-walled CNT/PPy nanocomposites [419]. Recently, the supercapacitor electrodes based on pyrrole treated single-walled CNT were developed [420]. High values of capacitance ($350\,F\,g^{-1}$), power density ($4.8\,kW\,kg^{-1}$), and energy density ($3.3\,kJ\,kg^{-1}$) were obtained in KOH aqueous solution, and the capacitance was almost seven times that ($55\,F\,g^{-1}$) of the untreated buckypaper. While buckypaper was predominantly composed of micropores, pyrrole treated samples were mainly composed of meso- and macropores. Correlating the capacitance with surface area, pore size, and pore size distribution, the macropore made a significant contribution to the capacitance performance of these materials.

The electrochemical quartz crystal microbalance was utilized to elucidate the deposition and intercalation behavior of PPy-CNT and PPy-chloride thin layer [421]. Interestingly, the capacitive charging current was observed on the massogram at the positive switching potential, and this result confirmed the redox pseudocapacitive behavior of PPy.

Carbon nanofiber has significant advantages in terms of low manufacture cost and mass production compared with CNT. Recently, PANI-coated carbon nanofiber was readily fabricated using one-step VDP [422]. The specific capacitance of PANI-coated carbon nanofiber was susceptible to the thickness of the PANI layer. The maximum specific capacitance was 264 F g^{-1} when the thickness of the PANI layer was ca. 20 nm.

9.5
Photovoltaic Cell

Conducting polymers have been considered as the alternative of inorganic semiconductors for low-cost electronics and optical devices owing to easy synthesis and mechanical flexibility. From this viewpoint, PT has been applied to solar cells and photovoltaics as the type of pure polymer or polymer-inorganic hybrids [424–429]. It was demonstrated that the morphological properties of each working layer in a hybrid solar cell dramatically affected the device performance [424]. The active layer commonly contained an electron donor phase (e.g., conducting polymers) and an electron acceptor phase (e.g., CdSe nanocrystals and fullerene). Consequently, it was significant to improve the homogeneity of the active layer consisting of two different phases for high-efficiency cells. Recently, it was reported that P3HT with an amino end-group enhanced the efficiency of P3HT-CdSe solar cells by increasing the dispersion of CdSe nanorods [426]. CdSe nanorods (diameter: ca. 7 nm, length: ca. 30 nm) could be well-dispersed in P3HT film through the coordination of an amino end-group with CdSe. PPV derivative/fullerene bulk heterojunction solar cells have been also explored: soluble fullerene and PPV derivative were employed as an electron acceptor and an electron donor, respectively [430–435].

9.6
Electrochromic Device

Electrochromism can be defined as a reversible color change in a material induced by an electric field. Early studies of electrochromism focused on inorganic (e.g., WO$_3$) compounds and organic molecules (e.g., bypyridiliums) [67]. However, recent researches have fairly focused on conducting polymers such as PPy, PANI, and PT [436]. Conducting polymers have advantages over the other electrochromic materials, including ease of processing, fast switching time, high contrast and coloration efficiency, tunable bandgap, and so forth. In particular, nanostructured conducting polymers can provide a much faster switching time compared with bulk counterparts. The electrochromic devices with fast switching speeds could be recently fabricated based on well-defined PEDOT nanotube arrays [437]. The thin walls (thickness: ca. 10–20 nm) of PEDOT nanotubes contributed to the rapid diffusion of counterions during the redox process, which allowed extremely fast switching speeds (below 10 ms). In addition, the micrometer-sized length of PEDOT nanotubes could produce a strong coloration.

Electrospun conducting polymer nanofibers (diameter: ca. 120 nm) also exhibited a desirable electrochromic property [438]. The electrospun conducting polymer nanofibers formed an interconnected network by solid/swollen-state oxidative crosslinking without significant perturbation of the morphology. Nanofibers showed relatively fast switching times of 2–3 s.

9.7
Field Emission Display

One-dimensional conducting polymer nanomaterials have been utilized as the field emission electron sources for flat panel displays [365–367]. Conducting polymer nanotubes or nanowires were mostly prepared by the electrochemical polymerization within the cylindrical pores of alumina membranes, and the field emission characteristics were evaluated. As a typical example, a field emission cell was composed of PEDOT nanowire (conductivity, 3.4×10^{-3} S cm^{-1}) tips (cathode) and ITO (anode). The turn-on field of PEDOT nanowire was 3.5–4.0 μA cm^{-2} at 10 V μm^{-1}, and the current density increased up to 100 μA cm^{-2} at 4.5 V μm^{-1}. The field enhancement factor of the PEDOT nanowire tips was $\sim$ 1200 and this value was comparable to that of CNT. PPy nanowire and PANI nanotube was also prepared using nanoporous template, and their field emission characteristics were investigated [365]. PPy nanowire and PANI nanotubes showed the turn-on fields of 3.5–4.0 and $\sim$ 5.0 μA cm^{-2} at 6 and 8 V μm^{-1}. These studies offered a great feasibility of conducting polymers as the building blocks for all-polymer field emission displays.

9.8
Actuator

Conducting polymers have been considered as the excellent material for actuator applications, because it undergoes reversible volume changes (expansion and contraction) during oxidation and reduction. Recent research trends concerning conducting polymer actuators have been reviewed by Smela [439]. PPy was comprised in a broad class of electrically actuated polymeric materials known as electroactive polymers, and exhibited several advantages including large strain (3% in-plane to > 30% out-of-plane), high strength, low voltage operation, biocompatibility, and so forth. Jager et al. have published a series of works, describing the novel fabrication of conducting polymer microactuators [440–442]. They developed microactuators, so-called micromuscles, based on a PPy-Au bilayer. The Au layer acted both as structural layer and an electrode. The micromuscles were used to lift plates, to open and close boxes [440]. In addition, the microrobotic arms could pick up, lift, move, and place micrometer-size objects within an area of about 250×100 μm^2 [441]. The controlled movement of the robot arm was due to individually controlled microactuators.

9.9
Optically Transparent Conducting Material

Over the last decade, significant advances have been made in the processing of conducting polymers, especially in the area of solution processing. These studies have led to transparent, highly conductive coatings and films. Now, with their ability to be processed into coatings and films coupled with their unique properties, conductive polymers show vast promise for industrial applications ranging from electrostatic dissipation to electrochromic displays. In traditional methods, there are two schools of thought on how to process conducting polymers into coatings and films: (1) modify or manipulate the chemistry of the conductive polymer to induce solubility in organic solvents, and (2) disperse and/or compatibilize the intractable conductive polymer in conventional film-forming matrices. Conducting polymer nanomaterials provide excellent transparent property owing to their nano-sized characteristic. The dispersion without losing the electrical path is a major concern in this application. Tailored nanostructures such as core-shell structures and surface charge repulsions can promote the transparent conductive property. To obtain a highly transparent film, conductive powders should have an average particle diameter, no greater than half of the shortest wavelength of visible light, and ultrathin networks of conductive fillers should be also formed in the transparent matrix. Recently, highly transparent conductive thin films were fabricated using PPy-PMMA core-shell nanoparticle as the conductive filler in a PMMA matrix [213]. As the core-shell nanoparticle was mixed with the PMMA matrix, the polymer nanocomposite film showed an excellent performance in both transparency and conductivity. PMMA shell prevented the aggregation of PPy nanoparticle in the nanocomposite film. Moreover, the nanometer-sized PMMA shell had a glass transition temperature (T_g) lower than that of a bulky one. Consequently, PPy nanoparticles could be effectively linked by annealing the film at over T_g, and the electrical paths were constructed without the loss of transparency.

9.10
Surface Protection

It is known that conducting polymers provide beneficial protection to various kinds of metals in a corrosive environment [443–445]. Over the past decades, conducting polymers have been considered as the potential candidate for effective corrosion protection without the use of heavy metals. It was found that PPy-clay nanocomposites with low clay loading (e.g., 1.0 wt %) provided an enhanced corrosion protection effect on cold-rolled steel compared with pristine PPy [446]. Similarly, PANI-clay nanocomposites exhibited a better anticorrosion performance than conventional PANI [447]. In the case of PT, there have been only a few attempts for corrosion control. The high oxidation potential of thiophene monomer gave rise to difficulty in synthesizing PT film on the surface of oxidizable metals. For example, the deposition of PT film could not be achieved on the copper surface [448]. This was because the intense dissolution of copper simultaneously occurred during the polymerization reaction. Consequently, the formation of stable PT film could be accomplished only after the coating of PPy thin film on the copper surface. Very adherent and smooth PT film could be obtained on nickel coated mild steel using cyclic voltammetry technique [449]. The PT film filled the role of effective physical barrier with low permeability on the metal surface.

9.11
Substituent for Carbon Nanomaterial

PPy consists of five-membered cyclic rings with crosslinking structures (α,α-/α,β-links). Accordingly, PPy has been shown to be an excellent precursor for fabricating various graphitic carbon nanomaterials such as a carbon nanosphere [146], carbon nanocapsule [84, 223], CNT [172, 249], mesoporous carbon [450], and fullerene [451].

A simple way to magnetic CNTs were developed using an iron-impregnated PPy nanotube precursor [249]. PPy nanotube precursor were prepared by chemical oxidation polymerization using a self-assembly method that was advantageous for large-scale synthesis. Then, subsequent carbonization of PPy nanotube was performed to generate magnetic CNT. The main advantage of this synthetic strategy was that the reverse microemulsion polymerization, followed by the carbonization process, made it possible to fabricate magnetic CNTs without using any specific template and encapsulation process. Magnetic CNT showed ferromagnetic behavior even at RT and the magnetic phase was γ-Fe_2O_3.

A very interesting route to fullerenes has been described by using PPy nanoparticles as the carbon precursor [451]. PPy nanoparticles with diameters of several nanometers could be transformed into fullerenes, in high yield, by mild-temperature carbonization ($\sim 1000\,°C$) under atmospheric pressure (Fig. 8). During the carbonization process of PPy, the carbonization reactions such as dehydrogenation and denitrogenation occurred and produced more compact polycondensed graphitic structures. The carbonization process initially affected the surface structure of the precursors thermodynamically. Consequently, the elimination and cyclization reactions for graphitization started at the surface of the polymer precursors. The graphitization was considered to proceed by mass-transfer from the core polymer chains to the surface graphitic materials, forming a hollow graphitic structure. Fullerenes have been largely fabricated by bottom-up approach in which tiny carbon molecules grow into larger units. In addition, the synthetic pro-

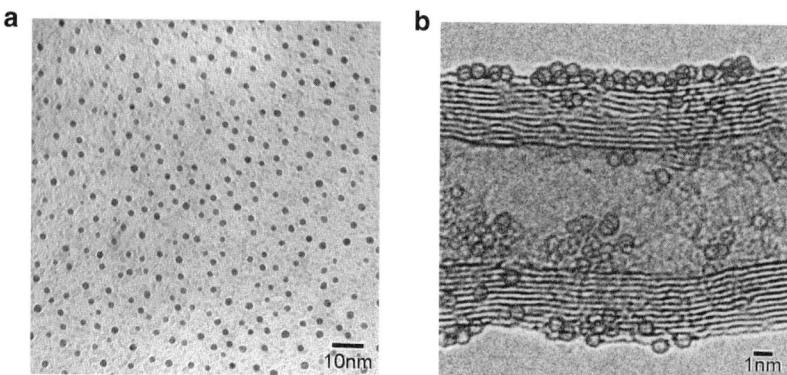

Fig. 8 **a** Normal TEM image of PPy nanoparticles with average diameter of 2 nm. **b** High resolution TEM image of the toluene extract of the carbonized PPy nanoparticles at 950 °C for 6 h (reproduced with permission from [451])

cess has been mainly conducted under vacuum pressure. However, the novel method adopts a top-down approach by mild-temperature carbonization (ca. 1000 °C) under atmosphere pressure and this is the first experimental evidence for polymer nanoparticle as a fullerene precursor in supercarbon science.

10
Conclusions

Currently, the fabrication of conducting polymer nanomaterials has been considered as a valuable and important topic in polymer material and science. In this review, recent fabrication methodologies of conducting polymer materials with the nano-sized dimension, namely, soft template method, hard template method, and template free method, have been presented. The fabrication technologies described here remove some of the obstacles to progress in the production of advanced conducting polymer materials. In addition, a variety of applications have been also focused on the well-established practical fields for PPy, PANI, PT, PEDOT, and PPV. It has been expected that research and development in conducting polymer nanomaterials proliferate novel performance and the enhancement of their physical and chemical properties compared to that in the bulk state. Moreover, new concepts toward nanotechnology will provide the emerging and promising applications of conducting polymers in future. This trend paves the way for the development of associated research fields with conducting nanomaterials. It is also the advent of a novel research area based on fusion technology such as the combination with information technology (IT), biotechnology (BT), and environmental technology (ET).

Acknowledgements This work has been financially supported by the Brain Korea 21 program of the Korean Ministry of Education and the Hyperstructured Organic Materials Research Center supported by Korea Science and Engineering Foundation.

References

1. Zhao QQ, Boxman A, Chowdhry U (2003) J Nanopart Res 5:567
2. Lazzari M, Lopez-Quintela MA (2003) Adv Mater 15:1583
3. Rao CNR, Chetham AK (2001) J Mater Chem 11:2887
4. Li W, Sha X, Dong W, Wang Z (2002) Chem Commun 2434
5. Li M, Mann S (2002) Adv Funct Mater 12:773
6. Barbe CJ, Graf R, Finnie S, Blackford M, Trautman R, Bartlett JR (2003) 26:457
7. Lee MS, Lee GD, Park SS, Hong SS (2003) J Ind Eng Chem (2003) 9:89
8. Anderson M, Osterlund L, Ljungstrom S, Palmgvist A (2002) J Phys Chem B 9:241
9. Mori Y, Okastu Y, Tsujimoto Y (2001) J Nanopart Res 3:219
10. Shchukin DG, Sukhorukov GB (2004) Adv Mater 16:671
11. Peng X, Manna L, Yang W, Wickjam, Scher E, Kadavanich A, Alvisatos AP (2000) Nature 404:59
12. Rao CNR, Kulkarui GU, Thomas PJ, Edwards PP (2000) Chem Soc Rev 29:27
13. Pileni MP (2000) Pure Appl Chem 72:53
14. Ohde H, Hunt F, Wai CM (2001) Chem Mater 13:4130
15. Hirai T, Sato T, Komasawa I (2001) J Phys Chem 105:9711
16. Dutta P, Fendler JH (2002) J Colloid Interface Sci 247:47
17. Chen DH, Wu SH (2000) Chem Mater 12:1354
18. Capek I (2004) Adv Colloid Interface Sci 110:49
19. Shirakawa H, Louis EJ, MacDiarmid AG, Chiang CK, Heeger AJ (1977) Chem Commun 578
20. Chiang CK, Fincher CR, Park YW, Heeger AJ, Shirakawa H, Louis EJ (1977) Phys Rev Lett 39:1098
21. Shirakawa H, Ito T, Ikeda S (1978) Makromol Chem 179:1565
22. Rabolt JF, Clarke TC, Street GB (1979) J Chem Phys 71:4614
23. Fincher CR, Ozaki M, Heeger AJ, MacDarmid AG (1979) Phys Rev B 19:4140
24. Harada I, Furukawa Y, Tasumi M, Shirakawa H, Ikeda S (1980) J Chem Phys 73:4746
25. Heeger AJ (2001) Angew Chem Int Ed 40:2591
26. Shirakawa H (2001) Angew Chem Int Ed 40:2574
27. Hörhold HH, Helbig M, Raabe D, Opfermann J, Scherf U, Stockmann R, Weiss D (1987) Z Chem 27:126
28. Burroughes JH, Bradley DDC, Brown AR, Marks RN, Mackay K, Friend RH, Burn PL, Holmes AB (1990) Nature 347:539
29. Braun D, Heeger AJ (1991) Appl Phys Lett 58:1982
30. Burn PL, Holmes AB, Kraft A, Bradley DDC, Brown AR, Friend RH, Gymer RW (1992) Nature 356:47
31. Gustafsson G, Cao Y, Treacy GM, Klavetter F, Colaneri N, Heeger AJ (1992) Nature 357:477
32. Greenham NC, Moratti SC, Bracley DDC, Friend RH, Holmes AB (1993) Nature 365:628
33. Miller JS (1993) Adv Mater 5:587
34. Miller JS (1993) Adv Mater 5:671
35. Toshima N, Hara S (1995) Prog Polym Sci 20:155

36. Baigent DR, Hamer PJ, Friend RH, Moratti SC, Holmes AB (1995) Synth Met 71:2175
37. Thorn-Csányi E, Kraxner P (1995) Macromol Chem Rapid Commun 16:147
38. Bao Z, Chan WK, Yu L (1995) J Am Chem Soc 117:12426
39. Lee C, Kim KY, Rhee SB (1995) Synth Met 84:221
40. Nishide H, Kaneko T, Nii T, Katoh K, Tsuchida E, Yamaguchi K (1995) J Am Chem Soc 117:548
41. Arroyo-Villan MI, Diaz-Quijada GA, Abdou MSA, Holdcroft S (1995) Macromolecules 28:975
42. Wu X, Chen TA, Rieke RD (1995) Macromolecules 28:2101
43. Sotzing GA, Reynolds JR (1995) J Chem Soc Chem Commun :703
44. Reddinger JL, Sotzing GA, Reynolds JR (1996) J Chem Soc Chem Commun 1777
45. Sapp SA, Sotzing GA, Reddinger JL, Reynolds JR (1996) Adv Mater 8:808
46. Feast WJ, Tsibouklis J, Pouwer KL, Groenendaal L, Meijer EW (1996) Polymer 37:5017
47. Yu G (1996) Synth Met 80:143
48. Cornil J, Beljonne D, dos Santos DA, Shuai Z, Brédas JL (1996) Synth Met 78:209
49. Yang Y, Pei Q, Heeger AJ (1996) J Appl Phys 79:934
50. Pelter A, Jenkins I, Jones DE (1997) Tetrahedron 53:10357
51. Roncali J (1997) Chem Rev 97:173
52. Pu L, Wagaman MW, Grubbs RH (1997) Macromolecules 30:3978
53. Middlecoff JS, Collard DM (1997) Synth Met 84:221
54. Sankaran B, Reynolds JR (1997) Macromolecules 30:2582
55. Kumar A, Welsh DM, Morvant MC, Piroux F, Abboud KA, Raynolds JR (1998) Chem Mater 10:896
56. Sotzing GA, Thomas CA, Reynolds JR, Steel PJ (1998) Macromolecules 31:3750
57. Saunders BR, Kumar N, Chandra S (1995) Chem Mater 7:1082
58. Nishio K, Fujimoto M, Ando O, Ono H, Murayama T (1996) J Appl Electrochem 26:425
59. Janata J, Josowicz M (2003) Nature Mater 2:19
60. Deronzier A, Moutet JC (1996) Coord Chem Rev 147:339
61. Groenendaal L, Peerlings HWI, van Dongen JLJ, Havinga EE, Vekemans JAJM, Meijer EW (1995) Macromolecules 28:116
62. MacDiarmid AG (1997) Synth Met 84:27
63. Wang L, Jing X, Wang F (1991) Synth Met 41–42:739
64. Reddinger JL, Reynolds JR (1999) Adv Polym Sci 145:57
65. Wessling B (1997) Synth Met 85:1313
66. Kumar D, Sharma RC (1998) Polym J 8:1053
67. Argun AA, Aubert PH, Thompson BC, Schwendeman I, Gaupp CL, Hwang J, Pinto NJ, Tanner DB, MacDiarmid AG, Reynolds JR (2004) Chem Mater 16:4401
68. Kelly TW, Baude PF, Gerlach C, Ender DE, Muyres D, Haase MA, Vogel DE, Thesis SD (2004) Chem Mater 16:4413
69. Newman CR, Frisbie CD, da Silva Filho DA, Brédas JL, Ewbank PC, Mann KR (2004) Chem Mater 16:4436
70. Coakley KM, McGehee MD (2004) Chem Mater 16:4533
71. Veres J, Ogier S, Lloyd G (2004) Chem Mater 16:4543
72. Kulkarni AP, Tonzola CJ, Babel A, Jenekhe SA (2004) Chem Mater 16:4556
73. Pron A, Rannou P (2002) Prog Polym Sci 27:135
74. Searson PC, Moffat TP (1994) Crit Rev Surf Chem 3:171
75. Toshima N, Hara S (1995) Prog Polym Sci 20:155
76. Malinauskas A (2001) Polymer 42:3957
77. Whang YE, Han JH, Nalwa HS, Watanabe T, Miyata S (1991) Synth Met 41:3043

78. Neoh KG, Kang ET (1989) J Appl Poly Sci 38:2009
79. Martina S, Enkelmann V, Wegner G, Schlüter AD (1992) Synth Met 51:299
80. Groendaal L, Peelings HWI, Van Dongen JLJ, Havinga EE, Vekemans JAJM, Meijer EW (1995) Macromolecules 28:116
81. Whang YE, Han JH, Nalwa HS, Watanabe T, Miyata S (1991) Synth Met 43:3043
82. Qian R (1993) In: Salantck WR, Lundström I, Ranby B (eds) Conjugated Polymers and Related Materials. Oxford University Press, London, p 161
83. Chandrasekhar P (1999) Conducting polymer, fundamentals and applications: a practical approach, ch 2. Kluwer, Norwell, MA
84. Jang J, Oh JH, Li XL (2004) J Mater Chem 14:2872
85. Stilwell DE, Park SM (1998) J Electrochem Soc 135:2254
86. Adams PN, Monkman AP (1997) Synth Met 87:165
87. Beadle PM, Nicolau YF, Banka E, Rannou P, Djurado D (1998) Synth Met 95:29
88. Stejskal J, Riede A, Hlavata D, Prokes J, Helmstedt M, Holler P (1998) Synth Met 96:55
89. Min G (2001) Synth Met 119:273
90. Morales GM, Miras MC, Barbero C (1999) Synth Met 101:686
91. Syed AA, Dinesan MK (1991) Talanta 38:815
92. Sun Z, Geng Y, Jing JL, Wang F (1997) Synth Met 84:99
93. Geng Y, Li J, Sun Z, Jing X, Wang F (1998) Synth Met 96:1
94. Kuramoto N, Tomita A (1997) Synth Met 88:147
95. Kuramoto N, Takahashi Y (1998) React Funct Polym 37:33
96. Stenger-Smith JD (1998) Prog Polym Sci 23:57
97. Tallman DE, Vang C, Bierwagon GP, Wallace GG (2002) Electrochem Soc 149:173
98. Akundy GS, Iroh JO (2001) Polymer 42:9665
99. Jérôme C, Demoustier-Champagne S, Legras R, Jérôme R (2000) Chem Eur J 6:3089
100. Choi SJ, Park SM (2000) Adv Mater 12:1547
101. Yang Y, Wan M (2001) J Mater Chem 11:2022
102. Lewis TW (2000) A study of the overoxidation of the conducting polymer polypyrrole. Dissertation, University of Wollongong
103. Jiang M, Wang J (2001) Electroanal Chem 500:584
104. Misoska V, Price WE, Ralph SF, Ogata N, Wallace GG (2001) Synth Met 123:279
105. Akundy GS, Rajagopalan R, Iroh JO (2002) J Appl Polym Sci 83:1970
106. Rajagopalan R, Iroh JO (2001) Electrochim Acta 46:2443
107. Sazou D (2001) Synth Met 118:133
108. Gao M, Huang S, Dai L, Wallace GG, Gao R, Wang Z (2000) Angew Chem 39:3664
109. Wang D, Caruso F (2001) Adv Mater 13:350
110. Bartlett PN, Birkin PR, Ghanem HA, Toh C-S (2001) J Mater Chem 11:849
111. He HX, Li CZ, Tao NJ (2001) J Appl Phys Lett 78:811
112. Norris ID, Kane-Magurie LAP, Wallace GG (2000) Macromolecules 33:3237
113. Zhou D, Innis PC, Wallace GG, Shimizu S, Maeda S (2000) Synth Met 114:287
114. Aboutanos V, Kane-Maguire LAP, Wallace GG (2000) Synth Met 114:313
115. Huang J, Virji S, Weiller B, Kaner RB (2003) J Am Chem Soc 125:314
116. Bates FS (1991) Science 251:898
117. Zhao D, Feng J, Huo Q, Melosh N, Fredrickson GH, Chmelka BF, Stucky GD (1998) Science 279:548
118. Raez J, Barjovanu R, Massey JA, Winnik MA, Manners I (2000) Angew Chem Int Ed 39:3862
119. Leclere P, Calderone A, Marsitzky D, Francke V, Geerts Y, Mullen K, Bredas JL, Lazaronic R (2000) Adv Mater 12:10742

120. Mannes I (1999) Chem Commun 857
121. Wei Z, Zhang Z, Wan M (2002) Langmuir 18:917
122. Qui H, Wan M (2001) J Polym Sci Part A Polym Chem 39:3485
123. Spector MS, Price RR, Schnur JM (1999) Adv Mater 11:337
124. Yao TM, Kim NE, Xia Y, Whitesides GM, Aksay IA (1997) Nature 390:674
125. Kresge CT, Leonowicz ME, Roth WJ, Vartuli JC, Beck JS (1992) Nature 359:710
126. Schnur JM (1993) Science 262:1669
127. Antonelli DM (1999) Adv Mater 11:487
128. Ghadiri M, Granja JR, Buehler L (1994) Nature 369:301
129. Michaelson JC, McEvoy AJ (1994) Chem Commun 79
130. Lindman J, Kronberg H (1998) Surfactants and polymers in aqueous solution. Wiley, Chichester, New York
131. Lovell PA, Al-Aasser MS (1997) Emulsion polymerization and emulsion polymers. Wiley, New York, p 700
132. Antonietti M, Basten R, Lohmann S (1995) Macromol Chem Phys 196:441
133. Candau F (1992) Polymerization in Organized Media. Gordon Science Publications, Philadelphia, p 215
134. Schork FJ, Poehlein GW, Wang S, Reimers J, Rodrigues J, Samer C (1999) Colloids Surf A 153:39
135. Chern CS, Chen TJ, Liou YC (1998) Polymer 39:3767
136. Landfester K (2000) Macromol Symp 150:171
137. Blythe PJ, Sudol ED, El-Aasser MS (2000) Macromol Symp 150:179
138. Jang J, Bae J (2004) Angew Chem Int Ed 43:3803
139. Yildiz U, Landfester K, Antomietti M (2003) Macromol Chem Phys 204:1966
140. Liu G, Yan X, Duncan S (2002) Macromolecules 35:9783
141. de Moel K, Alberda van Ekenstein GOR, Nijland H, Polushkin E, Brinke G, Maeki-Ontto R, Ikkala O (2001) Chem Mater 13:4580
142. Yang CS, Awschalom D, Stucky GD (2002) Chem Mater 14:1277
143. Kahlweit M (1985) Angew Chem Int Ed 24:654
144. Holliday BJ, Mirkin CA (2001) Angew Chem Int Ed 40:2022
145. Pavel FM, Mackay RA (2000) Langmuir 16:8568
146. Jang J, Oh JH, Stucky GD (2002) Angew Chem Int Ed 41:4016
147. Reynolds F, Jun K, Li Y (2001) Macromolecules 34:167
148. Jang J, Lee K (2002) Chem Commun 1098
149. Caruso F (2001) Adv Mater 13:11
150. Jang J, Ha H (2003) Chem Mater 15:2109
151. Zhang L, Wan M (2003) Adv Funct Mater 13:815
152. Duan X, Huang Y, Cui Y, Wang J, Lieber CM (2001) Nature 409:66
153. Jang J, Yoon H (2003) Chem Commun 720
154. Landfester K, Bechthold N, Tiarks F, Autonietti M (1999) Macromolecules 32:5222
155. Landfester K (2001) Adv Mater 13:765
156. Martin CR (1994) Science 266:1961
157. Gangopadhyay R (2004) Encyclo Nanosci Nanotech 2:105
158. Martin CR (1995) Acc Chem Res 28:61
159. Wan M (2004) Encyclo Nanosci Nanotech 2:153
160. Martin CR, Skotheim TA, Elsenbaumer RL, Reynolds JR (1998) Handbook of Conducting Polymers, ch 16, 2nd ed. Marcel Dekker, New York
161. Parthasarathy R, Martin CR (1994) Nature 369:298
162. Martin CR, Parthasarathy R, Menon V (1993) Synth Met 55:1165
163. Lei J, Cai Z, Martin CR (1992) Synth Met 46:53

164. Parthasarathy R, Martin CR (1994) Chem Mater 6:1627
165. Menon VP, Lei J, Martin CR (1996) Chem Mater 8:2382
166. Omerz-Kaifer G, Reddy PA, Gutsche CD, Echegoyen L (1998) J Am Chem Soc 120:2486
167. Marinakos SM, Brouseau LC, Jones A, Feldheim DL (1998) Chem Mater 10:1214
168. Jang J, Lim B, Lee J, Hyeon T (2001) Chem Commun 83
169. Hong BH, Lee JT, Lee CW, Kim JC, Bae SC, Kim KS (2001) J Am Chem Soc 12:10748
170. Delvaux M, Duchet J, Stavaux PY, Legras R, Demoustier-Champagne S (2000) Synth Met 113:275
171. Chem JH, Huang ZP, Wang DZ, Yang SX, Li WZ, Wen JG, Ren ZF (2002) Synth Met 125:289
172. Jang J, Oh JH (2004) Chem Commun 882
173. Langer JJ, Czajkowski I (1997) Adv Mater Opts Electronics 7:149
174. Lang JJ (1999) Adv Mater Opts Electronics 9:1
175. Langer JJ, Framski G, Joachimiak R (2001) Synth Met 121:1281
176. Langer JJ, Framski G, Golczak S (2001) Synth Met 121:1319
177. Hassanien A, Gao M, Tokumoto M, Dai L (2001) Chem Phys 342:479
178. Liu J, Wan M (2001) J Mater Chem 11:404
179. Qiu H, Wan M, Matthews B, Dai L (2001) Macromolecules 34:675
180. Pillalamarri SK, Blum FD, Tokuhiro A, Story JG, Bertino MF (2005) Chem Mater 17:227
181. Armes SP, Skotheim TA, Elsenbaumer RL, Reynolds JR (1998) Handbook of conducting polymers, ch 17. Marcel Dekker, New York
182. Armes SP, Miller JF, Vincent B (1987) J Colloid Interface Sci 118:410
183. Ghosh P, Kar SB (2004) J Appl Polym Sci 91:3737
184. Myers RE (1986) J Electron Mater 2:61
185. Huang WS, Humphrey BD, MacDiarmid AG (1986) J Chem Soc Faraday Trans 82:2385
186. Gangopadhyay R, De A, Ghosh G (2001) Synth Met 123:21
187. Banerjee P, Mandal BM (1995) Synth Met 74:257
188. Ghosh P, Siddhanta SK, Haque SR, Chakrabarti A (2001) Synth Met 123:83
189. Partch R, Gangolli SG, Matijevic E, Cai W, Arajs S (1991) J Colloid Interface Sci 144:27
190. Lascelles SF, Armes SP (1995) Adv Mater 7:864
191. Eisazadeh H, Spinks G, Wallace GG (1994) Polymer 35:3801
192. DeArmitt C, Armes SP (1993) Langmuir 9:652
193. Luk SY, Keane M, Lineton W, DeArmitt C, Armes SP (1995) J Chem Soc Faraday Trans 91:905
194. Jonas F, Heywang G (1994) Electrochim Acta 39:1345
195. Francoism B, Olinga T (1993) Synth Met 57:3489
196. Widawski G, Rawiso M, Francois B (1992) J Chim Phys Phys Chem Biol 89:1331
197. Jang J, Bae J, Lee K (2005) Polymer 46:3677
198. Rencker DH, Chun I (1996) Nanotechnology 7:216
199. Huang ZM, Zhang YZ, Kotaki M, Ramakrishna S (2003) Comp Sci Tech 63:2223
200. Gonzalez R, Pinto NJ (2005) Synth Met 151:275
201. Wilkes GL (2002) Electrospinning. http://www.che.vt.edu/Wilkes/electrospinning/electrspinning.html, Cited 2005
202. Frenot A, Chronakis IS (2003) Curr Opin Colloid Interf Sci 8:64
203. Norris ID, Shaker MM, Ko FK, MacDiarmid AG (2000) Synth Met 114:109
204. Kahol PK, Pinto NJ (2004) Synth Met 140:269

205. MacDiarmid AG, Jone WE, Norris ID, Gao J, Johnson AT, Pinto NJ, Hone J, Han B, Ko FK, Llauno M (2001) Synth Met 119:27
206. Sadki S, Schottland P, Brodie N, Sabouraud G (2000) Chem Soc Rev 29:283
207. Yan F, Xue G, Zhou M (2000) J Appl Polym Sci 77:135
208. Jang J, Oh JH (2002) Chem Commun 2200
209. Jang J, Oh JH (2004) Langmuir 20:8419
210. Simons MR, Chaloner PA, Armes SP (1998) Langmuir 14:611
211. Zelenev A, Sonnenberg W, Matijević E (1998) Colloid Polym Sci 276:838
212. Men'shikova AY, Shabsel's BM, Evseeva TG (2003) Russ J Appl Chem 76:822
213. Jang J, Oh JH (2005) Adv Funct Mater 15:494
214. Jang J, Nam Y, Yoon H (2005) Adv Mater 17:1382
215. Jang J, Oh JH (2003) Adv Mater 15:977
216. Cho G, Fung BM, Glatzhofer DT, Lee JS, Shul YG (2001) Langmuir 17:456
217. Shin HJ, Hwang IW, Hwang YN, Kim D, Han SH, Lee JS, Cho G (2003) J Phys Chem B 107:4699
218. Deng J, Peng Y, He C, Long X, Li P, Chan ASC (2003) Polym Int 52:1182
219. Marinakos SM, Shultz DA, Feldheim DL (1999) Adv Mater 11:34
220. Marinakos SM, Novak JP, Brousseau III LC, House AB, Edeki EM, Feldhaus JC, Feldheim DL (1999) J Am Chem Soc 121:8518
221. Cheng D, Xia H, Chan HSO (2004) Langmuir 20:9909
222. Cheng D, Zhou X, Xia H, Chan HSO (2005) Chem Mater 17:3578
223. Jang J, Li XL, Oh JH (2004) Chem Commun 794
224. Hao LY, Zhu CL, Jiang WQ, Chen CN, Hu Y, Chen ZY (2004) J Mater Chem 14:2929
225. Cai Z, Martin CR (1989) J Am Chem Soc 111:4138
226. Vito SD, Martin CR (1998) Chem Mater 10:1738
227. Park S, Lim JH, Chung SW, Mirkin CA (2004) Science 303:348
228. Hernández RM, Richter L, Semancik S, Stranick S, Mallouk TE (2004) 16:3431
229. Mativetsky JM, Datars WR (2002) Solid State Commun 122:151
230. Johnson BJS, Wolf JH, Zalusky AS, Hillmyer MA (2004) Chem Mater 16:2909
231. Ikegame M, Tajima K, Aida T (2003) Angew Chem Int Ed 42:2154
232. Zhang X, Manohar SK (2004) J Am Chem Soc 126:12714
233. Zhang X, Zhang J, Liu Z, Robinson C (2004) Chem Commun 1852
234. Hatano T, Bae AH, Takeuchi M, Fujita N, Kaneko K, Ihara H, Takafuji M, Shinkai S (2004) 43:465
235. Goren M, Qi Z, Lennox RB (2000) Chem Mater 12:1222
236. Jérôme C, Jérôme R (1998) Angew Chem Int Ed 37:2488
237. Jérôme C, Labaye DE, Jérôme R (2004) Synth Met 142:207
238. Zhang X, Zhang J, Wang R, Zhu T, Liu Z (2004) Chem Phys Chem 5:998
239. Fan J, Wan M, Zhu D, Chang B, Pan Z, Xie S (1999) J Appl Polym Sci 74:2605
240. An KH, Jeong SY, Hwang HR, Lee YH (2004) Adv Mater 16:1005
241. Zhou C, Kumar S (2005) Chem Mater 17:1997
242. Hughes M, Shaffer MSP, Renouf AC, Singh C, Chen GZ, Fray DJ, Windle AH (2002) Adv Mater 14:382
243. Hughes M, Chen GZ, Shaffer MSP, Fray DJ, Windle AH (2002) Chem Mater 14:1610
244. Chen A, Wang H, Li X (2005) Chem Commun 1863
245. Huang J, Ichinose I, Kunitake T (2005) Chem Commun 1717
246. Zhang W, Wen X, Yang S (2003) Langmuir 19:4420
247. Demoustier-Champagne S, Stavaux PY (1999) Chem Mater 11:829
248. He Y, Yuan J, Shi G (2005) J Mater Chem 15:859
249. Jang J, Yoon H (2003) Adv Mater 15:2088

250. Cuenot S, Demoustier-Champagne S, Nysten B (2000) Phys Rev Lett 85:1690
251. Cuenot S, Frétigny C, Demoustier-Champagne S, Nysten B (2004) Phys Rev Lett B 69:165410
252. Saha SK (2002) Appl Phys Lett 81:3645
253. Saha SK, Su YK, Lin CL, Jaw DW (2004) Nanotechnology 15:66
254. Holdcroft S (2001) Adv Mater 13:1753
255. Malinauskas A (2001) Polymer 42:3957
256. Goren M, Lennox RB (2001) Nano Lett 1:735
257. Carswell ADW, O'Rear EA, Grady BP (2003) J Am Chem Soc 125:14793
258. Seo I, Pyo M, Cho G (2002) Langmuir 18:7253
259. Lim JH, Mirkin CA (2002) Adv Mater 14:1474
260. Bai J, Snively CM, Delgass WN, Lauterbach J (2002) Adv Mater 14:1546
261. Gangopadhyay R, De A (2000) Chem Mater 12:608
262. Roux S, Soler-Illia GJAA, Demoustier-Champagne S, Audebert P, Sanchez C (2003) Adv Mater 15:217
263. Chen A, Wang H, Zhao B, Li X (2004) Synth Met 145:153
264. Wang Z, Roberge C, Wan Y, Dao LH, Guidoin R, Zhang Z (2003) J Biomed Mater Res A 66A:738
265. Shi G, Rouabhia M, Wang Z, Dao LH, Zhang Z (2004) Biomaterials 25:2477
266. Lee ES, Park JH, Wallace GG, Bae YH (2004) Polym Int 53:400
267. Xu XJ, Gan LM, Siow KS, Wong MK (2004) J Appl Polym Sci 91:1360
268. Kim D, Choi J, Kim JY, Han YK, Shon D (2002) Macromolecules 35:5314
269. Han D, Chu Y, Yang L, Liu Y, Lv Z (2005) Colloid Surf A 259:1793
270. Gao H, Jiang T, Han B, Wang Y, Du J, Liu Z, Zhang J (2004) Polymer 45:3017
271. Dorey S, Vasilev C, Vidal L, Labbe C, Gospodinova N (2005) Polymer 46:1309
272. Cheng D, Ng SC, Chan HSO (2005) Thin Solid Films 477:19
273. Lv R, Zhang S, Shi Q, Kan J (2005) Synth Met 150:115
274. Tang Z, Liu S, Wang Z, Dong S, Wang E (2000) Electrochem Commun 2:32
275. Xia H, Wang Q (2002) Chem Mater 14:2158
276. Schnitzler DC, Meruvia MS, Hümmelgen, Zarbin AJG (2003) Chem Mater 15:4658
277. Li X, Chen W, Bian C, He J, Xu N, Xue G (2003) Appl Surf Sci 217:16
278. Chung FY, Yang SM (2005) Synth Met 152:361
279. Su SJ, Kuramoto N (2000) Synth Met 114:147
280. Mallick K, Witcomb MJ, Dinsmore A, Scurrell MS (2005) Macromol Rapid Commun 26:232
281. Pang S, Li G, Zhang Z (2005) Macromol Rapid Commun (2005) 26:1262
282. Lee CF, Tsai HH, Wang LY, Chen CF, Chiu WY (2005) J Polym Sci Part A Polym Chem 43:342
283. Kohut-Svelko N, Reynaud S, Dedryvère R, Martinez H, Gonbeau D, François J (2005) Langmuir 21:1575
284. Wei Z, Wan M (2002) Adv Mater 14:1314
285. Shi X, Briseno AL, Sanedrin RJ, Zhou F (2003) Macromolecules 36:4093
286. Huang J, Kaner RB (2004) J Am Chem Soc 126:851
287. He Y (2005) Mater Sci Eng B 122:76
288. Huang J, Kaner RB (2004) Angew Chem Int Ed 43:5817
289. Li W, Wang HL (2004) J Am Chem Soc 126:2278
290. Zhang X, Goux WJ, Manohar SK (2004) J Am Chem Soc 126:4502
291. Zhang X, Manohar SK (2004) Chem Commun 2360
292. Li G, Zhang Z (2004) Macromolecules 37:2683
293. Norris ID, Shaker MM, Ko Fk, MacDiamid AJ (2000) Synth Met 114:109

294. Cao H, Xu Z, Sang H, Sheng D, Tie C (2001) Adv Mater 13:121
295. Xiong S, Wang Q, Xia H (2004) Mater Res Bull 39:1569
296. Dong H, Prasad S, Nyame V, Jones WE Jr (2004) Chem Mater 16:371
297. Huang K, Wan M (2002) Chem Mater 14:3486
298. Zhang Z, Wei Z, Wan M (2002) Macromolecules 35:5937
299. Zhang L, Wan M (2005) Thin Solid Films 477:24
300. Long Y, Chen Z, Wang N, Ma Y, Zhang Z, Zhang L, Wan M (2003) Appl Phys Lett 83:1863
301. Pinto NJ, Carrión PL, Ayala AM, Ortiz-Marciales M (2005) Synth Met 148:271
302. Long YZ, Luo JL, Xu J, Chen Z, Zhang L, Li J, Wan M (2004) J Phys Condens Mat 16:1123
303. Porter TL, Thompson D, Bradley M (1996) Thin Solid Films 288:268
304. Dabke RB, Dhanabalan A, Major S, Talwar SS, Lal R, Contractor AQ (1998) Thin Solid Films 335:203
305. Sfex R, Zhong LD, Turyan I, Mandler D, Yitzchaik S (2001) Langmuir 17:2556
306. Josefowicz JY, Avlyanov JK, MacDiarmid AG (2001) Thin Solid Films 393:186
307. Grigore L, Petty MC (2003) J Mater Sci Mater Electron 14:389
308. Morales J, Olayo MG, Cruz GJ, Castillo-Ortega MM, Olayo R (2000) J Polym Sci Part B Polym Phys 15:3247
309. Cruz GJ, Morales J, Castillo-Ortea MM, Olayo R (1997) Synth Met 88:213
310. Misra SCK, Mathure P, Srivastava BK (2004) Sens Actuators A 114:30
311. Xu B, Caruso AN, Dowben PA (2003) Appl Phys A Mater Sci Process 77:155
312. Choi J, Chipara M, Xu B, Yang CS, Doudin B, Dowben PA (2001) Chem Phys Lett 343:193
313. Nekrasov AA, Ivanov VF, Vannikov AV (2001) Electrochim Acta 46:3301
314. Venancio EC, Motheo AJ, Amaral FA, Bocchi N (2001) J Power Sources 94:36
315. Neves S, Polo Fonesca C (2002) J Power Sources 107:13
316. Sotomayor PT, Raimundo IM JR, Zarbin AJG, Rohwedder JJR, Oliveira Neto G, Alves OL (2001) Sens Actuators B 74:157
317. Tseng RJ, Huang J, Ouyang J, Kaner RB, Yang Y (2005) Nano Lett 5:1077
318. Cuentas-Gallegos AK, Lira-Cantu M, Casan-Pastor N, Gomez-Romero P (2005) Adv Funct Mater 15:1125
319. Zhang L, Wan M (2003) J Phys Chem B 107:6748
320. Skotheim TA, Elsenbaumer RL, Reynolds JR (eds) (1998) Handbook of conducting polymers. Marcel Dekker, New York, p 225
321. McCullough RD (1998) Adv Mater 10:93
322. Yamamoto T, Sanechika K, Yamamoto A (1980) J Polym Sci Polym Lett Ed 18:9
323. Lin JWP, Dudek LP (1980) J Polym Sci Polym Chem Ed 18:2869
324. Ong BS, Wu Y, Liu P, Gardner S (2005) Adv Mater 17:1141
325. Tepavcevic S, Wroble AT, Bissen M, Wallace DJ, Choi Y, Hanley L (2005) J Phys Chem B 109:7134
326. Lu Y, Yang Y, Sellinger A, Lu M, Huang J, Fan H, Haddad R, Lopez G, Burns AR, Sasaki DY, Shelnutt J, Brinker CJ (2001) Nature 410:913
327. Wu CG, Bein T (1994) Science 266:1013
328. Li G, Bhosale S, Wang T, Zhang Y, Zhu H, Fuhrhop JH (2003) Angew Chem Int Ed 42:3818
329. Merlo JA, Frisbie CD (2004) J Phys Chem B 108:19169
330. Liu J, Sheina E, Kowalewski T, McCullough RD (2002) Angew Chem Int Ed 41:329
331. Fu M, Zhu Y, Tan R, Shi G (2001) Adv Mater 13:1874
332. Fu M, Chen F, Zhang J, Shi G (2002) J Mater Chem 12:2331

333. Caoa J, Suna J, Shi G, Chena H, Zhang Q, Wang D, Wang M (2003) Mater Chem Phys 82:44
334. Vriezema DM, Hoogboom J, Velonia K, Takazawa K, Christianen PCM, Maan JC, Rowan AE, Nolte RJM (2003) Angew Chem Int Ed 42:772
335. Zhai L, McCullough RD (2002) Adv Mater 14:901
336. Xia C, Fan X, Park M, Advincula RC (2001) Langmuir 17:7893
337. Choi Y, Tepavcevic S, Xu Z, Hanley L (2004) Chem Mater 16:1924
338. Tepavcevic S, Choi Y, Hanley L (2004) Langmuir 20:8754
339. Tepavcevic S, Choi Y, Hanley L (2003) J Am Chem Soc 125:2396
340. Derue G, Coppee S, Gabriele S, Surin M, Geskin V, Monteverde F, Leclere P, Lazzaroni R, Damman P (2005) J Am Chem Soc 127:8018
341. Coppee S, Geskin VM, Lazzaroni R, Damman P (2004) Macromolecules 37:244
342. Xia C, Fan X, Locklin R, Advincula RC, Gies A, Nonidez W (2004) J Am Chem Soc 126:8735
343. Ballav N, Biswas M (2003) Synth Met 132:213
344. Wolf MO (2001) Adv Mater 13:545
345. Angelici RJ (2002) Organometallics 20:1259
346. Cravino A, Zerza G, Maggini M, Bucella S, Svensson M, Andersson MR, Neugebauer H, Sariciftcia NS (2000) Chem Commun 2487
347. Kingsborough RP, Swager TM (2000) Chem Mater 12:872
348. Takahashi K, Asano M, Imoto K, Yamaguchi T, Komura K, Nakamura J, Murata K (2003) J Phys Chem B 107:1646
349. Katz HE, Bao Z (2000) J Phys Chem B 104:671
350. Saxena V, Malhotra BD (2003) Current Appl Phys 3:293
351. Facchetti A, Letizia J, Yoon MH, Mushrush M, Katz HE, Marks TJ (2004) Chem Mater 16:4715
352. Facchetti A, Yoon MH, Stern CL, Katz HE, Marks TJ (2003) Angew Chem Int Ed 42:3900
353. Pielartzik H, Reynolds JR (2000) Adv Mater 12:481
354. Kirkchmeyer S, Reuter K (2005) J Mater Chem 15:2077
355. Choi JW, Han MG, Kim SY, Oh SG, Im SS (2004) Synth Met 141:293
356. Khan MA, Armes SP (1999) Langmuir 15:3469
357. Khan MA, Armes SP (2000) Adv Mater 12:671
358. Han MG, Armes SP (2003) Langmuir 19:4523
359. Han MG, Foulger SH (2004) Adv Mater 16:231
360. Han MG, Foulger SH (2004) Chem Commun 2154
361. Granström M, Inganös O (1995) Polymer 36:2867
362. Granström M, Inganös O (1996) Synth Met 76:141
363. Duvail JL, Rétho P, Garreau S, Louarn G, Godon C, Demoustier-Champagne S (2002) Synth Met 131:123
364. Duvail JL, Rétho P, Godon C, Marhic C, Louarn G, Chauvet O, Cuenot S, Nysten B, Dauginet-De Pra L, Demoustier-Champagne S (2003) Synth Met 135–136:329
365. Kim BH, Park DH, Joo J, Yu SG, Lee SH (2005) Synth Met 150:279
366. Kim BH, Kim MS, Park KT, Lee JK, Park DH, Joo J (2003) Appl Phys Lett 83:539
367. Joo J, Park KT, Kim BH, Kim MS, Lee SY, Jeong CK, Lee JK, Park DH, Yi WK, Lee SH, Ryu KS (2003) Synth Met 135–136:7
368. Han MG, Foulger SH (2005) Chem Comm 3092:1
369. Jang J, Chang M, Yoon H (2005) Adv Mater 17:1616
370. Zhang F, Myberg T, Inganäs O (2002) Nano Lett 2:1373
371. Samitsu S, Shimomura T, Ito K, Fujimori M, Heike S, Hashizume T (2005) Appl Phys Lett 86:233103

372. Hatano T, Takeuchi M, Ikeda A, Shinkai S (2003) Org Lett 5:1395
373. Wessling RA (1968) US Patent 3,401,152
374. Glich HG, Wheelwright WL (1966) J Polym Sci Part A-1 4:1337
375. Skaff H, Sill K, Emrick T (2004) J Am Chem Soc 126:11322
376. Luo YH, Liu HW, Xi F, Li L, Jin XG, Han CC, Chan CM (2003) J Am Chem Soc 125:6447
377. Ajayaghosh A, George SJ (2001) J Am Chem Soc 123:5148
378. Guo L, Wu Z, Liang Y (2004) Chem Commun 1664
379. Kim K, Jin JI (2001) Nano Lett 1:631
380. Babel A, Li D, Xia Y, Jenekhe SA (2005) Macromolecules 38:4705
381. Li D, Babel A, Xia Y (2004) Adv Mater 16:2062
382. Madhugiri S, Dalton A, Gutierrez J, Ferraris JP, Balkus Jr KJ (2003) J Am Chem Soc 125:14531
383. Ameer Q, Adeloju SB (2005) Sens Actuators B 541:552
384. Lellouche JP, Govindaraji S, Joseph A, Jang J, Lee KJ (2005) Chem Commun 4357
385. Dong B, Krutschke M, Zhang X, Chi L, Fuchs H (2005) Small 1:520
386. Tian Y, Wang J, Wang Z, Wang S (2004) Synth Met 143:309
387. Tian Y, Wang J, Wang Z, Wang S (2005) Sens Actuators B 104:23
388. Yang HH, Zhang SQ, Tan F, Zhuang ZX, Wang XR (2005) J Am Chem Soc 127:1378
389. Kros A, Nolte RJM, Sommerdijk NAJM (2002) Adv Mater 14:1779
390. Gao M, Dai L, Wallace GG (2003) Synth Met 137:1393
391. Wang J, Musameh M (2005) Anal Chim Acta 539:209
392. Ramanathan K, Bangar MA, Yun M, Chen W, Mulchandani A, Myung NV (2004) Nano Lett 7:1237
393. Ramanathan K, Bangar MA, Yun M, Chen W, Myung NV, Mulchandani A (2005) J Am Chem Soc 127:496
394. Huang J, Virji S, Weiller BH, Kaner RB (2004) Chem Eur J 10:1314
395. Virji S, Huang J, Kaner RB, Weiller BH (2004) Nano Lett 4:491
396. Virji S, Kaner RB, Weiller BH (2005) Chem Mater 17:1256
397. Virji S, Fowler JD, Baker CO, Huang J, Kaner RB, Weiller BH (2005) Small 1:624
398. Liu H, Kameoka J, Czaplewski DA, Craighead HG (2004) Nano Lett 4:671
399. Wang J, Chan S, Carlson RR, Luo Y, Ge G, Ries RS, Heath JR, Tseng HR (2004) Nano Lett 4:1693
400. Ho HA, Leclerc M (2003) J Am Chem Soc 125:4412
401. Floch FL, Ho HA, Harding-Lepage P, Bedard M, Neagu-Plesu R, Leclerc M (2005) Adv Mater 17:1251
402. Yamato H, Ohwa M Wernet W (1995) J Electroanal Chem 397:163
403. Kros A, van Hövell SWFM, Sommerdijk NAJM, Nolte RJM (2001) Adv Mater 13:1555
404. Kros A, Sommerdijk NAJM, Nolte RJM (2005) Sens Actuators B 106:289
405. Bae AH, Hatano T, Nakashima N, Murakami H, Shinkai S (2004) Org Biomol Chem 2:1139
406. Xu Y, Berger PR (2004) J Appl Phys 95:1497
407. Chabinyc ML, Lu JP, Street RA, Wu Y, Liu P, Ong BS (2004) J Appl Phys 96:2063
408. Hoshino S, Yoshida M, Uemura S, Kodzasa J, Takada N, Kamata K, Yase K (2004) J Appl Phys 95:5088
409. Chung HJ, Jung HH, Cho YS, Lee S, Ha JH, Choi JH, Kuk Y (2005) Appl Phys Lett 86:213113
410. Hu W, Nakashima H, Furukawa K, Kashimura Y, Ajito K, Liu Y, Zhu D, Torimitsu K (2005) J Am Chem Soc 127:2804
411. Kim K, Kim BH, Joo SH, Park JS, Joo J, Jin JI (2005) Adv Mater 17:464

412. Möller S, Perlov C, Jackson W, Taussig C, Forrest SR (2003) Nature 426:166
413. Ling Q, Song Y, Ding SJ, Zhu C, Chan DSH, Kwong DL, Kang ET, Neoh KG (2005) Adv Mater 17:455
414. Taylor DM, Mills CA (2001) J Appl Phys 90:306
415. Marsman AW, Hart CM, Gelinck GH, Geuns TCT, de Leeuw DM (2004) J Mater Res 19:2057
416. Vorotyntsev MA, Skompska M, Pousson E, Goux J, Moise C (2003) J Electroanal Chem 552:307
417. Botta C, Mercogliano C, Bolognesi A, Majumdar HS, Pala AJ (2004) Appl Phys Lett 85:2393
418. Jurewicz, K, Delpeux S, Bertagna V, Béguin F, Frackowiak E (2001) Chem Phys Lett 347:36
419. An KH, Jeon KK, Heo JK, Lim SC, Bae DJ, Lee YH (2002) J Electrochem Soc 149:A1058
420. Zhou C, Kumar S, Doyle CD, Tour JM (2005) Chem Mater 17:1997
421. Snook GA, Chen GZ, Fray DJ, Hughes M, Shaffer M (2004) J Electroanal Chem 568:135
422. Jang J, Bae J, Choi M, Yoon SH (2005) Carbon 43:2730
423. Park JH, Ko JM, Park OO, Kim DW (2002) J Power Sources 105:20
424. Roberson LB, Poggi MA, Kowalik J, Smestad GP, Bottomley LA, Tolbert LM (2004) Coordin Chem Rev 248:1491
425. Watanabe A, Kasuya A (2005) Thin Solid Films 483:358
426. Liu J, Tanaka T, Sivula L, Alivisatos AP, Frechet JMJ (2004) J Am Chem Soc 126:6550
427. Liu J, Kadnikova EN, Liu Y, McGehee MD, Frechet JMJ (2004) J Am Chem Soc 126:9486
428. Catellani M, Luzzati S, Lupsac NO, Mendichi R, Consonni R, Famulari A, Meille SV, Giacalone F, Segurac JL, Martin N (2004) J Mater Chem 14:67
429. Li ADQ, Li LS (2004) J Phys Chem B 108:12842
430. Shaheen SE, Brabec CJ, Sariciftci NS, Padinger F, Fromherz T, Hummelen JC (2001) Appl Phys Lett 78:841
431. Hoppe H, Glatzel T, Niggemann M, Hinsch A, Lux-Steiner MC, Sariciftci NS (2005) Nano Lett 5:269
432. Hoppe H, Niggemann M, Winder C, Kraut J, Hiesgen R, Hinsch A, Meissner D, Sariciftci NS (2004) Adv Funct Mater 14:1005
433. Yang X, van Duren JKJ, Janssen RAJ, Michels MAJ, Loos J (2004) Macromolecules 37:2151
434. Martens T, D'Haen J, Munters T, Beelen Z, Goris L, Manca J, D'Olieslaeger M, Vanderzande D, De Schepper L, Andriessen R (2003) Synth Met 138:243
435. van Hal, Wienk MM, Kroon JM, Verhees WJH, Slooff LH, van Gennip WJH, Jonkheijm P, Janssen RAJ (2003) Adv Mater 15:118
436. Argun AA, Cirpan A, Reynolds JR (2003) Adv Mater 15:1338
437. Cho SI, Kwon WJ, Choi SJ, Kim P, Park SA, Kim J, Son SJ, Xiao R, Kim SH, Lee SB (2005) Adv Mater 17:171
438. Jang SY, Seshadri V, Khil MS, Kumar A, Marquez M, Mather PT, Sotzing GA (2005) Adv Mater 17:2177
439. Smela E (2003) Adv Mater 15:481
440. Jager EWH, Smela E, Inganäs O (2000) Science 290:1540
441. Jager EWH, Inganäs O, Lundström I (2000) Science 288:2335
442. Jager EWH, Inganäs O, Lundström I (2001) Adv Mater 13:76
443. Tallman DE, Spinks G, Dominis A, Wallace GG (2002) J Solid State Electrochem 6:73

444. Spinks GM, Dominis AJ, Wallace GG, Tallman DE (2002) J Solid State Electrochem 6:85
445. Breslin CB, Fenelon AM, Conroy KG (2005) Mater Design 26:233
446. Yeh JM, Chin CP, Chang S (2003) J Appl Polym Sci 88:3264
447. Yeh JM, Liou SJ, Lai CY, Wu PC (2001) Chem Mater 13:1131
448. Tüken T, Yazici ME (2005) Prog Org Coat 53:38
449. Tüken T, Yazici ME (2005) Appl Surf Sci 239:398
450. Yang CM, Weidenthaler C, Spliethoff B, Mayanna M, Schüth F (2005) Chem Mater 17:355
451. Jang J, Oh JH (2004) Adv Mater 16:1650

Editor: Kwang-Sup Lee

Author Index Volumes 101–199

Author Index Volumes 1–100 see Volume 100

de Abajo, J. and *de la Campa, J. G.*: Processable Aromatic Polyimides. Vol. 140, pp. 23–60.
Abe, A., Furuya, H., Zhou, Z., Hiejima, T. and *Kobayashi, Y.*: Stepwise Phase Transitions of Chain Molecules: Crystallization/Melting via a Nematic Liquid-Crystalline Phase. Vol. 181, pp. 121–152.
Abetz, V. and *Simon, P. F. W.*: Phase Behaviour and Morphologies of Block Copolymers. Vol. 189, pp. 125–212.
Abetz, V. see Förster, S.: Vol. 166, pp. 173–210.
Adolf, D. B. see Ediger, M. D.: Vol. 116, pp. 73–110.
Advincula R.: Polymer Brushes by Anionic and Cationic Surface-Initiated Polymerization (SIP). Vol. 197, pp. 107–136.
Aharoni, S. M. and *Edwards, S. F.*: Rigid Polymer Networks. Vol. 118, pp. 1–231.
Akgun, B. see Brittain, W. J.: Vol. 198, pp. 125–147.
Alakhov, V. Y. see Kabanov, A. V.: Vol. 193, pp. 173–198.
Albertsson, A.-C. and *Varma, I. K.*: Aliphatic Polyesters: Synthesis, Properties and Applications. Vol. 157, pp. 99–138.
Albertsson, A.-C. see Edlund, U.: Vol. 157, pp. 53–98.
Albertsson, A.-C. see Söderqvist Lindblad, M.: Vol. 157, pp. 139–161.
Albertsson, A.-C. see Stridsberg, K. M.: Vol. 157, pp. 27–51.
Albertsson, A.-C. see Al-Malaika, S.: Vol. 169, pp. 177–199.
Allegra, G. and *Meille, S. V.*: Pre-Crystalline, High-Entropy Aggregates: A Role in Polymer Crystallization? Vol. 191, pp. 87–135.
Allen, S. see Ellis, J. S.: Vol. 193, pp. 123–172.
Al-Malaika, S.: Perspectives in Stabilisation of Polyolefins. Vol. 169, pp. 121–150.
Altstädt, V.: The Influence of Molecular Variables on Fatigue Resistance in Stress Cracking Environments. Vol. 188, pp. 105–152.
Améduri, B., Boutevin, B. and *Gramain, P.*: Synthesis of Block Copolymers by Radical Polymerization and Telomerization. Vol. 127, pp. 87–142.
Améduri, B. and *Boutevin, B.*: Synthesis and Properties of Fluorinated Telechelic Monodispersed Compounds. Vol. 102, pp. 133–170.
Ameduri, B. see Taguet, A.: Vol. 184, pp. 127–211.
Amir, R. J. and *Shabat, D.*: Domino Dendrimers. Vol. 192, pp. 59–94.
Amselem, S. see Domb, A. J.: Vol. 107, pp. 93–142.
Anantawaraskul, S., Soares, J. B. P. and *Wood-Adams, P. M.*: Fractionation of Semicrystalline Polymers by Crystallization Analysis Fractionation and Temperature Rising Elution Fractionation. Vol. 182, pp. 1–54.
Andrady, A. L.: Wavelenght Sensitivity in Polymer Photodegradation. Vol. 128, pp. 47–94.
Andreis, M. and *Koenig, J. L.*: Application of Nitrogen–15 NMR to Polymers. Vol. 124, pp. 191–238.
Angiolini, L. see Carlini, C.: Vol. 123, pp. 127–214.

Anjum, N. see Gupta, B.: Vol. 162, pp. 37–63.
Anseth, K. S., Newman, S. M. and *Bowman, C. N.*: Polymeric Dental Composites: Properties and Reaction Behavior of Multimethacrylate Dental Restorations. Vol. 122, pp. 177–218.
Antonietti, M. see Cölfen, H.: Vol. 150, pp. 67–187.
Aoki, H. see Ito, S.: Vol. 182, pp. 131–170.
Armitage, B. A. see O'Brien, D. F.: Vol. 126, pp. 53–58.
Arnal, M. L. see Müller, A. J.: Vol. 190, pp. 1–63.
Arndt, M. see Kaminski, W.: Vol. 127, pp. 143–187.
Arnold, A. and *Holm, C.*: Efficient Methods to Compute Long-Range Interactions for Soft Matter Systems. Vol. 185, pp. 59–109.
Arnold Jr., F. E. and *Arnold, F. E.*: Rigid-Rod Polymers and Molecular Composites. Vol. 117, pp. 257–296.
Arora, M. see Kumar, M. N. V. R.: Vol. 160, pp. 45–118.
Arshady, R.: Polymer Synthesis via Activated Esters: A New Dimension of Creativity in Macromolecular Chemistry. Vol. 111, pp. 1–42.
Aseyev, V. O., Tenhu, H. and *Winnik, F. M.*: Temperature Dependence of the Colloidal Stability of Neutral Amphiphilic Polymers in Water. Vol. 196, pp. 1–86.
Auer, S. and *Frenkel, D.*: Numerical Simulation of Crystal Nucleation in Colloids. Vol. 173, pp. 149–208.
Auriemma, F., de Rosa, C. and *Corradini, P.*: Solid Mesophases in Semicrystalline Polymers: Structural Analysis by Diffraction Techniques. Vol. 181, pp. 1–74.

Bahar, I., Erman, B. and *Monnerie, L.*: Effect of Molecular Structure on Local Chain Dynamics: Analytical Approaches and Computational Methods. Vol. 116, pp. 145–206.
Baietto-Dubourg, M. C. see Chateauminois, A.: Vol. 188, pp. 153–193.
Ballauff, M. see Dingenouts, N.: Vol. 144, pp. 1–48.
Ballauff, M. see Holm, C.: Vol. 166, pp. 1–27.
Ballauff, M. see Rühe, J.: Vol. 165, pp. 79–150.
Balsamo, V. see Müller, A. J.: Vol. 190, pp. 1–63.
Baltá-Calleja, F. J., González Arche, A., Ezquerra, T. A., Santa Cruz, C., Batallón, F., Frick, B. and *López Cabarcos, E.*: Structure and Properties of Ferroelectric Copolymers of Poly(vinylidene) Fluoride. Vol. 108, pp. 1–48.
Baltussen, J. J. M. see Northolt, M. G.: Vol. 178, pp. 1–108.
Barnes, M. D. see Otaigbe, J. U.: Vol. 154, pp. 1–86.
Barnes, C. M. see Satchi-Fainaro, R.: Vol. 193, pp. 1–65.
Barsett, H. see Paulsen, S. B.: Vol. 186, pp. 69–101.
Barshtein, G. R. and *Sabsai, O. Y.*: Compositions with Mineralorganic Fillers. Vol. 101, pp. 1–28.
Barton, J. see Hunkeler, D.: Vol. 112, pp. 115–134.
Baschnagel, J., Binder, K., Doruker, P., Gusev, A. A., Hahn, O., Kremer, K., Mattice, W. L., Müller-Plathe, F., Murat, M., Paul, W., Santos, S., Sutter, U. W. and *Tries, V.*: Bridging the Gap Between Atomistic and Coarse-Grained Models of Polymers: Status and Perspectives. Vol. 152, pp. 41–156.
Bassett, D. C.: On the Role of the Hexagonal Phase in the Crystallization of Polyethylene. Vol. 180, pp. 1–16.
Batallán, F. see Baltá-Calleja, F. J.: Vol. 108, pp. 1–48.
Batog, A. E., Pet'ko, I. P. and *Penczek, P.*: Aliphatic-Cycloaliphatic Epoxy Compounds and Polymers. Vol. 144, pp. 49–114.
Batrakova, E. V. see Kabanov, A. V.: Vol. 193, pp. 173–198.

Baughman, T. W. and *Wagener, K. B.*: Recent Advances in ADMET Polymerization. Vol. 176, pp. 1–42.
Baum, M. see Brittain, W. J.: Vol. 198, pp. 125–147.
Becker, O. and *Simon, G. P.*: Epoxy Layered Silicate Nanocomposites. Vol. 179, pp. 29–82.
Bell, C. L. and *Peppas, N. A.*: Biomedical Membranes from Hydrogels and Interpolymer Complexes. Vol. 122, pp. 125–176.
Bellon-Maurel, A. see Calmon-Decriaud, A.: Vol. 135, pp. 207–226.
Bennett, D. E. see O'Brien, D. F.: Vol. 126, pp. 53–84.
Bergbreiter, D. E. and *Kippenberger, A. M.*: Hyperbranched Surface Graft Polymerizations. Vol. 198, pp. 1–49.
Berry, G. C.: Static and Dynamic Light Scattering on Moderately Concentraded Solutions: Isotropic Solutions of Flexible and Rodlike Chains and Nematic Solutions of Rodlike Chains. Vol. 114, pp. 233–290.
Bershtein, V. A. and *Ryzhov, V. A.*: Far Infrared Spectroscopy of Polymers. Vol. 114, pp. 43–122.
Bhargava, R., Wang, S.-Q. and *Koenig, J. L*: FTIR Microspectroscopy of Polymeric Systems. Vol. 163, pp. 137–191.
Bhat, R. R., Tomlinson, M. R., Wu, T. and *Genzer, J.*: Surface-Grafted Polymer Gradients: Formation, Characterization, and Applications. Vol. 198, pp. 51–124.
Biesalski, M. see Rühe, J.: Vol. 165, pp. 79–150.
Bigg, D. M.: Thermal Conductivity of Heterophase Polymer Compositions. Vol. 119, pp. 1–30.
Binder, K.: Phase Transitions in Polymer Blends and Block Copolymer Melts: Some Recent Developments. Vol. 112, pp. 115–134.
Binder, K.: Phase Transitions of Polymer Blends and Block Copolymer Melts in Thin Films. Vol. 138, pp. 1–90.
Binder, K. see Baschnagel, J.: Vol. 152, pp. 41–156.
Binder, K., Müller, M., Virnau, P. and *González MacDowell, L.*: Polymer+Solvent Systems: Phase Diagrams, Interface Free Energies, and Nucleation. Vol. 173, pp. 1–104.
Bird, R. B. see Curtiss, C. F.: Vol. 125, pp. 1–102.
Biswas, M. and *Mukherjee, A.*: Synthesis and Evaluation of Metal-Containing Polymers. Vol. 115, pp. 89–124.
Biswas, M. and *Sinha Ray, S.*: Recent Progress in Synthesis and Evaluation of Polymer-Montmorillonite Nanocomposites. Vol. 155, pp. 167–221.
Blankenburg, L. see Klemm, E.: Vol. 177, pp. 53–90.
Blickle, C. see Brittain, W. J.: Vol. 198, pp. 125–147.
Blumen, A. see Gurtovenko, A. A.: Vol. 182, pp. 171–282.
Bogdal, D., Penczek, P., Pielichowski, J. and *Prociak, A.*: Microwave Assisted Synthesis, Crosslinking, and Processing of Polymeric Materials. Vol. 163, pp. 193–263.
Bohrisch, J., Eisenbach, C. D., Jaeger, W., Mori, H., Müller, A. H. E., Rehahn, M., Schaller, C., Traser, S. and *Wittmeyer, P.*: New Polyelectrolyte Architectures. Vol. 165, pp. 1–41.
Bolze, J. see Dingenouts, N.: Vol. 144, pp. 1–48.
Bosshard, C.: see Gubler, U.: Vol. 158, pp. 123–190.
Boutevin, B. and *Robin, J. J.*: Synthesis and Properties of Fluorinated Diols. Vol. 102, pp. 105–132.
Boutevin, B. see Améduri, B.: Vol. 102, pp. 133–170.
Boutevin, B. see Améduri, B.: Vol. 127, pp. 87–142.
Boutevin, B. see Guida-Pietrasanta, F.: Vol. 179, pp. 1–27.
Boutevin, B. see Taguet, A.: Vol. 184, pp. 127–211.
Bowman, C. N. see Anseth, K. S.: Vol. 122, pp. 177–218.

Boyd, R. H.: Prediction of Polymer Crystal Structures and Properties. Vol. 116, pp. 1–26.
Boyes, S. G. see *Brittain, W. J.*: Vol. 198, pp. 125–147.
Bracco, S. see *Sozzani, P.*: Vol. 181, pp. 153–177.
Briber, R. M. see *Hedrick, J. L.*: Vol. 141, pp. 1–44.
Brittain, W. J., Boyes, S. G., Granville, A. M., Baum, M., Mirous, B. K., Akgun, B., Zhao, B., Blickle, C. and *Foster, M. D.*: Surface Rearrangement of Diblock Copolymer Brushes— Stimuli Responsive Films. Vol. 198, pp. 125–147.
Bronnikov, S. V., Vettegren, V. I. and *Frenkel, S. Y.*: Kinetics of Deformation and Relaxation in Highly Oriented Polymers. Vol. 125, pp. 103–146.
Brown, H. R. see *Creton, C.*: Vol. 156, pp. 53–135.
Bruza, K. J. see *Kirchhoff, R. A.*: Vol. 117, pp. 1–66.
Buchmeiser M. R.: Metathesis Polymerization To and From Surfaces. Vol. 197, pp. 137–171.
Buchmeiser, M. R.: Regioselective Polymerization of 1-Alkynes and Stereoselective Cyclopolymerization of a, w-Heptadiynes. Vol. 176, pp. 89–119.
Budkowski, A.: Interfacial Phenomena in Thin Polymer Films: Phase Coexistence and Segregation. Vol. 148, pp. 1–112.
Bunz, U. H. F.: Synthesis and Structure of PAEs. Vol. 177, pp. 1–52.
Burban, J. H. see *Cussler, E. L.*: Vol. 110, pp. 67–80.
Burchard, W.: Solution Properties of Branched Macromolecules. Vol. 143, pp. 113–194.
Butté, A. see *Schork, F. J.*: Vol. 175, pp. 129–255.

Calmon-Decriaud, A., Bellon-Maurel, V., Silvestre, F.: Standard Methods for Testing the Aerobic Biodegradation of Polymeric Materials. Vol. 135, pp. 207–226.
Cameron, N. R. and *Sherrington, D. C.*: High Internal Phase Emulsions (HIPEs)-Structure, Properties and Use in Polymer Preparation. Vol. 126, pp. 163–214.
de la Campa, J. G. see *de Abajo, J.*: Vol. 140, pp. 23–60.
Candau, F. see *Hunkeler, D.*: Vol. 112, pp. 115–134.
Canelas, D. A. and *DeSimone, J. M.*: Polymerizations in Liquid and Supercritical Carbon Dioxide. Vol. 133, pp. 103–140.
Canva, M. and *Stegeman, G. I.*: Quadratic Parametric Interactions in Organic Waveguides. Vol. 158, pp. 87–121.
Capek, I.: Kinetics of the Free-Radical Emulsion Polymerization of Vinyl Chloride. Vol. 120, pp. 135–206.
Capek, I.: Radical Polymerization of Polyoxyethylene Macromonomers in Disperse Systems. Vol. 145, pp. 1–56.
Capek, I. and *Chern, C.-S.*: Radical Polymerization in Direct Mini-Emulsion Systems. Vol. 155, pp. 101–166.
Cappella, B. see *Munz, M.*: Vol. 164, pp. 87–210.
Carlesso, G. see *Prokop, A.*: Vol. 160, pp. 119–174.
Carlini, C. and *Angiolini, L.*: Polymers as Free Radical Photoinitiators. Vol. 123, pp. 127–214.
Carter, K. R. see *Hedrick, J. L.*: Vol. 141, pp. 1–44.
Casas-Vazquez, J. see *Jou, D.*: Vol. 120, pp. 207–266.
Chan, C.-M. and *Li, L.*: Direct Observation of the Growth of Lamellae and Spherulites by AFM. Vol. 188, pp. 1–41.
Chandrasekhar, V.: Polymer Solid Electrolytes: Synthesis and Structure. Vol. 135, pp. 139–206.
Chang, J. Y. see *Han, M. J.*: Vol. 153, pp. 1–36.
Chang, T.: Recent Advances in Liquid Chromatography Analysis of Synthetic Polymers. Vol. 163, pp. 1–60.

Charleux, B. and *Faust, R.*: Synthesis of Branched Polymers by Cationic Polymerization. Vol. 142, pp. 1–70.
Chateauminois, A. and *Baietto-Dubourg, M. C.*: Fracture of Glassy Polymers Within Sliding Contacts. Vol. 188, pp. 153–193.
Chen, P. see Jaffe, M.: Vol. 117, pp. 297–328.
Chern, C.-S. see Capek, I.: Vol. 155, pp. 101–166.
Chevolot, Y. see Mathieu, H. J.: Vol. 162, pp. 1–35.
Chim, Y. T. A. see Ellis, J. S.: Vol. 193, pp. 123–172.
Choe, E.-W. see Jaffe, M.: Vol. 117, pp. 297–328.
Chow, P. Y. and *Gan, L. M.*: Microemulsion Polymerizations and Reactions. Vol. 175, pp. 257–298.
Chow, T. S.: Glassy State Relaxation and Deformation in Polymers. Vol. 103, pp. 149–190.
Chujo, Y. see Uemura, T.: Vol. 167, pp. 81–106.
Chung, S.-J. see Lin, T.-C.: Vol. 161, pp. 157–193.
Chung, T.-S. see Jaffe, M.: Vol. 117, pp. 297–328.
Clarke, N.: Effect of Shear Flow on Polymer Blends. Vol. 183, pp. 127–173.
Coenjarts, C. see Li, M.: Vol. 190, pp. 183–226.
Cölfen, H. and *Antonietti, M.*: Field-Flow Fractionation Techniques for Polymer and Colloid Analysis. Vol. 150, pp. 67–187.
Colmenero, J. see Richter, D.: Vol. 174, pp. 1–221.
Comanita, B. see Roovers, J.: Vol. 142, pp. 179–228.
Comotti, A. see Sozzani, P.: Vol. 181, pp. 153–177.
Connell, J. W. see Hergenrother, P. M.: Vol. 117, pp. 67–110.
Corradini, P. see Auriemma, F.: Vol. 181, pp. 1–74.
Creton, C., Kramer, E. J., Brown, H. R. and *Hui, C.-Y.*: Adhesion and Fracture of Interfaces Between Immiscible Polymers: From the Molecular to the Continuum Scale. Vol. 156, pp. 53–135.
Criado-Sancho, M. see Jou, D.: Vol. 120, pp. 207–266.
Curro, J. G. see Schweizer, K. S.: Vol. 116, pp. 319–378.
Curtiss, C. F. and *Bird, R. B.*: Statistical Mechanics of Transport Phenomena: Polymeric Liquid Mixtures. Vol. 125, pp. 1–102.
Cussler, E. L., Wang, K. L. and *Burban, J. H.*: Hydrogels as Separation Agents. Vol. 110, pp. 67–80.
Czub, P. see Penczek, P.: Vol. 184, pp. 1–95.

Dalton, L.: Nonlinear Optical Polymeric Materials: From Chromophore Design to Commercial Applications. Vol. 158, pp. 1–86.
Dautzenberg, H. see Holm, C.: Vol. 166, pp. 113–171.
Davidson, J. M. see Prokop, A.: Vol. 160, pp. 119–174.
Davies, M. C. see Ellis, J. S.: Vol. 193, pp. 123–172.
Den Decker, M. G. see Northolt, M. G.: Vol. 178, pp. 1–108.
Desai, S. M. and *Singh, R. P.*: Surface Modification of Polyethylene. Vol. 169, pp. 231–293.
DeSimone, J. M. see Canelas, D. A.: Vol. 133, pp. 103–140.
DeSimone, J. M. see Kennedy, K. A.: Vol. 175, pp. 329–346.
Dhal, P. K., Holmes-Farley, S. R., Huval, C. C. and *Jozefiak, T. H.*: Polymers as Drugs. Vol. 192, pp. 9–58.
DiMari, S. see Prokop, A.: Vol. 136, pp. 1–52.
Dimonie, M. V. see Hunkeler, D.: Vol. 112, pp. 115–134.
Dingenouts, N., Bolze, J., Pötschke, D. and *Ballauf, M.*: Analysis of Polymer Latexes by Small-Angle X-Ray Scattering. Vol. 144, pp. 1–48.

Dodd, L. R. and *Theodorou, D. N.*: Atomistic Monte Carlo Simulation and Continuum Mean Field Theory of the Structure and Equation of State Properties of Alkane and Polymer Melts. Vol. 116, pp. 249–282.
Doelker, E.: Cellulose Derivatives. Vol. 107, pp. 199–266.
Dolden, J. G.: Calculation of a Mesogenic Index with Emphasis Upon LC-Polyimides. Vol. 141, pp. 189–245.
Domb, A. J., Amselem, S., Shah, J. and *Maniar, M.*: Polyanhydrides: Synthesis and Characterization. Vol. 107, pp. 93–142.
Domb, A. J. see Kumar, M. N. V. R.: Vol. 160, pp. 45–118.
Doruker, P. see Baschnagel, J.: Vol. 152, pp. 41–156.
Dubois, P. see Mecerreyes, D.: Vol. 147, pp. 1–60.
Dubrovskii, S. A. see Kazanskii, K. S.: Vol. 104, pp. 97–134.
Dudowicz, J. see Freed, K. F.: Vol. 183, pp. 63–126.
Duncan, R., Ringsdorf, H. and *Satchi-Fainaro, R.*: Polymer Therapeutics: Polymers as Drugs, Drug and Protein Conjugates and Gene Delivery Systems: Past, Present and Future Opportunities. Vol. 192, pp. 1–8.
Duncan, R. see Satchi-Fainaro, R.: Vol. 193, pp. 1–65.
Dunkin, I. R. see Steinke, J.: Vol. 123, pp. 81–126.
Dunson, D. L. see McGrath, J. E.: Vol. 140, pp. 61–106.
Dyer D. J.: Photoinitiated Synthesis of Grafted Polymers. Vol. 197, pp. 47–65.
Dziezok, P. see Rühe, J.: Vol. 165, pp. 79–150.

Eastmond, G. C.: Poly(e-caprolactone) Blends. Vol. 149, pp. 59–223.
Ebringerová, A., Hromádková, Z. and *Heinze, T.*: Hemicellulose. Vol. 186, pp. 1–67.
Economy, J. and *Goranov, K.*: Thermotropic Liquid Crystalline Polymers for High Performance Applications. Vol. 117, pp. 221–256.
Ediger, M. D. and *Adolf, D. B.*: Brownian Dynamics Simulations of Local Polymer Dynamics. Vol. 116, pp. 73–110.
Edlund, U. and *Albertsson, A.-C.*: Degradable Polymer Microspheres for Controlled Drug Delivery. Vol. 157, pp. 53–98.
Edwards, S. F. see Aharoni, S. M.: Vol. 118, pp. 1–231.
Eisenbach, C. D. see Bohrisch, J.: Vol. 165, pp. 1–41.
Ellis, J. S., Allen, S., Chim, Y. T. A., Roberts, C. J., Tendler, S. J. B. and *Davies, M. C.*: Molecular-Scale Studies on Biopolymers Using Atomic Force Microscopy. Vol. 193, pp. 123–172.
Endo, T. see Yagci, Y.: Vol. 127, pp. 59–86.
Engelhardt, H. and *Grosche, O.*: Capillary Electrophoresis in Polymer Analysis. Vol. 150, pp. 189–217.
Engelhardt, H. and *Martin, H.*: Characterization of Synthetic Polyelectrolytes by Capillary Electrophoretic Methods. Vol. 165, pp. 211–247.
Eriksson, P. see Jacobson, K.: Vol. 169, pp. 151–176.
Erman, B. see Bahar, I.: Vol. 116, pp. 145–206.
Eschner, M. see Spange, S.: Vol. 165, pp. 43–78.
Estel, K. see Spange, S.: Vol. 165, pp. 43–78.
Estevez, R. and *Van der Giessen, E.*: Modeling and Computational Analysis of Fracture of Glassy Polymers. Vol. 188, pp. 195–234.
Ewen, B. and *Richter, D.*: Neutron Spin Echo Investigations on the Segmental Dynamics of Polymers in Melts, Networks and Solutions. Vol. 134, pp. 1–130.
Ezquerra, T. A. see Baltá-Calleja, F. J.: Vol. 108, pp. 1–48.

Fatkullin, N. see Kimmich, R.: Vol. 170, pp. 1–113.
Faust, R. see Charleux, B.: Vol. 142, pp. 1–70.

Faust, R. see Kwon, Y.: Vol. 167, pp. 107–135.
Fekete, E. see Pukánszky, B.: Vol. 139, pp. 109–154.
Fendler, J. H.: Membrane-Mimetic Approach to Advanced Materials. Vol. 113, pp. 1–209.
Fetters, L. J. see Xu, Z.: Vol. 120, pp. 1–50.
Fontenot, K. see Schork, F. J.: Vol. 175, pp. 129–255.
Förster, S., Abetz, V. and *Müller, A. H. E.*: Polyelectrolyte Block Copolymer Micelles. Vol. 166, pp. 173–210.
Förster, S. and *Schmidt, M.*: Polyelectrolytes in Solution. Vol. 120, pp. 51–134.
Foster, M. D. see Brittain, W. J.: Vol. 198, pp. 125–147.
Freed, K. F. and *Dudowicz, J.*: Influence of Monomer Molecular Structure on the Miscibility of Polymer Blends. Vol. 183, pp. 63–126.
Freire, J. J.: Conformational Properties of Branched Polymers: Theory and Simulations. Vol. 143, pp. 35–112.
Frenkel, D. see Hu, W.: Vol. 191, pp. 1–35.
Frenkel, S. Y. see Bronnikov, S. V.: Vol. 125, pp. 103–146.
Frick, B. see Baltá-Calleja, F. J.: Vol. 108, pp. 1–48.
Fridman, M. L.: see Terent'eva, J. P.: Vol. 101, pp. 29–64.
Fuchs, G. see Trimmel, G.: Vol. 176, pp. 43–87.
Fuhrmann-Lieker, T. see Pudzich, R.: Vol. 199, pp. 83–142.
Fukuda, T. see Tsujii, Y.: Vol. 197, pp. 1–47.
Fukui, K. see Otaigbe, J. U.: Vol. 154, pp. 1–86.
Funke, W.: Microgels-Intramolecularly Crosslinked Macromolecules with a Globular Structure. Vol. 136, pp. 137–232.
Furusho, Y. see Takata, T.: Vol. 171, pp. 1–75.
Furuya, H. see Abe, A.: Vol. 181, pp. 121–152.

Galina, H.: Mean-Field Kinetic Modeling of Polymerization: The Smoluchowski Coagulation Equation. Vol. 137, pp. 135–172.
Gan, L. M. see Chow, P. Y.: Vol. 175, pp. 257–298.
Ganesh, K. see Kishore, K.: Vol. 121, pp. 81–122.
Gaw, K. O. and *Kakimoto, M.*: Polyimide-Epoxy Composites. Vol. 140, pp. 107–136.
Geckeler, K. E. see Rivas, B.: Vol. 102, pp. 171–188.
Geckeler, K. E.: Soluble Polymer Supports for Liquid-Phase Synthesis. Vol. 121, pp. 31–80.
Gedde, U. W. and *Mattozzi, A.*: Polyethylene Morphology. Vol. 169, pp. 29–73.
Gehrke, S. H.: Synthesis, Equilibrium Swelling, Kinetics Permeability and Applications of Environmentally Responsive Gels. Vol. 110, pp. 81–144.
Geil, P. H., Yang, J., Williams, R. A., Petersen, K. L., Long, T.-C. and *Xu, P.*: Effect of Molecular Weight and Melt Time and Temperature on the Morphology of Poly(tetrafluorethylene). Vol. 180, pp. 89–159.
de Gennes, P.-G.: Flexible Polymers in Nanopores. Vol. 138, pp. 91–106.
Genzer, J. see Bhat, R. R.: Vol. 198, pp. 51–124.
Georgiou, S.: Laser Cleaning Methodologies of Polymer Substrates. Vol. 168, pp. 1–49.
Geuss, M. see Munz, M.: Vol. 164, pp. 87–210.
Giannelis, E. P., Krishnamoorti, R. and *Manias, E.*: Polymer-Silicate Nanocomposites: Model Systems for Confined Polymers and Polymer Brushes. Vol. 138, pp. 107–148.
Van der Giessen, E. see Estevez, R.: Vol. 188, pp. 195–234.
Godovsky, D. Y.: Device Applications of Polymer-Nanocomposites. Vol. 153, pp. 163–205.
Godovsky, D. Y.: Electron Behavior and Magnetic Properties Polymer-Nanocomposites. Vol. 119, pp. 79–122.
Gohy, J.-F.: Block Copolymer Micelles. Vol. 190, pp. 65–136.

González Arche, A. see Baltá-Calleja, F. J.: Vol. 108, pp. 1–48.
Goranov, K. see Economy, J.: Vol. 117, pp. 221–256.
Goto, A. see Tsujii, Y.: Vol. 197, pp. 1–47.
Gramain, P. see Améduri, B.: Vol. 127, pp. 87–142.
Granville, A. M. see Brittain, W. J.: Vol. 198, pp. 125–147.
Grein, C.: Toughness of Neat, Rubber Modified and Filled β-Nucleated Polypropylene: From Fundamentals to Applications. Vol. 188, pp. 43–104.
Greish, K. see Maeda, H.: Vol. 193, pp. 103–121.
Grest, G. S.: Normal and Shear Forces Between Polymer Brushes. Vol. 138, pp. 149–184.
Grigorescu, G. and *Kulicke, W.-M.*: Prediction of Viscoelastic Properties and Shear Stability of Polymers in Solution. Vol. 152, p. 1–40.
Grimsdale, A. C. and *Müllen, K.*: Polyphenylene-type Emissive Materials: Poly(*para*-phenylene)s, Polyfluorenes, and Ladder Polymers. Vol. 199, pp. 1–82.
Gröhn, F. see Rühe, J.: Vol. 165, pp. 79–150.
Grosberg, A. Y. and *Khokhlov, A. R.*: After-Action of the Ideas of I. M. Lifshitz in Polymer and Biopolymer Physics. Vol. 196, pp. 189–210.
Grosberg, A. and *Nechaev, S.*: Polymer Topology. Vol. 106, pp. 1–30.
Grosche, O. see Engelhardt, H.: Vol. 150, pp. 189–217.
Grubbs, R., *Risse, W.* and *Novac, B.*: The Development of Well-defined Catalysts for Ring-Opening Olefin Metathesis. Vol. 102, pp. 47–72.
Gubler, U. and *Bosshard, C.*: Molecular Design for Third-Order Nonlinear Optics. Vol. 158, pp. 123–190.
Guida-Pietrasanta, F. and *Boutevin, B.*: Polysilalkylene or Silarylene Siloxanes Said Hybrid Silicones. Vol. 179, pp. 1–27.
van Gunsteren, W. F. see Gusev, A. A.: Vol. 116, pp. 207–248.
Gupta, B. and *Anjum, N.*: Plasma and Radiation-Induced Graft Modification of Polymers for Biomedical Applications. Vol. 162, pp. 37–63.
Gurtovenko, A. A. and *Blumen, A.*: Generalized Gaussian Structures: Models for Polymer Systems with Complex Topologies. Vol. 182, pp. 171–282.
Gusev, A. A., *Müller-Plathe, F.*, *van Gunsteren, W. F.* and *Suter, U. W.*: Dynamics of Small Molecules in Bulk Polymers. Vol. 116, pp. 207–248.
Gusev, A. A. see Baschnagel, J.: Vol. 152, pp. 41–156.
Guillot, J. see Hunkeler, D.: Vol. 112, pp. 115–134.
Guyot, A. and *Tauer, K.*: Reactive Surfactants in Emulsion Polymerization. Vol. 111, pp. 43–66.

Hadjichristidis, N., *Pispas, S.*, *Pitsikalis, M.*, *Iatrou, H.* and *Vlahos, C.*: Asymmetric Star Polymers Synthesis and Properties. Vol. 142, pp. 71–128.
Hadjichristidis, N., *Pitsikalis, M.* and *Iatrou, H.*: Synthesis of Block Copolymers. Vol. 189, pp. 1–124.
Hadjichristidis, N. see Xu, Z.: Vol. 120, pp. 1–50.
Hadjichristidis, N. see Pitsikalis, M.: Vol. 135, pp. 1–138.
Hahn, O. see Baschnagel, J.: Vol. 152, pp. 41–156.
Hakkarainen, M.: Aliphatic Polyesters: Abiotic and Biotic Degradation and Degradation Products. Vol. 157, pp. 1–26.
Hakkarainen, M. and *Albertsson, A.-C.*: Environmental Degradation of Polyethylene. Vol. 169, pp. 177–199.
Halary, J. L. see Monnerie, L.: Vol. 187, pp. 35–213.
Halary, J. L. see Monnerie, L.: Vol. 187, pp. 215–364.
Hall, H. K. see Penelle, J.: Vol. 102, pp. 73–104.
Hamley, I. W.: Crystallization in Block Copolymers. Vol. 148, pp. 113–138.

Hammouda, B.: SANS from Homogeneous Polymer Mixtures: A Unified Overview. Vol. 106, pp. 87–134.
Han, M. J. and *Chang, J. Y.*: Polynucleotide Analogues. Vol. 153, pp. 1–36.
Harada, A.: Design and Construction of Supramolecular Architectures Consisting of Cyclodextrins and Polymers. Vol. 133, pp. 141–192.
Haralson, M. A. see Prokop, A.: Vol. 136, pp. 1–52.
Harding, S. E.: Analysis of Polysaccharides by Ultracentrifugation. Size, Conformation and Interactions in Solution. Vol. 186, pp. 211–254.
Hasegawa, N. see Usuki, A.: Vol. 179, pp. 135–195.
Hassan, C. M. and *Peppas, N. A.*: Structure and Applications of Poly(vinyl alcohol) Hydrogels Produced by Conventional Crosslinking or by Freezing/Thawing Methods. Vol. 153, pp. 37–65.
Hawker, C. J.: Dentritic and Hyperbranched Macromolecules Precisely Controlled Macromolecular Architectures. Vol. 147, pp. 113–160.
Hawker, C. J. see Hedrick, J. L.: Vol. 141, pp. 1–44.
He, G. S. see Lin, T.-C.: Vol. 161, pp. 157–193.
Hedrick, J. L., Carter, K. R., Labadie, J. W., Miller, R. D., Volksen, W., Hawker, C. J., Yoon, D. Y., Russell, T. P., McGrath, J. E. and *Briber, R. M.*: Nanoporous Polyimides. Vol. 141, pp. 1–44.
Hedrick, J. L., Labadie, J. W., Volksen, W. and *Hilborn, J. G.*: Nanoscopically Engineered Polyimides. Vol. 147, pp. 61–112.
Hedrick, J. L. see Hergenrother, P. M.: Vol. 117, pp. 67–110.
Hedrick, J. L. see Kiefer, J.: Vol. 147, pp. 161–247.
Hedrick, J. L. see McGrath, J. E.: Vol. 140, pp. 61–106.
Heine, D. R., Grest, G. S. and *Curro, J. G.*: Structure of Polymer Melts and Blends: Comparison of Integral Equation theory and Computer Sumulation. Vol. 173, pp. 209–249.
Heinrich, G. and *Klüppel, M.*: Recent Advances in the Theory of Filler Networking in Elastomers. Vol. 160, pp. 1–44.
Heinze, T. see Ebringerová, A.: Vol. 186, pp. 1–67.
Heinze, T. see El Seoud, O. A.: Vol. 186, pp. 103–149.
Heller, J.: Poly (Ortho Esters). Vol. 107, pp. 41–92.
Helm, C. A. see Möhwald, H.: Vol. 165, pp. 151–175.
Hemielec, A. A. see Hunkeler, D.: Vol. 112, pp. 115–134.
Hergenrother, P. M., Connell, J. W., Labadie, J. W. and *Hedrick, J. L.*: Poly(arylene ether)s Containing Heterocyclic Units. Vol. 117, pp. 67–110.
Hernández-Barajas, J. see Wandrey, C.: Vol. 145, pp. 123–182.
Hervet, H. see Léger, L.: Vol. 138, pp. 185–226.
Hiejima, T. see Abe, A.: Vol. 181, pp. 121–152.
Hikosaka, M., Watanabe, K., Okada, K. and *Yamazaki, S.*: Topological Mechanism of Polymer Nucleation and Growth – The Role of Chain Sliding Diffusion and Entanglement. Vol. 191, pp. 137–186.
Hilborn, J. G. see Hedrick, J. L.: Vol. 147, pp. 61–112.
Hilborn, J. G. see Kiefer, J.: Vol. 147, pp. 161–247.
Hillborg, H. see Vancso, G. J.: Vol. 182, pp. 55–129.
Hillmyer, M. A.: Nanoporous Materials from Block Copolymer Precursors. Vol. 190, pp. 137–181.
Hiramatsu, N. see Matsushige, M.: Vol. 125, pp. 147–186.
Hirasa, O. see Suzuki, M.: Vol. 110, pp. 241–262.
Hirotsu, S.: Coexistence of Phases and the Nature of First-Order Transition in Poly-N-isopropylacrylamide Gels. Vol. 110, pp. 1–26.
Höcker, H. see Klee, D.: Vol. 149, pp. 1–57.

Holm, C. see Arnold, A.: Vol. 185, pp. 59–109.
Holm, C., Hofmann, T., Joanny, J. F., Kremer, K., Netz, R. R., Reineker, P., Seidel, C., Vilgis, T. A. and *Winkler, R. G.*: Polyelectrolyte Theory. Vol. 166, pp. 67–111.
Holm, C., Rehahn, M., Oppermann, W. and *Ballauff, M.*: Stiff-Chain Polyelectrolytes. Vol. 166, pp. 1–27.
Holmes-Farley, S. R. see Dhal, P. K.: Vol. 192, pp. 9–58.
Hornsby, P.: Rheology, Compounding and Processing of Filled Thermoplastics. Vol. 139, pp. 155–216.
Houbenov, N. see Rühe, J.: Vol. 165, pp. 79–150.
Hromádková, Z. see Ebringerová, A.: Vol. 186, pp. 1–67.
Hu, W. and *Frenkel, D.*: Polymer Crystallization Driven by Anisotropic Interactions. Vol. 191, pp. 1–35.
Huber, K. see Volk, N.: Vol. 166, pp. 29–65.
Hugenberg, N. see Rühe, J.: Vol. 165, pp. 79–150.
Hui, C.-Y. see Creton, C.: Vol. 156, pp. 53–135.
Hult, A., Johansson, M. and *Malmström, E.*: Hyperbranched Polymers. Vol. 143, pp. 1–34.
Hünenberger, P. H.: Thermostat Algorithms for Molecular-Dynamics Simulations. Vol. 173, pp. 105–147.
Hunkeler, D., Candau, F., Pichot, C., Hemielec, A. E., Xie, T. Y., Barton, J., Vaskova, V., Guillot, J., Dimonie, M. V. and *Reichert, K. H.*: Heterophase Polymerization: A Physical and Kinetic Comparision and Categorization. Vol. 112, pp. 115–134.
Hunkeler, D. see Macko, T.: Vol. 163, pp. 61–136.
Hunkeler, D. see Prokop, A.: Vol. 136, pp. 1–52; 53–74.
Hunkeler, D. see Wandrey, C.: Vol. 145, pp. 123–182.
Huval, C. C. see Dhal, P. K.: Vol. 192, pp. 9–58.

Iatrou, H. see Hadjichristidis, N.: Vol. 142, pp. 71–128.
Iatrou, H. see Hadjichristidis, N.: Vol. 189, pp. 1–124.
Ichikawa, T. see Yoshida, H.: Vol. 105, pp. 3–36.
Ihara, E. see Yasuda, H.: Vol. 133, pp. 53–102.
Ikada, Y. see Uyama, Y.: Vol. 137, pp. 1–40.
Ikehara, T. see Jinnuai, H.: Vol. 170, pp. 115–167.
Ilavsky, M.: Effect on Phase Transition on Swelling and Mechanical Behavior of Synthetic Hydrogels. Vol. 109, pp. 173–206.
Imai, M. see Kaji, K.: Vol. 191, pp. 187–240.
Imai, Y.: Rapid Synthesis of Polyimides from Nylon-Salt Monomers. Vol. 140, pp. 1–23.
Inomata, H. see Saito, S.: Vol. 106, pp. 207–232.
Inoue, S. see Sugimoto, H.: Vol. 146, pp. 39–120.
Irie, M.: Stimuli-Responsive Poly(N-isopropylacrylamide), Photo- and Chemical-Induced Phase Transitions. Vol. 110, pp. 49–66.
Ise, N. see Matsuoka, H.: Vol. 114, pp. 187–232.
Ishikawa, T.: Advances in Inorganic Fibers. Vol. 178, pp. 109–144.
Ito, H.: Chemical Amplification Resists for Microlithography. Vol. 172, pp. 37–245.
Ito, K. and *Kawaguchi, S.*: Poly(macronomers), Homo- and Copolymerization. Vol. 142, pp. 129–178.
Ito, K. see Kawaguchi, S.: Vol. 175, pp. 299–328.
Ito, S. and *Aoki, H.*: Nano-Imaging of Polymers by Optical Microscopy. Vol. 182, pp. 131–170.
Ito, Y. see Suginome, M.: Vol. 171, pp. 77–136.
Ivanov, A. E. see Zubov, V. P.: Vol. 104, pp. 135–176.

Jacob, S. and *Kennedy, J.*: Synthesis, Characterization and Properties of OCTA-ARM Polyisobutylene-Based Star Polymers. Vol. 146, pp. 1–38.
Jacobson, K., Eriksson, P., Reitberger, T. and *Stenberg, B.*: Chemiluminescence as a Tool for Polyolefin. Vol. 169, pp. 151–176.
Jaeger, W. see Bohrisch, J.: Vol. 165, pp. 1–41.
Jaffe, M., Chen, P., Choe, E.-W., Chung, T.-S. and *Makhija, S.*: High Performance Polymer Blends. Vol. 117, pp. 297–328.
Jancar, J.: Structure-Property Relationships in Thermoplastic Matrices. Vol. 139, pp. 1–66.
Jang, J.: Conducting Polymer Nanomaterials and Their Applications. Vol. 199, pp. 189–260.
Jen, A. K.-Y. see Kajzar, F.: Vol. 161, pp. 1–85.
Jerome, R. see Mecerreyes, D.: Vol. 147, pp. 1–60.
de Jeu, W. H. see Li, L.: Vol. 181, pp. 75–120.
Jiang, M., Li, M., Xiang, M. and *Zhou, H.*: Interpolymer Complexation and Miscibility and Enhancement by Hydrogen Bonding. Vol. 146, pp. 121–194.
Jin, J. see Shim, H.-K.: Vol. 158, pp. 191–241.
Jinnai, H., Nishikawa, Y., Ikehara, T. and *Nishi, T.*: Emerging Technologies for the 3D Analysis of Polymer Structures. Vol. 170, pp. 115–167.
Jo, W. H. and *Yang, J. S.*: Molecular Simulation Approaches for Multiphase Polymer Systems. Vol. 156, pp. 1–52.
Joanny, J.-F. see Holm, C.: Vol. 166, pp. 67–111.
Joanny, J.-F. see Thünemann, A. F.: Vol. 166, pp. 113–171.
Johannsmann, D. see Rühe, J.: Vol. 165, pp. 79–150.
Johansson, M. see Hult, A.: Vol. 143, pp. 1–34.
Joos-Müller, B. see Funke, W.: Vol. 136, pp. 137–232.
Jou, D., Casas-Vazquez, J. and *Criado-Sancho, M.*: Thermodynamics of Polymer Solutions under Flow: Phase Separation and Polymer Degradation. Vol. 120, pp. 207–266.
Jozefiak, T. H. see Dhal, P. K.: Vol. 192, pp. 9–58.

Kabanov, A. V., Batrakova, E. V., Sherman, S. and *Alakhov, V. Y.*: Polymer Genomics. Vol. 193, pp. 173–198.
Kaetsu, I.: Radiation Synthesis of Polymeric Materials for Biomedical and Biochemical Applications. Vol. 105, pp. 81–98.
Kaji, K., Nishida, K., Kanaya, T., Matsuba, G., Konishi, T. and *Imai, M.*: Spinodal Crystallization of Polymers: Crystallization from the Unstable Melt. Vol. 191, pp. 187–240.
Kaji, K. see Kanaya, T.: Vol. 154, pp. 87–141.
Kajzar, F., Lee, K.-S. and *Jen, A. K.-Y.*: Polymeric Materials and their Orientation Techniques for Second-Order Nonlinear Optics. Vol. 161, pp. 1–85.
Kakimoto, M. see Gaw, K. O.: Vol. 140, pp. 107–136.
Kaminski, W. and *Arndt, M.*: Metallocenes for Polymer Catalysis. Vol. 127, pp. 143–187.
Kammer, H. W., Kressler, H. and *Kummerloewe, C.*: Phase Behavior of Polymer Blends – Effects of Thermodynamics and Rheology. Vol. 106, pp. 31–86.
Kanaya, T. and *Kaji, K.*: Dynamcis in the Glassy State and Near the Glass Transition of Amorphous Polymers as Studied by Neutron Scattering. Vol. 154, pp. 87–141.
Kanaya, T. see Kaji, K.: Vol. 191, pp. 187–240.
Kandyrin, L. B. and *Kuleznev, V. N.*: The Dependence of Viscosity on the Composition of Concentrated Dispersions and the Free Volume Concept of Disperse Systems. Vol. 103, pp. 103–148.
Kaneko, M. see Ramaraj, R.: Vol. 123, pp. 215–242.
Kaneko, M. see Yagi, M.: Vol. 199, pp. 143–188.
Kang, E. T., Neoh, K. G. and *Tan, K. L.*: X-Ray Photoelectron Spectroscopic Studies of Electroactive Polymers. Vol. 106, pp. 135–190.

Kaplan, D. L. see Singh, A.: Vol. 194, pp. 211–224.
Kaplan, D. L. see Xu, P.: Vol. 194, pp. 69–94.
Karlsson, S. see Söderqvist Lindblad, M.: Vol. 157, pp. 139–161.
Karlsson, S.: Recycled Polyolefins. Material Properties and Means for Quality Determination. Vol. 169, pp. 201–229.
Kataoka, K. see Nishiyama, N.: Vol. 193, pp. 67–101.
Kato, K. see Uyama, Y.: Vol. 137, pp. 1–40.
Kato, M. see Usuki, A.: Vol. 179, pp. 135–195.
Kausch, H.-H. and *Michler, G. H.*: The Effect of Time on Crazing and Fracture. Vol. 187, pp. 1–33.
Kausch, H.-H. see Monnerie, L. Vol. 187, pp. 215–364.
Kautek, W. see Krüger, J.: Vol. 168, pp. 247–290.
Kawaguchi, S. see Ito, K.: Vol. 142, pp. 129–178.
Kawaguchi, S. and *Ito, K.*: Dispersion Polymerization. Vol. 175, pp. 299–328.
Kawata, S. see Sun, H.-B.: Vol. 170, pp. 169–273.
Kazanskii, K. S. and *Dubrovskii, S. A.*: Chemistry and Physics of Agricultural Hydrogels. Vol. 104, pp. 97–134.
Kennedy, J. P. see Jacob, S.: Vol. 146, pp. 1–38.
Kennedy, J. P. see Majoros, I.: Vol. 112, pp. 1–113.
Kennedy, K. A., Roberts, G. W. and *DeSimone, J. M.*: Heterogeneous Polymerization of Fluoroolefins in Supercritical Carbon Dioxide. Vol. 175, pp. 329–346.
Khalatur, P. G. and *Khokhlov, A. R.*: Computer-Aided Conformation-Dependent Design of Copolymer Sequences. Vol. 195, pp. 1–100.
Khokhlov, A., Starodybtzev, S. and *Vasilevskaya, V.*: Conformational Transitions of Polymer Gels: Theory and Experiment. Vol. 109, pp. 121–172.
Khokhlov, A. R. see Grosberg, A. Y.: Vol. 196, pp. 189–210.
Khokhlov, A. R. see Khalatur, P. G.: Vol. 195, pp. 1–100.
Khokhlov, A. R. see Kuchanov, S. I.: Vol. 196, pp. 129–188.
Khokhlov, A. R. see Okhapkin, I. M.: Vol. 195, pp. 177–210.
Kiefer, J., Hedrick, J. L. and *Hiborn, J. G.*: Macroporous Thermosets by Chemically Induced Phase Separation. Vol. 147, pp. 161–247.
Kihara, N. see Takata, T.: Vol. 171, pp. 1–75.
Kilian, H. G. and *Pieper, T.*: Packing of Chain Segments. A Method for Describing X-Ray Patterns of Crystalline, Liquid Crystalline and Non-Crystalline Polymers. Vol. 108, pp. 49–90.
Kim, J. see Quirk, R. P.: Vol. 153, pp. 67–162.
Kim, K.-S. see Lin, T.-C.: Vol. 161, pp. 157–193.
Kimmich, R. and *Fatkullin, N.*: Polymer Chain Dynamics and NMR. Vol. 170, pp. 1–113.
Kippelen, B. and *Peyghambarian, N.*: Photorefractive Polymers and their Applications. Vol. 161, pp. 87–156.
Kippenberger, A. M. see Bergbreiter, D. E.: Vol. 198, pp. 1–49.
Kirchhoff, R. A. and *Bruza, K. J.*: Polymers from Benzocyclobutenes. Vol. 117, pp. 1–66.
Kishore, K. and *Ganesh, K.*: Polymers Containing Disulfide, Tetrasulfide, Diselenide and Ditelluride Linkages in the Main Chain. Vol. 121, pp. 81–122.
Kitamaru, R.: Phase Structure of Polyethylene and Other Crystalline Polymers by Solid-State 13C/MNR. Vol. 137, pp. 41–102.
Klapper, M. see Rusanov, A. L.: Vol. 179, pp. 83–134.
Klee, D. and *Höcker, H.*: Polymers for Biomedical Applications: Improvement of the Interface Compatibility. Vol. 149, pp. 1–57.
Klemm, E., Pautzsch, T. and *Blankenburg, L.*: Organometallic PAEs. Vol. 177, pp. 53–90.

Klier, J. see Scranton, A. B.: Vol. 122, pp. 1–54.
v. Klitzing, R. and *Tieke, B.*: Polyelectrolyte Membranes. Vol. 165, pp. 177–210.
Kloeckner, J. see Wagner, E.: Vol. 192, pp. 135–173.
Klüppel, M.: The Role of Disorder in Filler Reinforcement of Elastomers on Various Length Scales. Vol. 164, pp. 1–86.
Klüppel, M. see Heinrich, G.: Vol. 160, pp. 1–44.
Knuuttila, H., Lehtinen, A. and *Nummila-Pakarinen, A.*: Advanced Polyethylene Technologies – Controlled Material Properties. Vol. 169, pp. 13–27.
Kobayashi, S. and *Ohmae, M.*: Enzymatic Polymerization to Polysaccharides. Vol. 194, pp. 159–210.
Kobayashi, S. see Uyama, H.: Vol. 194, pp. 51–67.
Kobayashi, S. see Uyama, H.: Vol. 194, pp. 133–158.
Kobayashi, S., Shoda, S. and *Uyama, H.*: Enzymatic Polymerization and Oligomerization. Vol. 121, pp. 1–30.
Kobayashi, T. see Abe, A.: Vol. 181, pp. 121–152.
Köhler, W. and *Schäfer, R.*: Polymer Analysis by Thermal-Diffusion Forced Rayleigh Scattering. Vol. 151, pp. 1–59.
Koenig, J. L. see Bhargava, R.: Vol. 163, pp. 137–191.
Koenig, J. L. see Andreis, M.: Vol. 124, pp. 191–238.
Koike, T.: Viscoelastic Behavior of Epoxy Resins Before Crosslinking. Vol. 148, pp. 139–188.
Kokko, E. see Löfgren, B.: Vol. 169, pp. 1–12.
Kokufuta, E.: Novel Applications for Stimulus-Sensitive Polymer Gels in the Preparation of Functional Immobilized Biocatalysts. Vol. 110, pp. 157–178.
Konishi, T. see Kaji, K.: Vol. 191, pp. 187–240.
Konno, M. see Saito, S.: Vol. 109, pp. 207–232.
Konradi, R. see Rühe, J.: Vol. 165, pp. 79–150.
Kopecek, J. see Putnam, D.: Vol. 122, pp. 55–124.
Koßmehl, G. see Schopf, G.: Vol. 129, pp. 1–145.
Kostoglodov, P. V. see Rusanov, A. L.: Vol. 179, pp. 83–134.
Kozlov, E. see Prokop, A.: Vol. 160, pp. 119–174.
Kramer, E. J. see Creton, C.: Vol. 156, pp. 53–135.
Kremer, K. see Baschnagel, J.: Vol. 152, pp. 41–156.
Kremer, K. see Holm, C.: Vol. 166, pp. 67–111.
Kressler, J. see Kammer, H. W.: Vol. 106, pp. 31–86.
Kricheldorf, H. R.: Liquid-Crystalline Polyimides. Vol. 141, pp. 83–188.
Krishnamoorti, R. see Giannelis, E. P.: Vol. 138, pp. 107–148.
Krüger, J. and *Kautek, W.*: Ultrashort Pulse Laser Interaction with Dielectrics and Polymers, Vol. 168, pp. 247–290.
Kuchanov, S. I.: Modern Aspects of Quantitative Theory of Free-Radical Copolymerization. Vol. 103, pp. 1–102.
Kuchanov, S. I. and *Khokhlov, A. R.*: Role of Physical Factors in the Process of Obtaining Copolymers. Vol. 196, pp. 129–188.
Kuchanov, S. I.: Principles of Quantitive Description of Chemical Structure of Synthetic Polymers. Vol. 152, pp. 157–202.
Kudaibergennow, S. E.: Recent Advances in Studying of Synthetic Polyampholytes in Solutions. Vol. 144, pp. 115–198.
Kuleznev, V. N. see Kandyrin, L. B.: Vol. 103, pp. 103–148.
Kulichkhin, S. G. see Malkin, A. Y.: Vol. 101, pp. 217–258.
Kulicke, W.-M. see Grigorescu, G.: Vol. 152, pp. 1–40.
Kumar, M. N. V. R., Kumar, N., Domb, A. J. and *Arora, M.*: Pharmaceutical Polymeric Controlled Drug Delivery Systems. Vol. 160, pp. 45–118.

Kumar, N. see Kumar, M. N. V. R.: Vol. 160, pp. 45–118.
Kummerloewe, C. see Kammer, H. W.: Vol. 106, pp. 31–86.
Kuznetsova, N. P. see Samsonov, G. V.: Vol. 104, pp. 1–50.
Kwon, Y. and *Faust, R.*: Synthesis of Polyisobutylene-Based Block Copolymers with Precisely Controlled Architecture by Living Cationic Polymerization. Vol. 167, pp. 107–135.

Labadie, J. W. see Hergenrother, P. M.: Vol. 117, pp. 67–110.
Labadie, J. W. see Hedrick, J. L.: Vol. 141, pp. 1–44.
Labadie, J. W. see Hedrick, J. L.: Vol. 147, pp. 61–112.
Lamparski, H. G. see O'Brien, D. F.: Vol. 126, pp. 53–84.
Laschewsky, A.: Molecular Concepts, Self-Organisation and Properties of Polysoaps. Vol. 124, pp. 1–86.
Laso, M. see Leontidis, E.: Vol. 116, pp. 283–318.
Lauprêtre, F. see Monnerie, L.: Vol. 187, pp. 35–213.
Lazár, M. and *Rychl, R.*: Oxidation of Hydrocarbon Polymers. Vol. 102, pp. 189–222.
Lechowicz, J. see Galina, H.: Vol. 137, pp. 135–172.
Léger, L., Raphaël, E. and *Hervet, H.*: Surface-Anchored Polymer Chains: Their Role in Adhesion and Friction. Vol. 138, pp. 185–226.
Lenz, R. W.: Biodegradable Polymers. Vol. 107, pp. 1–40.
Leontidis, E., de Pablo, J. J., Laso, M. and *Suter, U. W.*: A Critical Evaluation of Novel Algorithms for the Off-Lattice Monte Carlo Simulation of Condensed Polymer Phases. Vol. 116, pp. 283–318.
Lee, B. see Quirk, R. P.: Vol. 153, pp. 67–162.
Lee, K.-S. see Kajzar, F.: Vol. 161, pp. 1–85.
Lee, Y. see Quirk, R. P.: Vol. 153, pp. 67–162.
Lehtinen, A. see Knuuttila, H.: Vol. 169, pp. 13–27.
Leónard, D. see Mathieu, H. J.: Vol. 162, pp. 1–35.
Lesec, J. see Viovy, J.-L.: Vol. 114, pp. 1–42.
Levesque, D. see Weis, J.-J.: Vol. 185, pp. 163–225.
Li, L. and *de Jeu, W. H.*: Flow-induced mesophases in crystallizable polymers. Vol. 181, pp. 75–120.
Li, L. see Chan, C.-M.: Vol. 188, pp. 1–41.
Li, M., Coenjarts, C. and *Ober, C. K.*: Patternable Block Copolymers. Vol. 190, pp. 183–226.
Li, M. see Jiang, M.: Vol. 146, pp. 121–194.
Liang, G. L. see Sumpter, B. G.: Vol. 116, pp. 27–72.
Lienert, K.-W.: Poly(ester-imide)s for Industrial Use. Vol. 141, pp. 45–82.
Likhatchev, D. see Rusanov, A. L.: Vol. 179, pp. 83–134.
Lin, J. and *Sherrington, D. C.*: Recent Developments in the Synthesis, Thermostability and Liquid Crystal Properties of Aromatic Polyamides. Vol. 111, pp. 177–220.
Lin, T.-C., Chung, S.-J., Kim, K.-S., Wang, X., He, G. S., Swiatkiewicz, J., Pudavar, H. E. and *Prasad, P. N.*: Organics and Polymers with High Two-Photon Activities and their Applications. Vol. 161, pp. 157–193.
Linse, P.: Simulation of Charged Colloids in Solution. Vol. 185, pp. 111–162.
Lippert, T.: Laser Application of Polymers. Vol. 168, pp. 51–246.
Liu, Y. see Söderqvist Lindblad, M.: Vol. 157, pp. 139–161.
Long, T.-C. see Geil, P. H.: Vol. 180, pp. 89–159.
López Cabarcos, E. see Baltá-Calleja, F. J.: Vol. 108, pp. 1–48.
Lotz, B.: Analysis and Observation of Polymer Crystal Structures at the Individual Stem Level. Vol. 180, pp. 17–44.
Löfgren, B., Kokko, E. and *Seppälä, J.*: Specific Structures Enabled by Metallocene Catalysis in Polyethenes. Vol. 169, pp. 1–12.

Löwen, H. see *Thünemann, A. F.*: Vol. 166, pp. 113–171.
Lozinsky V. I.: Approaches to Chemical Synthesis of Protein-Like Copolymers. Vol. 196, pp. 87–128.
Luo, Y. see *Schork, F. J.*: Vol. 175, pp. 129–255.

Macko, T. and *Hunkeler, D.*: Liquid Chromatography under Critical and Limiting Conditions: A Survey of Experimental Systems for Synthetic Polymers. Vol. 163, pp. 61–136.
Maeda, H., Greish, K. and *Fang, J.*: The EPR Effect and Polymeric Drugs: A Paradigm Shift for Cancer Chemotherapy in the 21st Century. Vol. 193, pp. 103–121.
Majoros, I., Nagy, A. and *Kennedy, J. P.*: Conventional and Living Carbocationic Polymerizations United. I. A Comprehensive Model and New Diagnostic Method to Probe the Mechanism of Homopolymerizations. Vol. 112, pp. 1–113.
Makhaeva, E. E. see *Okhapkin, I. M.*: Vol. 195, pp. 177–210.
Makhija, S. see *Jaffe, M.*: Vol. 117, pp. 297–328.
Malmström, E. see *Hult, A.*: Vol. 143, pp. 1–34.
Malkin, A. Y. and *Kulichkhin, S. G.*: Rheokinetics of Curing. Vol. 101, pp. 217–258.
Maniar, M. see *Domb, A. J.*: Vol. 107, pp. 93–142.
Manias, E. see *Giannelis, E. P.*: Vol. 138, pp. 107–148.
Martin, H. see *Engelhardt, H.*: Vol. 165, pp. 211–247.
Marty, J. D. and *Mauzac, M.*: Molecular Imprinting: State of the Art and Perspectives. Vol. 172, pp. 1–35.
Mashima, K., Nakayama, Y. and *Nakamura, A.*: Recent Trends in Polymerization of a-Olefins Catalyzed by Organometallic Complexes of Early Transition Metals. Vol. 133, pp. 1–52.
Mathew, D. see *Reghunadhan Nair, C. P.*: Vol. 155, pp. 1–99.
Mathieu, H. J., Chevolot, Y, Ruiz-Taylor, L. and *Leónard, D.*: Engineering and Characterization of Polymer Surfaces for Biomedical Applications. Vol. 162, pp. 1–35.
Matsuba, G. see *Kaji, K.*: Vol. 191, pp. 187–240.
Matsuda T.: Photoiniferter-Driven Precision Surface Graft Microarchitectures for Biomedical Applications. Vol. 197, pp. 67–106.
Matsumura S.: Enzymatic Synthesis of Polyesters via Ring-Opening Polymerization. Vol. 194, pp. 95–132.
Matsumoto, A.: Free-Radical Crosslinking Polymerization and Copolymerization of Multivinyl Compounds. Vol. 123, pp. 41–80.
Matsumoto, A. see *Otsu, T.*: Vol. 136, pp. 75–138.
Matsuoka, H. and *Ise, N.*: Small-Angle and Ultra-Small Angle Scattering Study of the Ordered Structure in Polyelectrolyte Solutions and Colloidal Dispersions. Vol. 114, pp. 187–232.
Matsushige, K., Hiramatsu, N. and *Okabe, H.*: Ultrasonic Spectroscopy for Polymeric Materials. Vol. 125, pp. 147–186.
Mattice, W. L. see *Rehahn, M.*: Vol. 131/132, pp. 1–475.
Mattice, W. L. see *Baschnagel, J.*: Vol. 152, pp. 41–156.
Mattozzi, A. see *Gedde, U. W.*: Vol. 169, pp. 29–73.
Mauzac, M. see *Marty, J. D.*: Vol. 172, pp. 1–35.
Mays, W. see *Xu, Z.*: Vol. 120, pp. 1–50.
Mays, J. W. see *Pitsikalis, M.*: Vol. 135, pp. 1–138.
McGrath, J. E. see *Hedrick, J. L.*: Vol. 141, pp. 1–44.
McGrath, J. E., Dunson, D. L. and *Hedrick, J. L.*: Synthesis and Characterization of Segmented Polyimide-Polyorganosiloxane Copolymers. Vol. 140, pp. 61–106.
McLeish, T. C. B. and *Milner, S. T.*: Entangled Dynamics and Melt Flow of Branched Polymers. Vol. 143, pp. 195–256.

Mecerreyes, D., Dubois, P. and *Jerome, R.*: Novel Macromolecular Architectures Based on Aliphatic Polyesters: Relevance of the Coordination-Insertion Ring-Opening Polymerization. Vol. 147, pp. 1–60.
Mecham, S. J. see McGrath, J. E.: Vol. 140, pp. 61–106.
Meille, S. V. see Allegra, G.: Vol. 191, pp. 87–135.
Menzel, H. see Möhwald, H.: Vol. 165, pp. 151–175.
Meyer, T. see Spange, S.: Vol. 165, pp. 43–78.
Michler, G. H. see Kausch, H.-H.: Vol. 187, pp. 1–33.
Mikos, A. G. see Thomson, R. C.: Vol. 122, pp. 245–274.
Milner, S. T. see McLeish, T. C. B.: Vol. 143, pp. 195–256.
Mirous, B. K. see Brittain, W. J.: Vol. 198, pp. 125–147.
Mison, P. and *Sillion, B.*: Thermosetting Oligomers Containing Maleimides and Nadiimides End-Groups. Vol. 140, pp. 137–180.
Miyasaka, K.: PVA-Iodine Complexes: Formation, Structure and Properties. Vol. 108, pp. 91–130.
Miller, R. D. see Hedrick, J. L.: Vol. 141, pp. 1–44.
Minko, S. see Rühe, J.: Vol. 165, pp. 79–150.
Möhwald, H., Menzel, H., Helm, C. A. and *Stamm, M.*: Lipid and Polyampholyte Monolayers to Study Polyelectrolyte Interactions and Structure at Interfaces. Vol. 165, pp. 151–175.
Monkenbusch, M. see Richter, D.: Vol. 174, pp. 1–221.
Monnerie, L., Halary, J. L. and *Kausch, H.-H.*: Deformation, Yield and Fracture of Amorphous Polymers: Relation to the Secondary Transitions. Vol. 187, pp. 215–364.
Monnerie, L., Lauprêtre, F. and *Halary, J. L.*: Investigation of Solid-State Transitions in Linear and Crosslinked Amorphous Polymers. Vol. 187, pp. 35–213.
Monnerie, L. see Bahar, I.: Vol. 116, pp. 145–206.
Moore, J. S. see Ray, C. R.: Vol. 177, pp. 99–149.
Mori, H. see Bohrisch, J.: Vol. 165, pp. 1–41.
Morishima, Y.: Photoinduced Electron Transfer in Amphiphilic Polyelectrolyte Systems. Vol. 104, pp. 51–96.
Morton, M. see Quirk, R. P.: Vol. 153, pp. 67–162.
Motornov, M. see Rühe, J.: Vol. 165, pp. 79–150.
Mours, M. see Winter, H. H.: Vol. 134, pp. 165–234.
Müllen, K. see Grimsdale, A. C.: Vol. 199, pp. 1–82.
Müllen, K. see Scherf, U.: Vol. 123, pp. 1–40.
Müller, A. H. E. see Bohrisch, J.: Vol. 165, pp. 1–41.
Müller, A. H. E. see Förster, S.: Vol. 166, pp. 173–210.
Müller, A. J., Balsamo, V. and *Arnal, M. L.*: Nucleation and Crystallization in Diblock and Triblock Copolymers. Vol. 190, pp. 1–63.
Müller, M. and *Schmid, F.*: Incorporating Fluctuations and Dynamics in Self-Consistent Field Theories for Polymer Blends. Vol. 185, pp. 1–58.
Müller, M. see Thünemann, A. F.: Vol. 166, pp. 113–171.
Müller-Plathe, F. see Gusev, A. A.: Vol. 116, pp. 207–248.
Müller-Plathe, F. see Baschnagel, J.: Vol. 152, p. 41–156.
Mukerherjee, A. see Biswas, M.: Vol. 115, pp. 89–124.
Munz, M., Cappella, B., Sturm, H., Geuss, M. and *Schulz, E.*: Materials Contrasts and Nanolithography Techniques in Scanning Force Microscopy (SFM) and their Application to Polymers and Polymer Composites. Vol. 164, pp. 87–210.
Murat, M. see Baschnagel, J.: Vol. 152, p. 41–156.
Muthukumar, M.: Modeling Polymer Crystallization. Vol. 191, pp. 241–274.

Muzzarelli, C. see Muzzarelli, R. A. A.: Vol. 186, pp. 151–209.
Muzzarelli, R. A. A. and *Muzzarelli, C.*: Chitosan Chemistry: Relevance to the Biomedical Sciences. Vol. 186, pp. 151–209.
Mylnikov, V.: Photoconducting Polymers. Vol. 115, pp. 1–88.

Nagy, A. see Majoros, I.: Vol. 112, pp. 1–11.
Naji, A., Seidel, C. and *Netz, R. R.*: Theoretical Approaches to Neutral and Charged Polymer Brushes. Vol. 198, pp. 149–183.
Naka, K. see Uemura, T.: Vol. 167, pp. 81–106.
Nakamura, A. see Mashima, K.: Vol. 133, pp. 1–52.
Nakayama, Y. see Mashima, K.: Vol. 133, pp. 1–52.
Narasinham, B. and *Peppas, N. A.*: The Physics of Polymer Dissolution: Modeling Approaches and Experimental Behavior. Vol. 128, pp. 157–208.
Nechaev, S. see Grosberg, A.: Vol. 106, pp. 1–30.
Neoh, K. G. see Kang, E. T.: Vol. 106, pp. 135–190.
Netz, R. R. see Holm, C.: Vol. 166, pp. 67–111.
Netz, R. R. see Naji, A.: Vol. 198, pp. 149–183.
Netz, R. R. see Rühe, J.: Vol. 165, pp. 79–150.
Newman, S. M. see Anseth, K. S.: Vol. 122, pp. 177–218.
Nijenhuis, K. te: Thermoreversible Networks. Vol. 130, pp. 1–252.
Ninan, K. N. see Reghunadhan Nair, C. P.: Vol. 155, pp. 1–99.
Nishi, T. see Jinnai, H.: Vol. 170, pp. 115–167.
Nishida, K. see Kaji, K.: Vol. 191, pp. 187–240.
Nishikawa, Y. see Jinnai, H.: Vol. 170, pp. 115–167.
Nishiyama, N. and *Kataoka, K.*: Nanostructured Devices Based on Block Copolymer Assemblies for Drug Delivery: Designing Structures for Enhanced Drug Function. Vol. 193, pp. 67–101.
Noid, D. W. see Otaigbe, J. U.: Vol. 154, pp. 1–86.
Noid, D. W. see Sumpter, B. G.: Vol. 116, pp. 27–72.
Nomura, M., Tobita, H. and *Suzuki, K.*: Emulsion Polymerization: Kinetic and Mechanistic Aspects. Vol. 175, pp. 1–128.
Northolt, M. G., Picken, S. J., Den Decker, M. G., Baltussen, J. J. M. and *Schlatmann, R.*: The Tensile Strength of Polymer Fibres. Vol. 178, pp. 1–108.
Novac, B. see Grubbs, R.: Vol. 102, pp. 47–72.
Novikov, V. V. see Privalko, V. P.: Vol. 119, pp. 31–78.
Nummila-Pakarinen, A. see Knuuttila, H.: Vol. 169, pp. 13–27.

Ober, C. K. see Li, M.: Vol. 190, pp. 183–226.
O'Brien, D. F., Armitage, B. A., Bennett, D. E. and *Lamparski, H. G.*: Polymerization and Domain Formation in Lipid Assemblies. Vol. 126, pp. 53–84.
Ogasawara, M.: Application of Pulse Radiolysis to the Study of Polymers and Polymerizations. Vol. 105, pp. 37–80.
Ohmae, M. see Kobayashi, S.: Vol. 194, pp. 159–210.
Ohno, K. see Tsujii, Y.: Vol. 197, pp. 1–47.
Okabe, H. see Matsushige, K.: Vol. 125, pp. 147–186.
Okada, M.: Ring-Opening Polymerization of Bicyclic and Spiro Compounds. Reactivities and Polymerization Mechanisms. Vol. 102, pp. 1–46.
Okada, K. see Hikosaka, M.: Vol. 191, pp. 137–186.
Okano, T.: Molecular Design of Temperature-Responsive Polymers as Intelligent Materials. Vol. 110, pp. 179–198.

Okay, O. see Funke, W.: Vol. 136, pp. 137–232.
Okhapkin, I. M., Makhaeva, E. E. and *Khokhlov, A. R.*: Water Solutions of Amphiphilic Polymers: Nanostructure Formation and Possibilities for Catalysis. Vol. 195, pp. 177–210.
Onuki, A.: Theory of Phase Transition in Polymer Gels. Vol. 109, pp. 63–120.
Oppermann, W. see Holm, C.: Vol. 166, pp. 1–27.
Oppermann, W. see Volk, N.: Vol. 166, pp. 29–65.
Osad'ko, I. S.: Selective Spectroscopy of Chromophore Doped Polymers and Glasses. Vol. 114, pp. 123–186.
Osakada, K. and *Takeuchi, D.*: Coordination Polymerization of Dienes, Allenes, and Methylenecycloalkanes. Vol. 171, pp. 137–194.
Otaigbe, J. U., Barnes, M. D., Fukui, K., Sumpter, B. G. and *Noid, D. W.*: Generation, Characterization, and Modeling of Polymer Micro- and Nano-Particles. Vol. 154, pp. 1–86.
Otsu, T. and *Matsumoto, A.*: Controlled Synthesis of Polymers Using the Iniferter Technique: Developments in Living Radical Polymerization. Vol. 136, pp. 75–138.

de Pablo, J. J. see Leontidis, E.: Vol. 116, pp. 283–318.
Padias, A. B. see Penelle, J.: Vol. 102, pp. 73–104.
Pascault, J.-P. see Williams, R. J. J.: Vol. 128, pp. 95–156.
Pasch, H.: Analysis of Complex Polymers by Interaction Chromatography. Vol. 128, pp. 1–46.
Pasch, H.: Hyphenated Techniques in Liquid Chromatography of Polymers. Vol. 150, pp. 1–66.
Pasut, G. and *Veronese, F. M.*: PEGylation of Proteins as Tailored Chemistry for Optimized Bioconjugates. Vol. 192, pp. 95–134.
Paul, W. see Baschnagel, J.: Vol. 152, pp. 41–156.
Paulsen, S. B. and *Barsett, H.*: Bioactive Pectic Polysaccharides. Vol. 186, pp. 69–101.
Pautzsch, T. see Klemm, E.: Vol. 177, pp. 53–90.
Penczek, P., Czub, P. and *Pielichowski, J.*: Unsaturated Polyester Resins: Chemistry and Technology. Vol. 184, pp. 1–95.
Penczek, P. see Batog, A. E.: Vol. 144, pp. 49–114.
Penczek, P. see Bogdal, D.: Vol. 163, pp. 193–263.
Penelle, J., Hall, H. K., Padias, A. B. and *Tanaka, H.*: Captodative Olefins in Polymer Chemistry. Vol. 102, pp. 73–104.
Peppas, N. A. see Bell, C. L.: Vol. 122, pp. 125–176.
Peppas, N. A. see Hassan, C. M.: Vol. 153, pp. 37–65.
Peppas, N. A. see Narasimhan, B.: Vol. 128, pp. 157–208.
Petersen, K. L. see Geil, P. H.: Vol. 180, pp. 89–159.
Pet'ko, I. P. see Batog, A. E.: Vol. 144, pp. 49–114.
Pheyghambarian, N. see Kippelen, B.: Vol. 161, pp. 87–156.
Pichot, C. see Hunkeler, D.: Vol. 112, pp. 115–134.
Picken, S. J. see Northolt, M. G.: Vol. 178, pp. 1–108.
Pielichowski, J. see Bogdal, D.: Vol. 163, pp. 193–263.
Pielichowski, J. see Penczek, P.: Vol. 184, pp. 1–95.
Pieper, T. see Kilian, H. G.: Vol. 108, pp. 49–90.
Pispas, S. see Pitsikalis, M.: Vol. 135, pp. 1–138.
Pispas, S. see Hadjichristidis, N.: Vol. 142, pp. 71–128.
Pitsikalis, M., Pispas, S., Mays, J. W. and *Hadjichristidis, N.*: Nonlinear Block Copolymer Architectures. Vol. 135, pp. 1–138.
Pitsikalis, M. see Hadjichristidis, N.: Vol. 142, pp. 71–128.
Pitsikalis, M. see Hadjichristidis, N.: Vol. 189, pp. 1–124.
Pleul, D. see Spange, S.: Vol. 165, pp. 43–78.

Plummer, C. J. G.: Microdeformation and Fracture in Bulk Polyolefins. Vol. 169, pp. 75–119.
Pötschke, D. see *Dingenouts, N.*: Vol. 144, pp. 1–48.
Pokrovskii, V. N.: The Mesoscopic Theory of the Slow Relaxation of Linear Macromolecules. Vol. 154, pp. 143–219.
Pospíšil, J.: Functionalized Oligomers and Polymers as Stabilizers for Conventional Polymers. Vol. 101, pp. 65–168.
Pospíšil, J.: Aromatic and Heterocyclic Amines in Polymer Stabilization. Vol. 124, pp. 87–190.
Powers, A. C. see *Prokop, A.*: Vol. 136, pp. 53–74.
Prasad, P. N. see *Lin, T.-C.*: Vol. 161, pp. 157–193.
Priddy, D. B.: Recent Advances in Styrene Polymerization. Vol. 111, pp. 67–114.
Priddy, D. B.: Thermal Discoloration Chemistry of Styrene-co-Acrylonitrile. Vol. 121, pp. 123–154.
Privalko, V. P. and *Novikov, V. V.*: Model Treatments of the Heat Conductivity of Heterogeneous Polymers. Vol. 119, pp. 31–78.
Prociak, A. see *Bogdal, D.*: Vol. 163, pp. 193–263.
Prokop, A., Hunkeler, D., DiMari, S., Haralson, M. A. and *Wang, T. G.*: Water Soluble Polymers for Immunoisolation I: Complex Coacervation and Cytotoxicity. Vol. 136, pp. 1–52.
Prokop, A., Hunkeler, D., Powers, A. C., Whitesell, R. R. and *Wang, T. G.*: Water Soluble Polymers for Immunoisolation II: Evaluation of Multicomponent Microencapsulation Systems. Vol. 136, pp. 53–74.
Prokop, A., Kozlov, E., Carlesso, G. and *Davidsen, J. M.*: Hydrogel-Based Colloidal Polymeric System for Protein and Drug Delivery: Physical and Chemical Characterization, Permeability Control and Applications. Vol. 160, pp. 119–174.
Pruitt, L. A.: The Effects of Radiation on the Structural and Mechanical Properties of Medical Polymers. Vol. 162, pp. 65–95.
Pudavar, H. E. see *Lin, T.-C.*: Vol. 161, pp. 157–193.
Pudzich, R., Fuhrmann-Lieker, T. and *Salbeck, J.*: Spiro Compounds for Organic Electroluminescence and Related Applications. Vol. 199, pp. 83–142.
Pukánszky, B. and *Fekete, E.*: Adhesion and Surface Modification. Vol. 139, pp. 109–154.
Putnam, D. and *Kopecek, J.*: Polymer Conjugates with Anticancer Acitivity. Vol. 122, pp. 55–124.
Putra, E. G. R. see *Ungar, G.*: Vol. 180, pp. 45–87.

Quirk, R. P., Yoo, T., Lee, Y., M., Kim, J. and *Lee, B.*: Applications of 1,1-Diphenylethylene Chemistry in Anionic Synthesis of Polymers with Controlled Structures. Vol. 153, pp. 67–162.

Ramaraj, R. and *Kaneko, M.*: Metal Complex in Polymer Membrane as a Model for Photosynthetic Oxygen Evolving Center. Vol. 123, pp. 215–242.
Rangarajan, B. see *Scranton, A. B.*: Vol. 122, pp. 1–54.
Ranucci, E. see *Söderqvist Lindblad, M.*: Vol. 157, pp. 139–161.
Raphaël, E. see *Léger, L.*: Vol. 138, pp. 185–226.
Rastogi, S. and *Terry, A. E.*: Morphological implications of the interphase bridging crystalline and amorphous regions in semi-crystalline polymers. Vol. 180, pp. 161–194.
Ray, C. R. and *Moore, J. S.*: Supramolecular Organization of Foldable Phenylene Ethynylene Oligomers. Vol. 177, pp. 99–149.
Reddinger, J. L. and *Reynolds, J. R.*: Molecular Engineering of p-Conjugated Polymers. Vol. 145, pp. 57–122.
Reghunadhan Nair, C. P., Mathew, D. and *Ninan, K. N.*: Cyanate Ester Resins, Recent Developments. Vol. 155, pp. 1–99.

Reichert, K. H. see Hunkeler, D.: Vol. 112, pp. 115–134.
Reihmann, M. and *Ritter, H.*: Synthesis of Phenol Polymers Using Peroxidases. Vol. 194, pp. 1–49.
Rehahn, M., Mattice, W. L. and *Suter, U. W.*: Rotational Isomeric State Models in Macromolecular Systems. Vol. 131/132, pp. 1–475.
Rehahn, M. see Bohrisch, J.: Vol. 165, pp. 1–41.
Rehahn, M. see Holm, C.: Vol. 166, pp. 1–27.
Reineker, P. see Holm, C.: Vol. 166, pp. 67–111.
Reitberger, T. see Jacobson, K.: Vol. 169, pp. 151–176.
Ritter, H. see Reihmann, M.: Vol. 194, pp. 1–49.
Reynolds, J. R. see Reddinger, J. L.: Vol. 145, pp. 57–122.
Richter, D. see Ewen, B.: Vol. 134, pp. 1–130.
Richter, D., Monkenbusch, M. and *Colmenero, J.*: Neutron Spin Echo in Polymer Systems. Vol. 174, pp. 1–221.
Riegler, S. see Trimmel, G.: Vol. 176, pp. 43–87.
Ringsdorf, H. see Duncan, R.: Vol. 192, pp. 1–8.
Risse, W. see Grubbs, R.: Vol. 102, pp. 47–72.
Rivas, B. L. and *Geckeler, K. E.*: Synthesis and Metal Complexation of Poly(ethyleneimine) and Derivatives. Vol. 102, pp. 171–188.
Roberts, C. J. see Ellis, J. S.: Vol. 193, pp. 123–172.
Roberts, G. W. see Kennedy, K. A.: Vol. 175, pp. 329–346.
Robin, J. J.: The Use of Ozone in the Synthesis of New Polymers and the Modification of Polymers. Vol. 167, pp. 35–79.
Robin, J. J. see Boutevin, B.: Vol. 102, pp. 105–132.
Rodríguez-Pérez, M. A.: Crosslinked Polyolefin Foams: Production, Structure, Properties, and Applications. Vol. 184, pp. 97–126.
Roe, R.-J.: MD Simulation Study of Glass Transition and Short Time Dynamics in Polymer Liquids. Vol. 116, pp. 111–114.
Roovers, J. and *Comanita, B.*: Dendrimers and Dendrimer-Polymer Hybrids. Vol. 142, pp. 179–228.
Rothon, R. N.: Mineral Fillers in Thermoplastics: Filler Manufacture and Characterisation. Vol. 139, pp. 67–108.
de Rosa, C. see Auriemma, F.: Vol. 181, pp. 1–74.
Rozenberg, B. A. see Williams, R. J. J.: Vol. 128, pp. 95–156.
Rühe, J., Ballauff, M., Biesalski, M., Dziezok, P., Gröhn, F., Johannsmann, D., Houbenov, N., Hugenberg, N., Konradi, R., Minko, S., Motornov, M., Netz, R. R., Schmidt, M., Seidel, C., Stamm, M., Stephan, T., Usov, D. and *Zhang, H.*: Polyelectrolyte Brushes. Vol. 165, pp. 79–150.
Ruckenstein, E.: Concentrated Emulsion Polymerization. Vol. 127, pp. 1–58.
Ruiz-Taylor, L. see Mathieu, H. J.: Vol. 162, pp. 1–35.
Rusanov, A. L.: Novel Bis (Naphtalic Anhydrides) and Their Polyheteroarylenes with Improved Processability. Vol. 111, pp. 115–176.
Rusanov, A. L., Likhatchev, D., Kostoglodov, P. V., Müllen, K. and *Klapper, M.*: Proton-Exchanging Electrolyte Membranes Based on Aromatic Condensation Polymers. Vol. 179, pp. 83–134.
Russel, T. P. see Hedrick, J. L.: Vol. 141, pp. 1–44.
Russum, J. P. see Schork, F. J.: Vol. 175, pp. 129–255.
Rychly, J. see Lazár, M.: Vol. 102, pp. 189–222.
Ryner, M. see Stridsberg, K. M.: Vol. 157, pp. 27–51.
Ryzhov, V. A. see Bershtein, V. A.: Vol. 114, pp. 43–122.

Sabsai, O. Y. see Barshtein, G. R.: Vol. 101, pp. 1–28.
Saburov, V. V. see Zubov, V. P.: Vol. 104, pp. 135–176.
Saito, S., Konno, M. and *Inomata, H.*: Volume Phase Transition of N-Alkylacrylamide Gels. Vol. 109, pp. 207–232.
Salbeck, J. see Pudzich, R.: Vol. 199, pp. 83–142.
Samsonov, G. V. and *Kuznetsova, N. P.*: Crosslinked Polyelectrolytes in Biology. Vol. 104, pp. 1–50.
Santa Cruz, C. see Baltá-Calleja, F. J.: Vol. 108, pp. 1–48.
Santos, S. see Baschnagel, J.: Vol. 152, p. 41–156.
Satchi-Fainaro, R., Duncan, R. and *Barnes, C. M.*: Polymer Therapeutics for Cancer: Current Status and Future Challenges. Vol. 193, pp. 1–65.
Satchi-Fainaro, R. see Duncan, R.: Vol. 192, pp. 1–8.
Sato, T. and *Teramoto, A.*: Concentrated Solutions of Liquid-Christalline Polymers. Vol. 126, pp. 85–162.
Schaller, C. see Bohrisch, J.: Vol. 165, pp. 1–41.
Schäfer, R. see Köhler, W.: Vol. 151, pp. 1–59.
Scherf, U. and *Müllen, K.*: The Synthesis of Ladder Polymers. Vol. 123, pp. 1–40.
Sherman, S. see Kabanov, A. V.: Vol. 193, pp. 173–198.
Schlatmann, R. see Northolt, M. G.: Vol. 178, pp. 1–108.
Schmid, F. see Müller, M.: Vol. 185, pp. 1–58.
Schmidt, M. see Förster, S.: Vol. 120, pp. 51–134.
Schmidt, M. see Rühe, J.: Vol. 165, pp. 79–150.
Schmidt, M. see Volk, N.: Vol. 166, pp. 29–65.
Scholz, M.: Effects of Ion Radiation on Cells and Tissues. Vol. 162, pp. 97–158.
Schönherr, H. see Vancso, G. J.: Vol. 182, pp. 55–129.
Schopf, G. and *Koßmehl, G.*: Polythiophenes – Electrically Conductive Polymers. Vol. 129, pp. 1–145.
Schork, F. J., Luo, Y., Smulders, W., Russum, J. P., Butté, A. and *Fontenot, K.*: Miniemulsion Polymerization. Vol. 175, pp. 127–255.
Schulz, E. see Munz, M.: Vol. 164, pp. 97–210.
Schwahn, D.: Critical to Mean Field Crossover in Polymer Blends. Vol. 183, pp. 1–61.
Seppälä, J. see Löfgren, B.: Vol. 169, pp. 1–12.
Sturm, H. see Munz, M.: Vol. 164, pp. 87–210.
Schweizer, K. S.: Prism Theory of the Structure, Thermodynamics, and Phase Transitions of Polymer Liquids and Alloys. Vol. 116, pp. 319–378.
Scranton, A. B., Rangarajan, B. and *Klier, J.*: Biomedical Applications of Polyelectrolytes. Vol. 122, pp. 1–54.
Sefton, M. V. and *Stevenson, W. T. K.*: Microencapsulation of Live Animal Cells Using Polycrylates. Vol. 107, pp. 143–198.
Seidel, C. see Holm, C.: Vol. 166, pp. 67–111.
Seidel, C. see Naji, A.: Vol. 198, pp. 149–183.
Seidel, C. see Rühe, J.: Vol. 165, pp. 79–150.
El Seoud, O. A. and *Heinze, T.*: Organic Esters of Cellulose: New Perspectives for Old Polymers. Vol. 186, pp. 103–149.
Shabat, D. see Amir, R. J.: Vol. 192, pp. 59–94.
Shamanin, V. V.: Bases of the Axiomatic Theory of Addition Polymerization. Vol. 112, pp. 135–180.
Shcherbina, M. A. see Ungar, G.: Vol. 180, pp. 45–87.
Sheiko, S. S.: Imaging of Polymers Using Scanning Force Microscopy: From Superstructures to Individual Molecules. Vol. 151, pp. 61–174.

Sherrington, D. C. see *Cameron, N. R.:* Vol. 126, pp. 163–214.
Sherrington, D. C. see *Lin, J.:* Vol. 111, pp. 177–220.
Sherrington, D. C. see *Steinke, J.:* Vol. 123, pp. 81–126.
Shibayama, M. see *Tanaka, T.:* Vol. 109, pp. 1–62.
Shiga, T.: Deformation and Viscoelastic Behavior of Polymer Gels in Electric Fields. Vol. 134, pp. 131–164.
Shim, H.-K. and *Jin, J.:* Light-Emitting Characteristics of Conjugated Polymers. Vol. 158, pp. 191–241.
Shoda, S. see *Kobayashi, S.:* Vol. 121, pp. 1–30.
Siegel, R. A.: Hydrophobic Weak Polyelectrolyte Gels: Studies of Swelling Equilibria and Kinetics. Vol. 109, pp. 233–268.
de Silva, D. S. M. see *Ungar, G.:* Vol. 180, pp. 45–87.
Silvestre, F. see *Calmon-Decriaud, A.:* Vol. 207, pp. 207–226.
Sillion, B. see *Mison, P.:* Vol. 140, pp. 137–180.
Simon, F. see *Spange, S.:* Vol. 165, pp. 43–78.
Simon, G. P. see *Becker, O.:* Vol. 179, pp. 29–82.
Simon, P. F. W. see *Abetz, V.:* Vol. 189, pp. 125–212.
Simonutti, R. see *Sozzani, P.:* Vol. 181, pp. 153–177.
Singh, A. and *Kaplan, D. L.:* In Vitro Enzyme-Induced Vinyl Polymerization. Vol. 194, pp. 211–224.
Singh, A. see *Xu, P.:* Vol. 194, pp. 69–94.
Singh, R. P. see *Sivaram, S.:* Vol. 101, pp. 169–216.
Singh, R. P. see *Desai, S. M.:* Vol. 169, pp. 231–293.
Sinha Ray, S. see *Biswas, M.:* Vol. 155, pp. 167–221.
Sivaram, S. and *Singh, R. P.:* Degradation and Stabilization of Ethylene-Propylene Copolymers and Their Blends: A Critical Review. Vol. 101, pp. 169–216.
Slugovc, C. see *Trimmel, G.:* Vol. 176, pp. 43–87.
Smulders, W. see *Schork, F. J.:* Vol. 175, pp. 129–255.
Soares, J. B. P. see *Anantawaraskul, S.:* Vol. 182, pp. 1–54.
Sozzani, P., Bracco, S., Comotti, A. and *Simonutti, R.:* Motional Phase Disorder of Polymer Chains as Crystallized to Hexagonal Lattices. Vol. 181, pp. 153–177.
Söderqvist Lindblad, M., Liu, Y., Albertsson, A.-C., Ranucci, E. and *Karlsson, S.:* Polymer from Renewable Resources. Vol. 157, pp. 139–161.
Spange, S., Meyer, T., Voigt, I., Eschner, M., Estel, K., Pleul, D. and *Simon, F.:* Poly(Vinylformamide-co-Vinylamine)/Inorganic Oxid Hybrid Materials. Vol. 165, pp. 43–78.
Stamm, M. see *Möhwald, H.:* Vol. 165, pp. 151–175.
Stamm, M. see *Rühe, J.:* Vol. 165, pp. 79–150.
Starodybtzev, S. see *Khokhlov, A.:* Vol. 109, pp. 121–172.
Stegeman, G. I. see *Canva, M.:* Vol. 158, pp. 87–121.
Steinke, J., Sherrington, D. C. and *Dunkin, I. R.:* Imprinting of Synthetic Polymers Using Molecular Templates. Vol. 123, pp. 81–126.
Stelzer, F. see *Trimmel, G.:* Vol. 176, pp. 43–87.
Stenberg, B. see *Jacobson, K.:* Vol. 169, pp. 151–176.
Stenzenberger, H. D.: Addition Polyimides. Vol. 117, pp. 165–220.
Stephan, T. see *Rühe, J.:* Vol. 165, pp. 79–150.
Stevenson, W. T. K. see *Sefton, M. V.:* Vol. 107, pp. 143–198.
Stridsberg, K. M., Ryner, M. and *Albertsson, A.-C.:* Controlled Ring-Opening Polymerization: Polymers with Designed Macromoleculars Architecture. Vol. 157, pp. 27–51.
Sturm, H. see *Munz, M.:* Vol. 164, pp. 87–210.

Suematsu, K.: Recent Progress of Gel Theory: Ring, Excluded Volume, and Dimension. Vol. 156, pp. 136–214.
Sugimoto, H. and *Inoue, S.*: Polymerization by Metalloporphyrin and Related Complexes. Vol. 146, pp. 39–120.
Suginome, M. and *Ito, Y.*: Transition Metal-Mediated Polymerization of Isocyanides. Vol. 171, pp. 77–136.
Sumpter, B. G., Noid, D. W., Liang, G. L. and *Wunderlich, B.*: Atomistic Dynamics of Macromolecular Crystals. Vol. 116, pp. 27–72.
Sumpter, B. G. see Otaigbe, J. U.: Vol. 154, pp. 1–86.
Sun, H.-B. and *Kawata, S.*: Two-Photon Photopolymerization and 3D Lithographic Microfabrication. Vol. 170, pp. 169–273.
Suter, U. W. see Gusev, A. A.: Vol. 116, pp. 207–248.
Suter, U. W. see Leontidis, E.: Vol. 116, pp. 283–318.
Suter, U. W. see Rehahn, M.: Vol. 131/132, pp. 1–475.
Suter, U. W. see Baschnagel, J.: Vol. 152, pp. 41–156.
Suzuki, A.: Phase Transition in Gels of Sub-Millimeter Size Induced by Interaction with Stimuli. Vol. 110, pp. 199–240.
Suzuki, A. and *Hirasa, O.*: An Approach to Artifical Muscle by Polymer Gels due to Micro-Phase Separation. Vol. 110, pp. 241–262.
Suzuki, K. see Nomura, M.: Vol. 175, pp. 1–128.
Swiatkiewicz, J. see Lin, T.-C.: Vol. 161, pp. 157–193.

Tagawa, S.: Radiation Effects on Ion Beams on Polymers. Vol. 105, pp. 99–116.
Taguet, A., Ameduri, B. and *Boutevin, B.*: Crosslinking of Vinylidene Fluoride-Containing Fluoropolymers. Vol. 184, pp. 127–211.
Takata, T., Kihara, N. and *Furusho, Y.*: Polyrotaxanes and Polycatenanes: Recent Advances in Syntheses and Applications of Polymers Comprising of Interlocked Structures. Vol. 171, pp. 1–75.
Takeuchi, D. see Osakada, K.: Vol. 171, pp. 137–194.
Tan, K. L. see Kang, E. T.: Vol. 106, pp. 135–190.
Tanaka, H. and *Shibayama, M.*: Phase Transition and Related Phenomena of Polymer Gels. Vol. 109, pp. 1–62.
Tanaka, T. see Penelle, J.: Vol. 102, pp. 73–104.
Tauer, K. see Guyot, A.: Vol. 111, pp. 43–66.
Tendler, S. J. B. see Ellis, J. S.: Vol. 193, pp. 123–172.
Tenhu, H. see Aseyev, V. O.: Vol. 196, pp. 1–86.
Teramoto, A. see Sato, T.: Vol. 126, pp. 85–162.
Terent'eva, J. P. and *Fridman, M. L.*: Compositions Based on Aminoresins. Vol. 101, pp. 29–64.
Terry, A. E. see Rastogi, S.: Vol. 180, pp. 161–194.
Theodorou, D. N. see Dodd, L. R.: Vol. 116, pp. 249–282.
Thomson, R. C., Wake, M. C., Yaszemski, M. J. and *Mikos, A. G.*: Biodegradable Polymer Scaffolds to Regenerate Organs. Vol. 122, pp. 245–274.
Thünemann, A. F., Müller, M., Dautzenberg, H., Joanny, J.-F. and *Löwen, H.*: Polyelectrolyte complexes. Vol. 166, pp. 113–171.
Tieke, B. see v. Klitzing, R.: Vol. 165, pp. 177–210.
Tobita, H. see Nomura, M.: Vol. 175, pp. 1–128.
Tokita, M.: Friction Between Polymer Networks of Gels and Solvent. Vol. 110, pp. 27–48.
Tomlinson, M. R. see Bhat, R. R.: Vol. 198, pp. 51–124.
Traser, S. see Bohrisch, J.: Vol. 165, pp. 1–41.
Tries, V. see Baschnagel, J.: Vol. 152, p. 41–156.

Trimmel, G., Riegler, S., Fuchs, G., Slugovc, C. and *Stelzer, F.*: Liquid Crystalline Polymers by Metathesis Polymerization. Vol. 176, pp. 43–87.
Tsujii, Y., Ohno, K., Yamamoto, S., Goto, A. and *Fukuda, T.*: Structure and Properties of High-Density Polymer Brushes Prepared by Surface-Initiated Living Radical Polymerization. Vol. 197, pp. 1–47.
Tsuruta, T.: Contemporary Topics in Polymeric Materials for Biomedical Applications. Vol. 126, pp. 1–52.

Uemura, T., Naka, K. and *Chujo, Y.*: Functional Macromolecules with Electron-Donating Dithiafulvene Unit. Vol. 167, pp. 81–106.
Ungar, G., Putra, E. G. R., de Silva, D. S. M., Shcherbina, M. A. and *Waddon, A. J.*: The Effect of Self-Poisoning on Crystal Morphology and Growth Rates. Vol. 180, pp. 45–87.
Usov, D. see Rühe, J.: Vol. 165, pp. 79–150.
Usuki, A., Hasegawa, N. and *Kato, M.*: Polymer-Clay Nanocomposites. Vol. 179, pp. 135–195.
Uyama, H. and *Kobayashi, S.*: Enzymatic Synthesis and Properties of Polymers from Polyphenols. Vol. 194, pp. 51–67.
Uyama, H. and *Kobayashi, S.*: Enzymatic Synthesis of Polyesters via Polycondensation. Vol. 194, pp. 133–158.
Uyama, H. see Kobayashi, S.: Vol. 121, pp. 1–30.
Uyama, Y.: Surface Modification of Polymers by Grafting. Vol. 137, pp. 1–40.

Vancso, G. J., Hillborg, H. and *Schönherr, H.*: Chemical Composition of Polymer Surfaces Imaged by Atomic Force Microscopy and Complementary Approaches. Vol. 182, pp. 55–129.
Varma, I. K. see Albertsson, A.-C.: Vol. 157, pp. 99–138.
Vasilevskaya, V. see Khokhlov, A.: Vol. 109, pp. 121–172.
Vaskova, V. see Hunkeler, D.: Vol. 112, pp. 115–134.
Verdugo, P.: Polymer Gel Phase Transition in Condensation-Decondensation of Secretory Products. Vol. 110, pp. 145–156.
Veronese, F. M. see Pasut, G.: Vol. 192, pp. 95–134.
Vettegren, V. I. see Bronnikov, S. V.: Vol. 125, pp. 103–146.
Vilgis, T. A. see Holm, C.: Vol. 166, pp. 67–111.
Viovy, J.-L. and *Lesec, J.*: Separation of Macromolecules in Gels: Permeation Chromatography and Electrophoresis. Vol. 114, pp. 1–42.
Vlahos, C. see Hadjichristidis, N.: Vol. 142, pp. 71–128.
Voigt, I. see Spange, S.: Vol. 165, pp. 43–78.
Volk, N., Vollmer, D., Schmidt, M., Oppermann, W. and *Huber, K.*: Conformation and Phase Diagrams of Flexible Polyelectrolytes. Vol. 166, pp. 29–65.
Volksen, W.: Condensation Polyimides: Synthesis, Solution Behavior, and Imidization Characteristics. Vol. 117, pp. 111–164.
Volksen, W. see Hedrick, J. L.: Vol. 141, pp. 1–44.
Volksen, W. see Hedrick, J. L.: Vol. 147, pp. 61–112.
Vollmer, D. see Volk, N.: Vol. 166, pp. 29–65.
Voskerician, G. and *Weder, C.*: Electronic Properties of PAEs. Vol. 177, pp. 209–248.

Waddon, A. J. see Ungar, G.: Vol. 180, pp. 45–87.
Wagener, K. B. see Baughman, T. W.: Vol. 176, pp. 1–42.
Wagner, E. and *Kloeckner, J.*: Gene Delivery Using Polymer Therapeutics. Vol. 192, pp. 135–173.
Wake, M. C. see Thomson, R. C.: Vol. 122, pp. 245–274.

Wandrey, C., Hernández-Barajas, J. and *Hunkeler, D.*: Diallyldimethylammonium Chloride and its Polymers. Vol. 145, pp. 123–182.

Wang, K. L. see Cussler, E. L.: Vol. 110, pp. 67–80.

Wang, S.-Q.: Molecular Transitions and Dynamics at Polymer/Wall Interfaces: Origins of Flow Instabilities and Wall Slip. Vol. 138, pp. 227–276.

Wang, S.-Q. see Bhargava, R.: Vol. 163, pp. 137–191.

Wang, T. G. see Prokop, A.: Vol. 136, pp. 1–52; 53–74.

Wang, X. see Lin, T.-C.: Vol. 161, pp. 157–193.

Watanabe, K. see Hikosaka, M.: Vol. 191, pp. 137–186.

Webster, O. W.: Group Transfer Polymerization: Mechanism and Comparison with Other Methods of Controlled Polymerization of Acrylic Monomers. Vol. 167, pp. 1–34.

Weder, C. see Voskerician, G.: Vol. 177, pp. 209–248.

Weis, J.-J. and *Levesque, D.*: Simple Dipolar Fluids as Generic Models for Soft Matter. Vol. 185, pp. 163–225.

Whitesell, R. R. see Prokop, A.: Vol. 136, pp. 53–74.

Williams, R. A. see Geil, P. H.: Vol. 180, pp. 89–159.

Williams, R. J. J., Rozenberg, B. A. and *Pascault, J.-P.*: Reaction Induced Phase Separation in Modified Thermosetting Polymers. Vol. 128, pp. 95–156.

Winkler, R. G. see Holm, C.: Vol. 166, pp. 67–111.

Winnik, F. M. see Aseyev, V. O.: Vol. 196, pp. 1–86.

Winter, H. H. and *Mours, M.*: Rheology of Polymers Near Liquid-Solid Transitions. Vol. 134, pp. 165–234.

Wittmeyer, P. see Bohrisch, J.: Vol. 165, pp. 1–41.

Wood-Adams, P. M. see Anantawaraskul, S.: Vol. 182, pp. 1–54.

Wu, C.: Laser Light Scattering Characterization of Special Intractable Macromolecules in Solution. Vol. 137, pp. 103–134.

Wu, C. see Zhang, G.: Vol. 195, pp. 101–176.

Wu, T. see Bhat, R. R.: Vol. 198, pp. 51–124.

Wunderlich, B. see Sumpter, B. G.: Vol. 116, pp. 27–72.

Xiang, M. see Jiang, M.: Vol. 146, pp. 121–194.

Xie, T. Y. see Hunkeler, D.: Vol. 112, pp. 115–134.

Xu, P., Singh, A. and *Kaplan, D. L.*: Enzymatic Catalysis in the Synthesis of Polyanilines and Derivatives of Polyanilines. Vol. 194, pp. 69–94.

Xu, P. see Geil, P. H.: Vol. 180, pp. 89–159.

Xu, Z., Hadjichristidis, N., Fetters, L. J. and *Mays, J. W.*: Structure/Chain-Flexibility Relationships of Polymers. Vol. 120, pp. 1–50.

Yagci, Y. and *Endo, T.*: N-Benzyl and N-Alkoxy Pyridium Salts as Thermal and Photochemical Initiators for Cationic Polymerization. Vol. 127, pp. 59–86.

Yagi, M. and *Kaneko, M.*: Charge Transport and Catalysis by Molecules Confined in Polymeric Materials and Application to Future Nanodevices for Energy Conversion. Vol. 199, pp. 143–188.

Yamaguchi, I. see Yamamoto, T.: Vol. 177, pp. 181–208.

Yamamoto, T.: Molecular Dynamics Modeling of the Crystal-Melt Interfaces and the Growth of Chain Folded Lamellae. Vol. 191, pp. 37–85.

Yamamoto, T., Yamaguchi, I. and *Yasuda, T.*: PAEs with Heteroaromatic Rings. Vol. 177, pp. 181–208.

Yamamoto, S. see Tsujii, Y.: Vol. 197, pp. 1–47.

Yamaoka, H.: Polymer Materials for Fusion Reactors. Vol. 105, pp. 117–144.

Yamazaki, S. see Hikosaka, M.: Vol. 191, pp. 137–186.
Yannas, I. V.: Tissue Regeneration Templates Based on Collagen-Glycosaminoglycan Copolymers. Vol. 122, pp. 219–244.
Yang, J. see Geil, P. H.: Vol. 180, pp. 89–159.
Yang, J. S. see Jo, W. H.: Vol. 156, pp. 1–52.
Yasuda, H. and *Ihara, E.*: Rare Earth Metal-Initiated Living Polymerizations of Polar and Nonpolar Monomers. Vol. 133, pp. 53–102.
Yasuda, T. see Yamamoto, T.: Vol. 177, pp. 181–208.
Yaszemski, M. J. see Thomson, R. C.: Vol. 122, pp. 245–274.
Yoo, T. see Quirk, R. P.: Vol. 153, pp. 67–162.
Yoon, D. Y. see Hedrick, J. L.: Vol. 141, pp. 1–44.
Yoshida, H. and *Ichikawa, T.*: Electron Spin Studies of Free Radicals in Irradiated Polymers. Vol. 105, pp. 3–36.

Zhang, G. and *Wu, C.*: Folding and Formation of Mesoglobules in Dilute Copolymer Solutions. Vol. 195, pp. 101–176.
Zhang, H. see Rühe, J.: Vol. 165, pp. 79–150.
Zhang, Y.: Synchrotron Radiation Direct Photo Etching of Polymers. Vol. 168, pp. 291–340.
Zhao, B. see Brittain, W. J.: Vol. 198, pp. 125–147.
Zheng, J. and *Swager, T. M.*: Poly(arylene ethynylene)s in Chemosensing and Biosensing. Vol. 177, pp. 151–177.
Zhou, H. see Jiang, M.: Vol. 146, pp. 121–194.
Zhou, Z. see Abe, A.: Vol. 181, pp. 121–152.
Zubov, V. P., Ivanov, A. E. and *Saburov, V. V.*: Polymer-Coated Adsorbents for the Separation of Biopolymers and Particles. Vol. 104, pp. 135–176.

Subject Index

Actuator 244
Alkylindenofluorene-based polymers 58
Amplified spontaneous emission 83
Arylamines 114
–, spiro-linked 115

Biosensor 236
Bis(phenanthroline) ligands, spiro-bridged 128
Blue emission, polyfluorenes 43
Bromoarylboronic acid 9

Carbazole 63
Carbazole-based ladder polymers 65
Carbon nanomaterial, substituent 246
Catalysis 143, 165, 176
Charge transport 83, 110, 118, 143
– –, redox molecules, polymeric solid materials 146
Chromaticity diagram 96
CO_2 reduction catalysts 176
Commission Internationale de l' Éclairage (CIE) 96
Conducting materials, optically transparent 245
Conducting polymer nanomaterials 199
Core-shell 209, 219
Cyclophane bisboronic acid 20

Data storage 241
Dibromobenzene 9
2,2'-Dibromo-9,9'-spirobifluorene 91
2,7-Dichlorocarbazole 63
2,2'-Diiodo-9,9'-spirobifluorene 91
Dioctylfluorene–caprolactone 72
2,2'-Diphenylvinylboronic acid 112

Effective conjugation lengths 7

Electrochemical properties 110
Electrochromic devices 243
Electroluminescence 1, 84, 110, 113, 116, 119, 125, 137

Field emission displays 244
Field-effect transistors 83
Fluorene–binaphthyl 56
Fluorene–carbazole 51
Fluorenes, alternating copolymers, arylenes 54
Fluorenone 25
Fluorophenyl–fluorene 56

Hexabromospirobifluorene 92
HOMO/LUMO 4, 94

Indenofluorene–anthracene 59

Ladder polymers 1
Lasers, organic 83
Light-emitting diodes (LEDs) 1, 3, 83, 85
Light-emitting electrochemical cell (LEC) 24
Lithium triflate 24
LPPPs, blends 24
–, defect emission 25
LPPP–PPP 29
Luminance 99
Luminous efficiency 99

Manganese complexes 166
Molecular catalysts, multielectron reactions 165
Molecular glasses 83

Nafion film 146
Nanocomposites 217
–, polyaniline 224

-, polythiophene   229
Nanofiber   227, 232, 235
-, polyaniline   213, 221
Nanohybrid, polythiophene   229
Nanomaterials, conducting polymer   199
-, core-shell   219
Nanoparticle   207, 226, 230, 235
-, polyaniline   218
Nanopattern   216
-, polyaniline   223
Nanoribbon   232, 235
Nanorod, polyaniline   221
Nanosphere, hollow   212, 220, 226
Nanotube   215, 227, 232, 235
-, polyaniline   222
Nanowire, polythiophene   227

OLEDs   99
Oligoaryls   100
Oligophenyls, spiro-linked   100
Oligothienylene, spirobifluorene   122
Organic lasers   83
Oxadiazole compounds   125

PDAFs   32
-, defect emission   35
Perfluorpropylperoxide   9
PFs, substituted alkyl side chains   34
Photoluminescence   1, 5
Photosynthesis, artificial   143, 183
Photovoltaic cells   243
Poly(N-alkyl-carbazole)s   63, 64
Poly[9,9-bis(2-ethylhexyl)fluorene]   32
Poly(2,7-dioctyl-4,5,9,10-tetrahydropyrene)   62
Poly(9,9-dialkylfluorene)s (PDAFs)   30
Poly(9,9-diarylfluorene)s   46
Poly(9,9-dihexyl-2,7-fluorene)   31
Poly(9,9-dioctylfluorene)   31
Poly(3,4-ethylenedioxythiophene) nanomaterials   230
Poly(ethylene oxide) (PEO)   24
Poly(indenofluorene)s   58
Poly(ladder-type pentaphenylene)s   61
Poly(methyl methacrylate) (PMMA)   24
Poly(p-phenylene)s   see PPPs
Poly(p-phenylene vinylene)   234
Poly(spirobifluorene oligophenylene)   56
Poly[tetra(4-alkylphenyl)indenofluorene]s   61

Poly(tetraarylindenofluorene)s   60
Poly(tetrahydropyrene)s   21, 62
Poly(N-vinylcarbazole) (PVK)   7
Polyacenes, ribbon-like   19
Polyaniline nanomaterials   218
Polycarbazoles   63
Polycarboxylate   15
Polyethylenedioxythiophene (PEDOT)   94
Polyfluorenes   3, 22, 25, 30
-, blue emission   43
-, dendronised   43
-, improved charge injection   48
-, water-based, biosensor applications   55
Polyfluorene–PMMA triblock   72
Polyfluorene–polyaniline block   70
Polyfluorene–poly(ethylene oxide)   67, 69
Polyfluorene–polystyrene   69, 71
Polyfluorenone, precursor route   53
Polyindenofluorenes   3, 22, 59
Polyionene   31
Polymeric solid materials   143
Polyphenylenes, 'stepladder'-type   62
Polypyrrole nanomaterials   207
Polyrotaxane, fluorescent, polycarboxylate   15
Polysaccharide solids, excess liquid   159
- -, ionic conductivity   164
Polystyrene (PS)   66
Polystyrenesulfonic acid (PSS)   94
Polythiophene nanomaterials   225
PPPs, coupling of benzene rings   6
-, cyclophane-substituted   14
-, ladder-type   19
-, -, methine bridges   20
-, perfluoropropylation   9
-, precursor routes   8
-, solubilising substituents   9
-, soluble, luminescent properties   12
-, stepladder-type   30
-, substituted   9
Proton reduction catalysts   173

Ruthenium complexes   169

Sensors, chemical   236
Solar cells   83, 99, 120
- -, dye-sensitized   143, 178

Subject Index

Spiro compounds   83
– –, arylamines   114
– –, chirality   135
– –, mixed chromophores   128
– –, oligoaryls   100
– –, stilbene/azobenzene units   112
Spiro polymers   136
Spirobifluorene units, polymers   56
Spiro-octahedric compounds   101
Spiro-oligophenyles, starburst   102
Starburst   102
Step-ladder polymers   3
Stilbene/azobenzene units   112
Supercapacitor   242
Surface protection   245
Switch   240

TBPSF   127
Template method   199, 203
Template-free method   205
Ter-(9,9-diarylfluorene)s   105
Tetrabromospirobifluorene   92, 112
Thin film   216
– –, polyaniline   223
Thiophene compounds   122
Transistor   240

Vinylbiphenyl   8
N-Vinylcarbazole   8

Water oxidation catalysts   165

ZnS powder   84

Printing: Krips bv, Meppel
Binding: Stürtz, Würzburg